Integration of Distributed Energy Resources in Power Systems

Implementation, Operation, and Control

Integration of Distributed Energy Resources in Power Systems

Implementation, Operation, and Control

Edited by

Toshihisa Funabashi

Institute of Materials and Systems for Sustainability (IMaSS)
Nagoya University, Nagoya, Japan

AMSTERDAM • BOSTON • HEIDELBERG • LONDON
NEW YORK • OXFORD • PARIS • SAN DIEGO
SAN FRANCISCO • SINGAPORE • SYDNEY • TOKYO

Academic Press is an Imprint of Elsevier

Academic Press is an imprint of Elsevier
125 London Wall, London EC2Y 5AS, UK
525 B Street, Suite 1800, San Diego, CA 92101-4495, USA
50 Hampshire Street, 5th Floor, Cambridge, MA 02139, USA
The Boulevard, Langford Lane, Kidlington, Oxford OX5 1GB, UK

Library of Congress Cataloging-in-Publication Data
A catalog record for this book is available from the Library of Congress

British Library Cataloguing-in-Publication Data
A catalogue record for this book is available from the British Library

ISBN: 978-0-12-803212-1

For information on all Academic Press publications
visit our website at http://www.elsevier.com/

Publisher: Joe Hayton
Acquisition Editor: Raquel Zanol
Editorial Project Manager: Mariana Kühl Leme
Editorial Project Manager Intern: Ana Claudia Garcia
Production Project Manager: Anusha Sambamoorthy
Designer: Maria Inês Cruz

Contents

List of Contributors....................xi

CHAPTER 1 Introduction 1

Toshihisa Funabashi

1.1 Introduction....................1
1.2 Distributed Generation Resources....................3
1.2.1 Reciprocating Engines....................3
1.2.2 Microturbine Generator (MTG) System....................4
1.2.3 Fuel Cells....................5
1.3 Renewable Energy Sources....................6
1.3.1 Wind Energy Conversion System....................6
1.3.2 PV Energy System....................7
1.3.3 Biomass....................8
1.3.4 Geothermal Energy....................9
1.3.5 Hydro Energy....................9
1.4 Energy Storage Systems....................9
1.4.1 Electric Double Layer Capacitor....................10
1.4.2 Battery Energy Storage System....................10
1.4.3 Superconducting Magnetic Energy Storage....................10
1.4.4 Flywheel....................11
1.4.5 Plug in Electric Vehicle....................11
1.5 Smart Grid....................11
References....................13

CHAPTER 2 Integration of Distributed Energy Resources in Distribution Power Systems 15

Alberto Borghetti, Carlo Alberto Nucci

2.1 Introduction....................15
2.2 Interconnection Issues and Countermeasures....................17
2.2.1 Volt–VAR Control....................19
2.2.2 Gossip-like VVC MAS Procedure....................21
2.3 Role of ICT in the Integration of Distributed Energy Resources....................25
2.3.1 Models of the Communication Networks....................27
2.3.2 Model of the Power Distribution Feeder....................32
2.3.3 Test Results....................33
2.4 Conclusions....................46
Acknowledgment....................47
References....................47

CHAPTER 3 Operational Aspects of Distribution Systems with Massive DER Penetrations 51

Tomonobu Senjyu, Abdul Motin Howlader

3.1 Introduction....51
3.2 Control Objectives....53
3.2.1 Importance of Distributed Generations....54
3.2.2 Challenges of Distributed Generations System....56
3.2.3 Overview of Control System....57
3.3 Control Method....58
3.3.1 The Objective Function and Constraints....58
3.4 Particle Swarm Optimization....60
3.4.1 PV Generator System....61
3.4.2 BESS at the Interconnection Point....62
3.4.3 Plug-in Electric Vehicle....63
3.5 Simulation Results....63
3.5.1 Dynamic Responses for the Without Optimization Approach....66
3.5.2 Dynamic Responses for the Comparison Method....66
3.5.3 Dynamic Responses for the Proposed Method....70
3.6 Conclusions....73
References....73

CHAPTER 4 Prediction of Photovoltaic Power Generation Output and Network Operation 77

Takeyoshi Kato

4.1 Needs for Forecasting Photovoltaic (PV) Power Output in Electric Power Systems....78
4.2 Power Output Fluctuation Characteristics....79
4.2.1 Fluctuation Characteristics of Irradiance at Single Point....81
4.2.2 Fluctuation Characteristics of Spatial Average Irradiance in Utility Service Area....84
4.3 Forecasting Methods....86
4.3.1 Overview....86
4.3.2 Accuracy Measures....88
4.3.3 NWP Models....89
4.3.4 Satellite Cloud Motion Vector Approach....94
4.3.5 All Sky Images....96
4.3.6 Statistical Models....96
4.4 Examples of Forecasted Results....98
4.5 Smoothing Effect on Forecast Accuracy....100

4.6 Power System Operation Considering PV Power Output Fluctuations 101
4.7 Energy Management Examples of Smart House with PV 104
4.7.1 United States/Japan Demonstration Smart Grid Project in Los Alamos 104
References 106

CHAPTER 5 Prediction of Wind Power Generation Output and Network Operation 109
Ryoichi Hara

5.1 Need for Forecasting Wind Power Output in Electric Power Systems 110
5.2 Power Output Fluctuation Characteristics 112
5.2.1 Fundamentals 112
5.2.2 Maximum Variation 115
5.2.3 Umbrella Curve 115
5.2.4 Standard Deviation 115
5.2.5 Power Spectral Density 118
5.3 Power Output Smoothing Control 120
5.3.1 Application of Energy-storage System 120
5.3.2 Kinetic Energy of Wind Turbines 121
5.3.3 Pitch Angle Control 121
5.4 Forecasting Methods 122
5.4.1 Difficulties 122
5.4.2 Physical Approach 123
5.4.3 Statistic Approach 123
5.4.4 Regional Forecasting 124
5.4.5 Probabilistic Forecast 125
5.5 Examples of Forecasted Results 125
5.6 Forecasting Applications 128
5.6.1 Scheduled Generation of Wind Farms and Solar Power Plants with Energy-storage Systems 128
5.6.2 Suppression of Ramp Variation of Wind Output 129
References 130

CHAPTER 6 Energy Management Systems for DERs 132
Atsushi Yona

6.1 Basic Concepts of Home Energy Management Systems 132
6.2 Control Strategies for Energy Storage Systems 137
6.3 Control Strategies for EVs as Storage 145
6.4 Use of Smart Meter Data 146
References 155

CHAPTER 7 Protection of DERs 157
Raza Haider, Chul-Hwan Kim
7.1 Introduction 157
7.2 Protection in Distribution System 160
7.2.1 General Protection 160
7.2.2 IEEE Standards for Protection 163
7.2.3 What to do Under Fault Conditions 164
7.2.4 Fault Currents Change 165
7.2.5 Smart Protection 166
7.3 Power System Disturbances 168
7.3.1 Power Quality Issues 168
7.3.2 Faults 170
7.3.3 Consequences of Electric Faults 172
7.3.4 Application Based Three-phase Fault Analysis 173
7.4 Impact of DER on Protection System 173
7.4.1 Protection Failure 174
7.4.2 Hosting Capacity 175
7.4.3 Loss of Coordination 176
7.4.4 Protection Issues of DER 178
7.5 Protection Schemes for Distribution Systems with DER 180
7.5.1 Islanded Operation 181
7.5.2 The Protection Equipment for DER Networks 182
7.5.3 Recent Technological Trends in DER Protection 188
7.6 Conclusions 190
References 191

CHAPTER 8 Lightning Protections of Renewable Energy Generation Systems 193
Shozo Sekioka
8.1 Introduction 193
8.2 Lightning Protection Principle 196
8.2.1 Reduction 196
8.2.2 Suppression 197
8.2.3 Shielding 199
8.2.4 Lightning Characteristics for Lightning Protection Design 201
8.2.5 IEC International Standard 206
8.3 Lightning Protection for Wind Power Generation Systems 209
8.3.1 Lightning Damage in Wind Power Generation System 209
8.3.2 Lightning Protection Methods for Wind Turbine 211
8.3.3 Grounding Resistance 212
8.3.4 Lightning Protection of Blade Using Receptor 215
8.3.5 Energy Coordination of Surge Arrester/Surge Protective Device 216

8.4 Lightning Protection of Wind Farms 218
8.5 Lightning Protection for Photovoltaic Power Generation Systems 222
8.5.1 Lightning Damage in Photovoltaic System 223
8.5.2 Lightning Protection Against Lightning Overvoltages in Photovoltaic Systems 223
8.5.3 Direct Lightning Flash to Photovoltaic System 224
References 226

CHAPTER 9 Distributed Energy Resources and Power Electronics 229
Masahide Hojo

9.1 Power Electronics in PV Power Generation Systems 231
9.2 Power Electronics in Wind Power Generation Systems 232
9.3 Power Electronics in Battery Energy Storage Systems 233
9.4 Power Quality Problems with Related to DERs 234

CHAPTER 10 AC/DC Microgrids 236
Kazuto Yukita

10.1 Basic Concept of AC Microgrids 237
10.1.1 Islanding Mode 238
10.1.2 Connected Mode 239
10.1.3 Backup Mode 239
10.1.4 Experiment Field and Device Specification 239
10.1.5 System Operation 240
10.1.6 Measurement of Power Quality 242
10.2 Battery Charge Pattern and Cost 244
10.2.1 Battery Charge Method 246
10.2.2 Test Results 246
10.3 Supply and Demand Control of Microgrids 251
10.3.1 Peak Cut/Peak Shift Mode Operation 251
10.3.2 Receiving Constant Power Mode Operation 253
10.4 Basic Concept of DC Microgrids 254
10.4.1 System Operation of DC Microgrid 255
10.5 Examples of Microgrids in the World 258
10.6 Conclusions 259
References 259

CHAPTER 11 Stability Problems of Distributed Generators 261
Toshihisa Funabashi, Jinjun Liu, Tomonobu Senjyu

11.1 Voltage Stability in Distribution Systems 262
11.1.1 Definitions 262
11.2 Stability Problem with DGs Connected to a Weak Power System 262
11.2.1 Voltage Stability Analysis 263

11.2.2 Voltage Stability Index 264
11.2.3 Battery Control Method 264
11.2.4 Simulation Results 267
11.3 Stability Problem with Power Electronics in DGs 267
11.3.1 Terminal Characteristics of Submodule 268
11.3.2 Stability Criteria....................269
11.3.3 Comparison between Stability Criteria....................271
11.3.4 Conclusions....................272
11.4 Stability Problems in Microgrids....................273
11.4.1 Microgrid Model....................274
11.4.2 Inverter Control Method 276
11.4.3 FRT Requirements....................277
11.4.4 Criteria of Power Quality....................277
11.4.5 Simulation and Results 277
11.4.6 Conclusions....................279
References....................279

CHAPTER 12 Virtual Synchronous Generators and their Applications in Microgrids....................282
Toshifumi Ise, Hassan Bevrani
12.1 Basic Concepts of Virtual Synchronous Generators 282
12.2 Control Schemes of Virtual Synchronous Generators....................284
12.2.1 VSYNC's VSG Design....................285
12.2.2 IEPE's VSG Topology 286
12.2.3 KHI's VSG 287
12.2.4 VSG System of Osaka University 287
12.3 Applications for Microgrids 290
References....................292

CHAPTER 13 Application of DERs in Electricity Market....................295
Yusuke Manabe
13.1 Basic Concept of Electricity Market and DERs 295
13.1.1 RPS 296
13.1.2 FIT 297
13.1.3 Effect of DER's Mass Penetration for Electricity Market 297
13.2 Electricity Market Reform and Virtual Power Plan 299
References....................302

Subject Index 303

List of contributors

Hassan Bevrani
Department of Electrical & Computer Engineering, University of Kurdistan, Kurdistan, Sanandaj, Iran

Alberto Borghetti
Department of Electrical, Electronic and Information Engineering, University of Bologna, Bologna, Italy

Toshihisa Funabashi
Institute of Materials and Systems for Sustainability (IMaSS), Nagoya University, Nagoya, Japan

Raza Haider
Department of Electrical Engineering, Balochistan University of Engineering and Technology, Khuzdar, Pakistan

Ryoichi Hara
Graduate School of Information Science and Technology, Hokkaido University, Hokkaido, Japan

Masahide Hojo
Department of Electrical and Electronic Engineering, Faculty of Engineering, Tokushima University, Japan

Abdul Motin Howlader
Postdoctoral Fellow University of Hawaii, Manoa Honolulu, Hawaii

Toshifumi Ise
Graduate School of Engineering, Osaka University, Suita, Osaka, Japan

Takeyoshi Kato
Institute of Materials and Systems for Sustainability (IMaSS), Nagoya University, Nagoya, Japan

Chul-Hwan Kim
School of Electronics and Electrical Engineering, Sungkyunkwan University, Republic of Korea

Jinjun Liu
Xi'an Jiatong University, Xi'an, China

Yusuke Manabe
Funded Research Division Energy Systems (Chubu Electric Power), Institute of Materials and Systems for Sustainability (IMaSS), Nagoya University, Nagoya, Japan

Carlo Alberto Nucci
Department of Electrical, Electronic and Information Engineering, University of Bologna, Bologna, Italy

Shozo Sekioka
Department of Electrical & Electronic Engineering, Shonan Institute of Technology, Japan

Tomonobu Senjyu
Department of Electrical and Electronics Engineering, University of the Ryukyus, Okinawa, Japan

Atsushi Yona
Department of Electrical and Electronics Engineering, Faculty of Engineering, University of the Ryukyus, Okinawa, Japan

Kazuto Yukita
Aichi Institute of Technology, Department of Electrical Engineering, Toyota, Japan

Chapter 1

Introduction

Toshihisa Funabashi
Institute of Materials and Systems for Sustainability (IMaSS), Nagoya University, Nagoya, Japan

CHAPTER OUTLINE

1.1 Introduction 1
1.2 Distributed generation resources 3
1.2.1 Reciprocating engines 3
1.2.2 Microturbine generator (MTG) system 4
1.2.3 Fuel cells 5
1.3 Renewable energy sources 6
1.3.1 Wind energy conversion system 6
1.3.2 PV energy system 7
1.3.3 Biomass 8
1.3.4 Geothermal energy 9
1.3.5 Hydro energy 9
1.4 Energy storage systems 9
1.4.1 Electric double layer capacitor 10
1.4.2 Battery energy storage system 10
1.4.3 Superconducting magnetic energy storage 10
1.4.4 Flywheel 11
1.4.5 Plug in electric vehicle 11
1.5 Smart grid 11
References 13

1.1 INTRODUCTION

World energy demand has been increasing exponentially. Conventional energy resources (eg, coal, oil, and gas) are exhaustible and limited in supply. Therefore, there is an urgent need to conserve what we have and explore alternative energy resources. Among various types of renewable energy resources, solar and wind energies are the most promising for humankind [1]. Because of the large amount of renewable energies, the renewable energy sources will be the backbone of the energy system in future [2]. Over time, renewable energy will gradually displace coal, oil, and gas from our energy consumption patterns. In order to integrate a large amount of renewable energies into the power system, it is required to reconfigure the existing energy systems. The intelligent power grid or smart grid (SG) is key to this transformation. In the future, SG systems will be composed of several elements such as distributed renewable energy sources, a strong power grid, a flexible consumption, and an intelligent power control system [3]. Distributed renewable energy sources (eg, wind turbine, photovoltaic, fuel cell, biomass, smart house, etc.) and energy storage devices (eg, battery, electric double layer capacitor, superconducting magnetic energy storage, etc.) are expected to play a vital role for the green SG system and to meet the future energy demand [4,5]. The distributed generations (DGs) locate generation close to the load, that is, on the distribution network or on the customer side of the meter [6]. DGs have great potential to improve distribution system performance and should be encouraged [7].

Rating of DGs: The maximum rating of the DG which can be connected to a distributed generation depends on the capacity of the distribution system that is interrelated with the voltage level of the distribution system. Hence, the capacity of DGs can vary widely. There are four different categories of DGs which are as follows [8]:

Micro. DG range: ~ 1 W < 5 kW;
Small. DG range: 5 kW < 5 MW;
Medium. DG range: 5 MW < 50 MW;
Large. DG range: 50 MW < ~ 300 MW.

Due to the various types of DGs, the generation electric current can be either direct current (DC) or alternating current (AC). Photovoltaic, fuel cell, and batteries generate the DC which is appropriate for DC loads and DC SG. On the other hand, the DC can be converted to the AC by using power electronics interface and then it can be connected to the AC loads and power grid. Other DGs such as wind turbine, micro turbine, and biomass deliver an AC which for some applications must be controlled by using modern power electronic equipments in order to acquire the regulated voltage [9].

Application of DGs: There are several applications of DG in the power system such as [9]:

- The DG can be scattered in different places. It can be utilized as a standby power source. If the grid power cuts off the sensitive loads, for example, process industries and hospitals, the DG can provide the emergency power for these loads.

- The DG can supply power for the isolated communities where areas are geographical obstacles and difficult to connect the main power grid. Therefore, the DG can improve the economic condition for isolated communities.
- The electric power cost depends on the electric load. When the load demand is high, the electric power price will be high and vice versa. The DG can supply the electric power to the load when the demand is high. As a result, the customer can reduce the electricity cost to pay time-of-use rates.
- The DG can supply power for the rural and remote applications which include lighting, heating, cooling, communication, and small industrial processes.
- Individual DG owner is usually used as a base load to provide part of the main required power and support the grid by enhancing the system voltage profile. The DG also helps to reduce the power losses and improving the system power quality.

1.2 DISTRIBUTED GENERATION RESOURCES

Photovoltaic (PV) energy, wind turbines, and other distributed generation plants are typically situated in remote areas, requiring the operation systems that are fully integrated into transmission and distribution network [10]. The aim of the SG is to integrate all generation plants reduce the cost and greenhouse gas emission. A detailed discussion about the distributed energy resources and SG system is considered next in this section.

The DG is also known as the local generation, on-site generation, or distributed energy which produces electricity from some small energy sources. The energy sources are directly connected to the medium voltage (MV) or low voltage (LV) distribution systems, rather than to the bulk power transmission systems. Different types of the DG resources are depicted in Fig. 1.1 [11].

The DG can be power supplied by conventional generation systems (eg, diesel and gas generators) and nonconventional generation systems (eg, fuel cells and renewable energy resources). Various types of energy storages are also considered as the DG resources.

1.2.1 Reciprocating engines

The reciprocating engine is also known as the piston engine. It is an internal combustion engine (ICE) and can burn a variety of fuels, including natural gas, diesel, biodiesel, biofuels, etc. The reciprocating engine, with its compact size, wide range of power outputs, and fuel preferences, is an

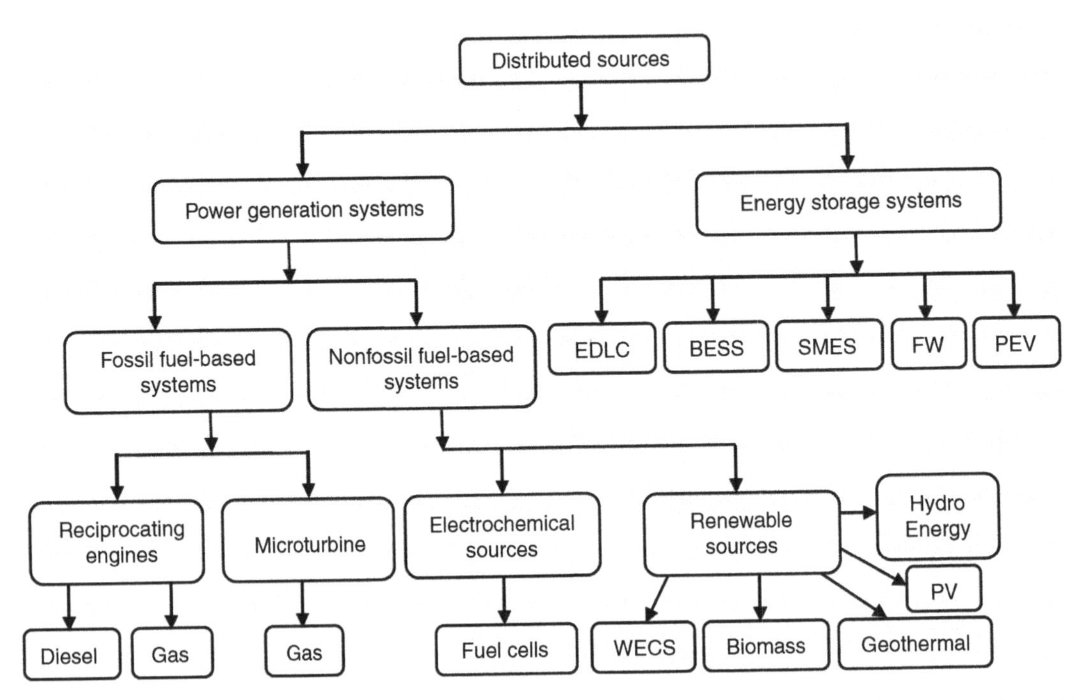

■ **FIGURE 1.1 DG sources.**

ideal prime mover for powering electricity generating sets used to deliver primary power in remote locations or more generally for providing mobile and emergency or stand-by electrical power. The power generation scales of the reciprocating engines are differed from the 1 kVA (small scale) to several tens of MVA (large scale) [11].

In case of the DG application, the reciprocating engine provides the lowest cost of all combined heat and power (CHP) systems, high efficiencies, short start-up times to full loads (10–15 s), and high reliability. But these types of engines generate the emission pollutants (eg, NO_X, CO, SO_X, etc.) which might be harmful for the environment.

1.2.2 Microturbine generator (MTG) system

The MTG is one of the best systems of the DG. The MTG has the advantages of being low (initial) cost, multifueled, reliable, and lightweight. In addition, the MTG offers the cogeneration system that generates heat energy as well as electric energy. This feature is suitable for the energy system of hotels, hospitals, supermarkets, etc. Although the MTG generates the electric power using the natural gas, it has an environmental benefit, that is, low nitrogen dioxide emission. But the energy efficiency of the MTG is lower than the reciprocating engines [12].

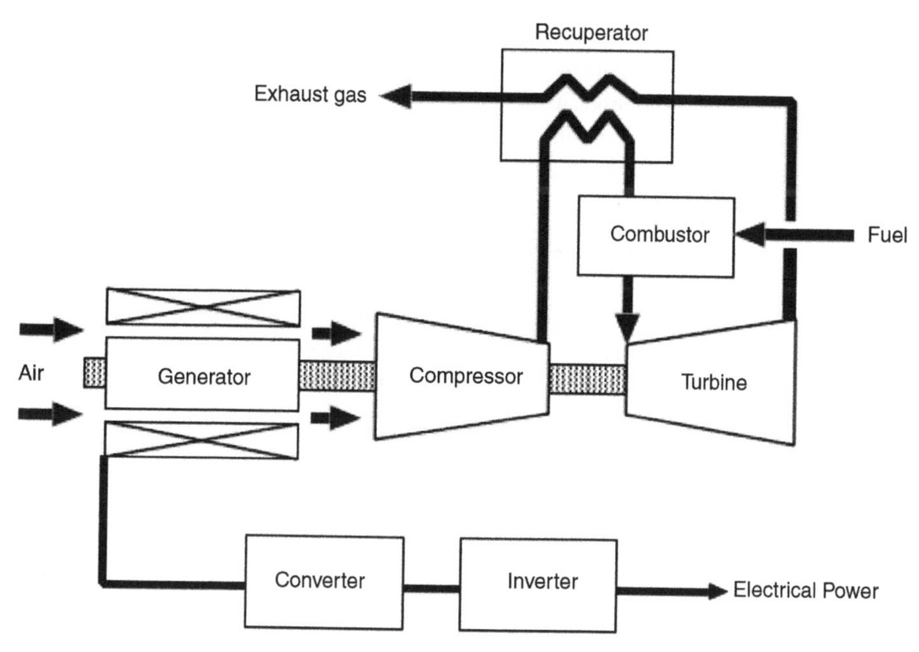

■ **FIGURE 1.2 System configuration of the MTG.**

Fig. 1.2 illustrates the system configuration of the MTG. The operation of this system is briefly explained as follows:

- The compressor compresses the outside air.
- The compressed air is heated by the exhausting gas in the recuperator.
- The heated air is combined with the natural gas. Then the mixed gas burns into the combustor.
- The combustion gas flows to the turbine, which generates the kinetic energy. The output power of the turbine is utilized for both generator and compressor.
- Since the output voltage of the generator is enclosed to the high frequency, it is converted into the DC voltage by the converter. The DC voltage is then converted into the AC voltage through the inverter and the output power sends to the consumers.

1.2.3 Fuel cells

A fuel cell can produce electricity by a chemical reaction. There are many types of fuel cells, but they all consist of an anode, a cathode, and an electrolyte that allows charges to move between the two sides of the fuel cell. Electrons are drained from the anode to the cathode through an external circuit, producing DC electricity. As the main difference among fuel cell types is the electrolyte, fuel cells are classified by the type of electrolyte. The major types of the fuel cells are alkaline fuel cell (AFC), proton exchange membrane fuel cell, phosphoric acid fuel cell, solid oxide fuel cell, and molten carbonate fuel cell (MCFC). Since the fuel cells generate the DC electricity, to obtain the AC electricity from fuel cells, power-conditioning equipment is required to handle the conversion from DC to AC that is obligatory to be included into the power distribution network [11].

The power generation capacity of fuel cells is diverged from 10 kW to 3 MW which depends on the types of fuel cells. In a fuel cell, hydrogen gas is used as a fuel, hence, virtually no harmful emissions are generated by the fuel cells. This results in power production that is almost entirely absent of nitrogen oxide (NO_X), sulfur dioxide (SO_X) or particulate matter. On the other hand, fuel cells are highly efficient, fuel flexible, and suitable for the CHP. There might be several disadvantages of the fuel cells such as MCFC requires a long start-up time and low-power density, AFC is sensitive to CO_2 in fuel and air.

1.3 RENEWABLE ENERGY SOURCES

Due to the crisis of exhausting fossil fuels and considering the greenhouse effect, it is predicted that over the next 20 years, fossil fuels will contribute 64% of the growth in energy. Renewables (eg, wind, solar, hydro, wave, biofuels, etc.) will account for 18% of the global energy by 2030. The rate at which renewables penetrate the global energy market is similar to the emergence of nuclear power in the 1970s and 1980s [13]. Renewable energy sources are emission free; hence, they will be vital part of the future SG system. Different types of renewable energy sources are described in the forthcoming sections.

1.3.1 Wind energy conversion system

A wind energy conversion system (WECS) is powered by wind energy and generates mechanical energy that sends energy to the electrical generator for making electricity. Fig. 1.3 shows the interconnection of a WECS. The generator of the wind turbine can be a permanent magnet synchronous generator (PMSG), doubly fed induction generator, induction generator, synchronous generator, etc. Wind energy acquired from the wind turbine is sent to the generator. To achieve maximum power from the WECS, the rotational speed of the generator is controlled by a pulse width modulation converter. The output power of the generator is supplied to the grid through a generator-side converter and a grid-side inverter. A wind farm can be distributed in onshore, offshore, seashore, or hilly areas. The WECS might be the most promising DG for future SG.

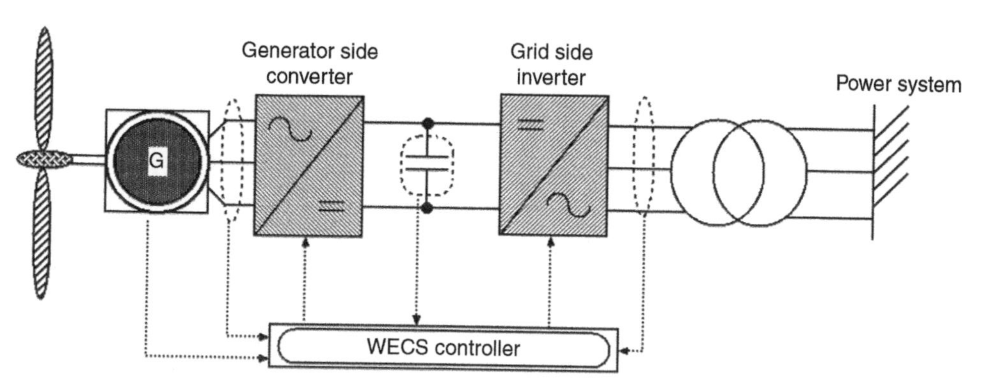

■ FIGURE 1.3 Wind energy conversion system.

Wind energy is an alternative to fossil fuels, it is plentiful, renewable, widely distributed, clean, low cost, produces no emissions during operation, and uses a tiny land area [14]. The effects on the environment are generally less problematic than those from other conventional power sources. Due to the variable wind speed, the output power of the WECS fluctuates and may create a frequency deviation of the power grid. To solve this problem, much research has already been conducted.

The world wind energy association (WWEA) published the key statistics of the World Wind Energy Report 2013.

The world wind energy capacity reached 318.5 GW by end of 2013 (this was 282.2 GW in 2012). In total, 103 countries are today using wind power on commercial basis. China was still by far the leading wind market with a new capacity of 16 GW and a total capacity of 91.3 GW. Wind power contributes close to 4% of the global electricity demand. For the year 2020, the WWEA predicts a wind capacity of more than 700 GW [15].

1.3.2 PV energy system

PV power systems convert sunlight directly into electricity. The PV system can be stand-alone (off-grid) or grid connected. Like WECS, PV is also a clean source of energy. However, the primary obstacle to increased use of PV systems is their high initial cost but continuous price reductions have been occurring. In some off-grid locations as short as one-quarter of a mile, PV systems can be cost-effective versus the costs of running power lines into the property and the subsequent continual electric charges. PV modules can be grouped together as an array of modules connected in series and parallel to provide any level of power requirements, from W to kW, and MW size. The PV energy generation system with battery pack is shown in Fig. 1.4. The PV array receives power from the sunlight and generates electricity, and a DC–DC converter converts the output voltage into a desired level. Depending on the PV system requirements, a battery energy storage

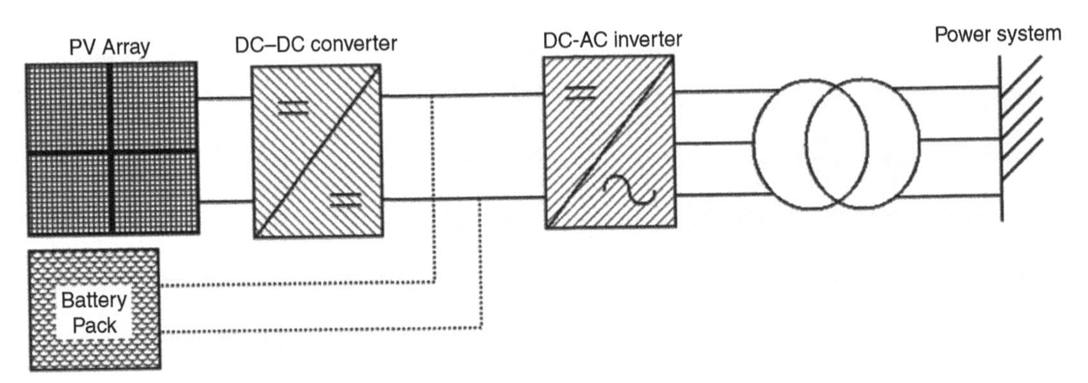

■ **FIGURE 1.4 PV energy system.**

system (BESS) stores the power into the battery and delivers the power to the PV system. A DC–AC inverter transports power to the power grid.

1.3.3 Biomass

Biomass fuels generate energy from things that once lived such as wood products, dried vegetation, crop residues, aquatic plants, and even garbage. When plants lived, they used a lot of the sun's energy to make their own food (photosynthesis). They stored the foods in the plants in a form of chemical energy. As the plants died, the energy became entrapped in the remains. This trapped energy is usually released by burning and can be converted into biomass energy. Wood remains the largest biomass energy source to date. Industrial biomass can be grown from numerous types of plants, including miscanthus, switchgrass, hemp, corn, poplar, willow, sorghum, sugarcane, bamboo, and a variety of tree species, ranging from eucalyptus to oil palm [16]. Burning materials like wood, waste, and other plant matters releases stored chemical energy in the form of heat, which can be used to turn shaft to produce electricity.

The use of biomass can be environmentally friendly because the biomass is reduced, recycled, and then reused. It is also a renewable resource because plants that can make biomass can be grown over and over. Biomass takes carbon out of the atmosphere while it is growing, and returns it as it is burned. If it is managed on a sustainable basis, biomass is harvested as part of a constantly replenished crop. This is either during woodland or arboricultural management or coppicing or as part of a continuous programmed of replanting with the new growth taking up CO_2 from the atmosphere at the same time as it is released by combustion of the previous harvest. The energy stored in biomass fuels came originally from the sun.

Biomass is such a widely utilized source of energy, probably due to its low cost and indigenous nature, that it accounts for almost 6.9% of the world's total energy supply.

1.3.4 Geothermal energy

Deep down in the Earth's coating, molten rock exists, and it is simply rocks that have melted into liquid form as a result of excessive heat under the Earth. This can be found about 1800 miles deep below the surface, but closer to the surface, the rock layers are hot enough to keep water and air spaces there at a temperature of about 10–16°C. Geothermal technology takes advantage of the hot close-to-Earth-surface temperatures to generate power.

In places with hotter "close-to-Earth-surface" temperatures, deep wells can be drilled and cold water pumped down. The water runs through splinters in the rocks and is heated up. It comebacks to the surface as hot water and steam, where its energy can be used to drive turbines and generates electricity. Geothermal energy is also known as a renewable energy source because the water is refilled by the rainfall, and the heat is continuously produced by the Earth. Geothermal systems are extremely environmentally friendly, unlike conventional power stations. Even though it may occasionally release some gases from deep down inside the Earth, which may be slightly harmful, these can be contained quite easily. Again the cost of the land on which to build a geothermal power plant is usually less expensive than the construction of an oil, gas, coal, or nuclear power plant. Geothermal energy accounts for more than 6 GW of generated energy, and is installed in 21 countries across the world [17].

1.3.5 Hydro energy

Moving water generates kinetic energy. This can be transferred into useful energy in different ways. Hydroelectric power schemes store water high up in dams. Water has gravitational potential energy which is released when it falls. The moving water turns the generator shaft, producing electricity. Hydro energy is also a renewable energy and there is no possibility to generate emission. Hydro energy provides almost 2.7% of the world's total energy supply.

1.4 ENERGY STORAGE SYSTEMS

Energy storage systems are an essential part of the renewable power generation system. The renewable power sources like solar, wind, and hydro are fluctuating resources. To supply a smooth output power to the power grid, energy storage systems are installed to the power generation system. Again the renewable sources (wind and solar) are unreliable, and in the case of the wind energy, the wind velocity sometimes drops below the power generation level, and sunlight may only be available 6–8 h per day to generate electricity. When the power generation becomes zero or the energy demand is high, the energy storage systems can deliver power to the consumers. Therefore, an energy storage system can be an important component of the SG system to improve the reliability of the power network. There are various types of energy storages, such as electric double layer capacitor (EDLC), BESS, superconducting magnetic energy storage (SMES), flywheel (FW), plug in electric vehicle (PEV), etc. A short description of these energy system storage systems is given in this section.

1.4.1 Electric double layer capacitor

EDLC is also known as super capacitor or ultra capacitor. The EDLC is an electrochemical capacitor employing conducting polymers as the electrodes [18,19]. The EDLC enables large power effects per weight having a goal up to 10 kW/kg but a storage capacity around 10 Wh/kg only. The storage time is short or typically up to 30–60 s. A 1-m^3 EDLC to rage may, in the future, yield a 1–5 MW power pulse and weight 100–500 kg [20]. The price is around 200–600 €/kWh and 50–150 €/Wh but in 5–10 years a price level of 10–15 €/Wh is predicted. The most important drawback of EDLCs is their high cost [21].

1.4.2 Battery energy storage system

Batteries are a well-established technology for storage the electricity. The power and capacity that are bound together through the electrode surface means that increasing the power level simultaneously increases the storage capacity. Many types of batteries are now mature technologies. In fact, research activities involving lead–acid batteries have been conducted for over 140 years. Notwithstanding, a tremendous effort is being carried out to turn technologies like nickel–cadmium (Ni–Cd), sodium–sulfur (NaS), and lithium ion (Li ion) batteries into cost effective options for higher power applications. The capital power costs of the lead–acid, Ni–Cd, NaS, and Li ion may vary from 50–100, 400–2400, 210–250, and 900–1300 $/kWh, respectively [22,23].

1.4.3 Superconducting magnetic energy storage

The SMES system is a relatively recent technology. Its operation is based on storing energy in a magnetic field, which is created by a DC current through a large superconducting coil at a cryogenic temperature. The energy stored is calculated as the product of the self-inductance of the coil and the square of the current flowing through it. The response time is very short. The SMES technology has been demonstrated but the price is still very high. According to [24], the power injection of a 1 MW/1 kWh SMES can be increased by 200 kW only in 20 ms and the capital power cost may vary between 1,000 and 10,000 $/kW.

1.4.4 Flywheel

In an FW the storage capacity is based on the kinetic energy of a rotating disc which depends on the square of the rotational speed. A mass rotates on two magnetic bearings in order to decrease friction at high speed, coupled with an electric machine. Energy is transferred to the FW when the machine operates as a motor (the FW accelerates), charging the energy storage device. The FW energy storage system (FESS) is discharged when the electric machine regenerates through the drive (slowing the FW). FESSs have long lifetimes, high energy density, and a large maximum output power. The energy efficiency of an FESS can be as high as 90%. Typical capacities range from 3–133 kWh.

1.4.5 Plug in electric vehicle

Recent PEVs have been increased extensively and usually include a BESS. PEVs may play an important part in balancing the energy on the grid by serving as distributed sources of stored energy, a concept called "vehicle-to-grid". By drawing on a large number of batteries plugged into the SG throughout its service region, a utility can potentially inject extra power into the grid during critical peak times, avoiding brownouts and rolling blackouts. Therefore, they can play a vital role to improve the power system reliability and the power quality of the SG. PEVs can drastically lessen the dependence on oil, and they emit nothing about air pollutants when running in all-electric modes. However, they do rely on power plants to charge their batteries, and conventional fossil-fueled power plants release pollution. To run a PEV as cleanly as possible, it needs to be charged in the hours of the morning when power demand is at its lowest and when wind power is typically at its peak. The SG technologies will help to meet this goal by interacting with the PEV to charge it at the most optimal time [25].

1.5 SMART GRID

The electrical grid was built in the 1890s and improved upon as technology advanced through each decade. This type of the traditional grid includes centralized power generation, and at distribution level one-directional power flow and weak market integration [25].

The world's electricity systems face a number of challenges, including ageing infrastructure, continued growth in demand, the integration of increasing numbers of variable renewable energy sources, electric vehicles and other distributed networks, the need to improve the security of supply, and the need to lower carbon emissions. To move forward, we need a new kind of electric grid, one that is built from the bottom up to handle the groundswell of digital and computerized equipment and technology dependent on it – and one that can automate and manage the increasing complexity and needs of electricity in the 21st century. SG technologies offer ways not just to meet these challenges but also to develop a cleaner energy supply that is more energy efficient, more affordable, and more sustainable. An SG is an electricity network that uses digital and other advanced technologies to monitor and manage the transport of electricity from all generation sources to meet the varying electricity demands of the user. SGs coordinate the needs and capabilities of all generators, grid operators, end-users, and electricity market stakeholders to operate all parts of the system as efficiently as possible, minimizing costs and environmental impacts while maximizing system reliability, resilience, and stability [26]. An SG includes centralized and distributed power generation produced substantially by renewable energy sources. It integrates distributed and active resources (ie, generation, demand, storage, and electricity vehicles) into energy markets and power systems. SGs can be characterized by a controllable multidirectional power flow. Following are the general features of a SG system [26]:

- Wide-area monitoring and control
- Information and communication technology integration
- Renewable and distributed generation integration
- Transmission enhancement
- Distribution grid management
- Advanced metering infrastructure
- Electric vehicle charging infrastructure
- Demand-side management systems

An emission-free SG system is shown in Fig. 1.5. The SG includes distributed renewable sources, PEV, energy demand, and control system. The integration of DG and flexible loads in a distribution network will benefit the network when managed appropriately. To lessen distribution losses, voltage regulation within the acceptable range, smooth power flow at the interconnection point, and stable power output to the power network are vital issues for the SG system, when DGs are integrated as power generation sources. The combination of DG and other active sources into a distribution system is needed in order to fully exploit the benefits of active resources in the network management. With proper management of active resources the overall system performance may be enhanced from presently used practices. One important control task in power systems is to maintain balance between power production and consumption which means keeping the power system's frequency at a proper level. This procedure is becoming more and more challenging due to an increase in the penetration level of intermittent renewable power production. In recent years, there have also been many serious frequency unsteadiness related to wide-area power system blackouts in Europe and the USA, and their costs, both economical and social, are high [27].

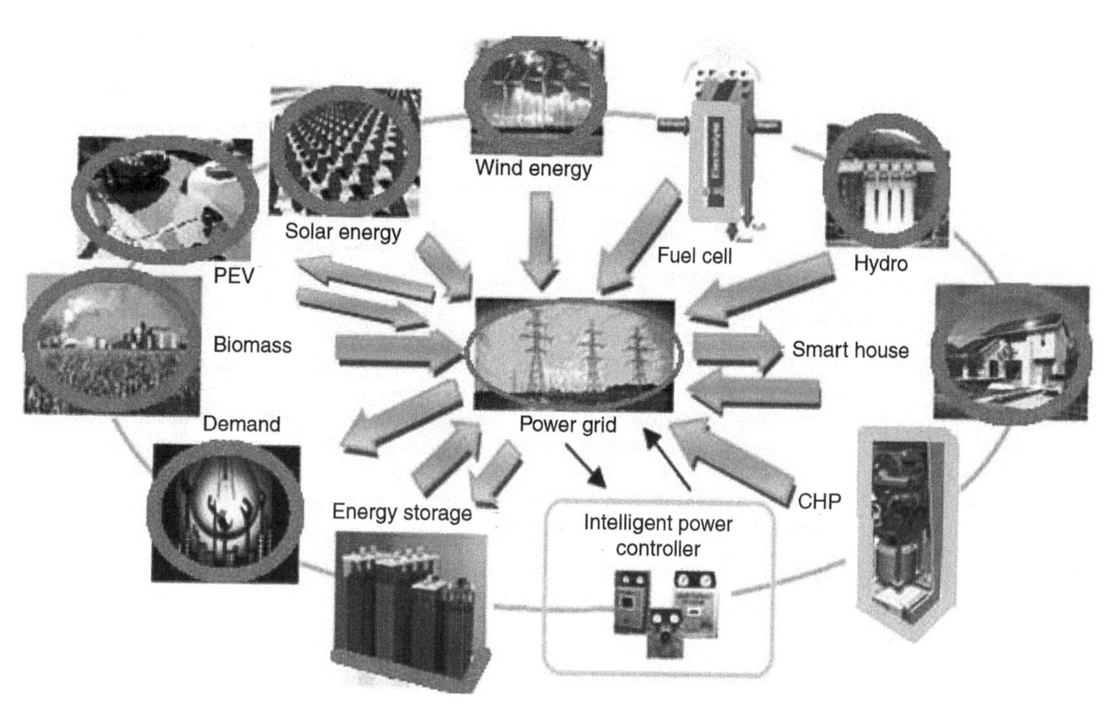

FIGURE 1.5 Schematic diagram of the zero-emission smart grid integrated with the distributed renewable energy sources, intelligent power controller, PEV, energy storage, and energy demand.

REFERENCES

[1] Howlader AM, Urasaki N, Yona A, Senjyu T, Saber AY, et al. Review of output power smoothing methods for wind energy conversion. Renew Sust Energ Rev 2013;26:135–46.

[2] Howlader AM, Senjyu T, Saber AY, et al. An integrated power smoothing control for a grid-interactive wind farm considering wake effects. IEEE Syst J 2014;9(3):1–12.

[3] Howlader AM, Urasaki N, Yona A, Senjyu T, Saber AY, et al. Control strategies for wind farm based smart grid system. In: IEEE 10th International Conference on the Power Electronics and Drive Systems. KitaKyushu, Japan; 2013.

[4] Ferrari ML, Traverso A, Pascenti M, Massardo AF, et al. Plant management tools tested with a small-scale distributed generation laboratory. Energ Convers Manage 2014;78:105–13.

[5] Howlader HOR, Matayoshi H, Senjyu T, et al. Distributed generation incorporated with the thermal generation for optimum operation of a smart grid considering forecast error. Energ Convers Manage 2014;96:303–14.

[6] El-Khattam W, Salama MMA. Distributed generation technologies, definitions, and benefits. Electr Pow Syst Res 2004;71:119–28.

[7] Barker, PP, de Mello, RW. Determining the impact of distributed generation on power systems. I. Radial distribution systems, IEEE Power Engineering Society Summer Meeting, Seattle, WA; 2000.

[8] Ackerman T, Anderson G, Soder L, et al. Distributed generation: a definition. Electr Pow Syst Res 2001;57:195–204.

[9] El-Khattam W, Salama MMA, et al. Distributed generation technologies, definitions and benefits. Electr Pow Syst Res 2004;71:119–28.

[10] IEEE Smart Grid. Integrating distributed generation into the smarter grid, http://smartgrid.ieee.org/december-2013/1013-integrating-distributed-generation-into-the-smarter-grid.
[11] Akorede MF, Hizam H, Pouresmaeil E, et al. Distributed energy resources and benefits to the environment. Renew Sust Energ Rev 2010;14(2):724–34.
[12] Urasaki N, Howlader AM, Senjyu T, Funabashi T, Saber AY, et al. High efficiency drive for micro-turbine generator based on current phase and revolving speed optimizations. Int J Emerg Elec Power Syst 2011;12(5).
[13] Howlader AM, Urasaki N, Yona A, Senjyu T, Saber AY, et al. Design and implement a digital H_∞ robust controller for a MW-class PMSG-based grid-interactive wind energy conversion system. Energies 2013;6(4):2084–109.
[14] Fthenakis V, Kim HC, et al. Land use and electricity generation: a life-cycle analysis. Renew Sust Energ Rev 2009;13(6–7):1465–76.
[15] World Wind Energy Association (WWEA). http://www.wwindea.org/home/index.php.
[16] Volk TA, Abrahamson LP, White EH, Neuhauser E, Gray E, Demeter C, Lindsey C, Jarnefeld J, Aneshansley DJ, Pellerin R, Edick S, et al. Developing a willow biomass crop enterprise for Bio-energy and Bio-products in the United States. Proceedings of Bioenergy 2000. Adam's Mark Hotel, Buffalo, New York, USA; 2000.
[17] Fridleifsson IB, et al. Geothermal energy for the benefit of the people. Renew Sust Energ Rev 2001;5:299–312.
[18] Rafik F, Gualous H, Gallay R, Crausaz A, Berthon A, et al. Frequency thermal and voltage super capacitor characterization and modelling. J Power Sources 2007;165:928–34.
[19] Sharma P, Bhatti TS, et al. A review on electrochemical double-layer capacitors. Energ Convers Manage 2010;51:2901–12.
[20] Schneuwly A., et al. Ultacapacitors, the new thinking in automotive world, http://www.maxwell.com .
[21] Kusko A, Dedad J, et al. Short-term and long-term energy storage methods for standby electric power systems. IEEE Ind Appl Mag 2007;13:66–72.
[22] Rabiee A, Khorram H, Aghaei J, et al. A review of energy storage systems in microgrids with wind turbines. Renew Sust Energ Rev 2013;18:316–26.
[23] Diaz-Gonzalez F, Sumper A, Gomis-Bellmunt O, Villafafila-Robles R, et al. A review of energy storage technologies for wind power applications. Renew Sust Energ Rev 2012;16:2154–71.
[24] Nielsen KE, Molinas M, et al. Superconducting magnetic energy storage (SMES) in power systems with renewable energy sources. In: IEEE international symposium on industrial electronics; 2010, pp. 2487–2492.
[25] What is smart grid. https://www.smartgrid.gov/the_smart_grid.
[26] International Energy Agency (IEA). Technology Road Map, http://www.iea.org/.
[27] Jarventausta P, Repo S, Rautiainen A, Partanen J, et al. Smart grid power system control in distributed generation environment. Ann Rev Control 2010;34(2):277–86.

Chapter 2

Integration of distributed energy resources in distribution power systems

Alberto Borghetti, Carlo Alberto Nucci

Department of Electrical, Electronic and Information Engineering, University of Bologna, Bologna, Italy

CHAPTER OUTLINE

2.1 Introduction 15
2.2 Interconnection issues and countermeasures 17
2.2.1 Volt–VAR control 19
2.2.2 Gossip-like VVC MAS procedure 21
2.3 Role of ICT in the integration of distributed energy resources 25
2.3.1 Models of the communication networks 27
2.3.2 Model of the power distribution feeder 32
2.3.3 Test results 33
2.4 Conclusions 46
Acknowledgment 47
References 47

2.1 INTRODUCTION

The motivation for achieving new solutions realizing appropriate interconnection of distributed energy resources (DERs) in distribution power systems can be justified by the observation of some modern distribution network configurations. Fig. 2.1 shows the geographic information system (GIS) view of the 15 kV distribution network of an Italian town (Imola) with a population of around 70,000 people with three substations each equipped with two or three 30 MVA transformers. The length of the total line is about 200 km. Each of the small circles represents one of the 700 active clients, that is, clients that own a generation setup. The total peak value of the power generation capability is 150 MW connected at a medium voltage (MV) level and 13 MW connected at a low voltage (LV) bus. Most of the generation is provided by photovoltaic (PV)

Integration of Distributed Energy Resources in Power Systems

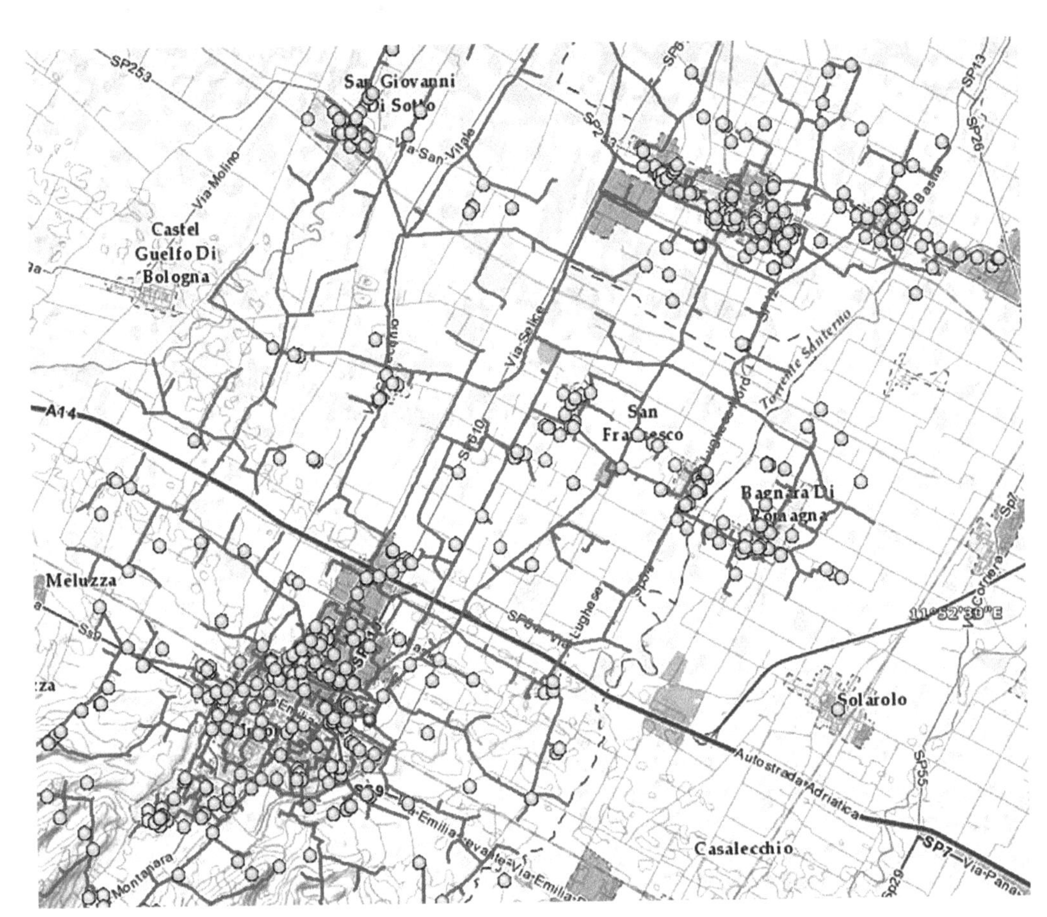

FIGURE 2.1 GIS view of the medium voltage distribution network of Imola in red. Gray (yellow in the web version) circles indicate clients with generation capability. *(Source: Data courtesy of Hera Multiutility Company.)*

units and it has been installed mainly in the last 10 years. This is a somewhat typical example of what is occurring in several places in the world.

Indeed, as described in Refs. [1,2], the worldwide increase of electricity production from renewable energy sources (RESs) is coupled with the growing installation of distributed generation (DG) and DERs, also including storage systems. A large share of the RES plants (wind parks, micro wind turbines, large and small PV, and medium or small scale hydro power plant), as well as other types of DG, such as small industrial and micro combined heat and power plants, or combined cooling, heat, and power plants, are connected to MV and LV distribution networks.

High percentage levels of power production from volatile RESs, for example PV and wind power plants, with respect to the load may cause serious problems for the operation of the high voltage (HV) transmission network, such as bottlenecks, which require appropriate congestions management, load curtailment, increasing reserve margins and balancing capacity in pool markets, steep up/down power ramps to conventional plants, etc.

Today, several MV distribution networks have an installed capacity of DG greater than the total peak demand and, more often than in the past, we find some HV/MV distribution transformers that operate with a reverse power flow for a significant number of hours per year. For the transmission system operator, in the periods characterized by high production from RES, these distribution networks behave as large generators. For distribution system operators (DSOs) the operation of these networks may represent a difficult task due the following causes:

- The point of connection of DG in an MV feeder may be at some distance from the buses of significant consumption; or
- The production from nondispatchable DG is in general not synchronized with the demand.

Without the implementation of an adequate control of DG outputs, these two causes may severely reduce the expected DG benefits, such as the reduction of peak loads, of power loss and congestions, the avoidance of network overcapacity, and the deferral of network reinforcements.

Traditionally, distribution network are usually planned in order to properly operate in all the possible combinations of load and generation levels (the fit-and-forget approach). Since peak load may occur in moments of very low DG production and vice versa, the presence of DG may even increase capital expenditures of the DSO for network reinforcement, although it reduces the utilization rate of network assets.

Also, power losses tend to increase in distribution networks with a large penetration of DG, while they can be reduced in the transmission network.

The following section of this chapter reviews the most typical connection issues and countermeasures with some references to international standard. The third section addresses the role of the increased use of information and communication technologies (ICT) in the integration of DERs, and to this aim we describe the example of a multiagent system (MAS) approach for the Volt-VAR control (VVC) tested using an ICT power cosimulation platform specifically developed for the purpose.

2.2 INTERCONNECTION ISSUES AND COUNTERMEASURES

In many countries, the legislation, for example Directive 2009/72/EC of the European Parliament, states that DSOs are obliged to enable the connection of both load and DG. DSOs may refuse access to the grid once they can prove the grid lacks the required network capacity.

In general, as described in Ref. [3] and in the references therein, the so-called hosting capacity of distribution systems is determined by the following:

- The minimum and maximum voltage limits with respect to the rated value;
- The maximum current allowed to circulate in the network branches in steady-state conditions; and
- The fault level, usually defined by the rating of existing switchgear in the vicinity of the point of connection.

Indeed DG tends to increase the bus voltages, may cause congestions in some overhead lines or cables, and could increase the maximum expected fault current value at the connection points, especially for the case of synchronous machines directly connected to the network.

Moreover, there are specific issues that need to be taken into account in the operation of distribution systems with DG. As described, for instance, in Ref. [4], DG may interfere with the operation of the automatic voltage regulator of on-load tap changers (OLTCs) installed in the HV/MV transformer. In particular, the adoption of the usual line drop compensation, which raises the reference voltage value at the MV terminal of the transformer in proportion to the load, appears unsuitable for distribution feeders with substantial presence of DG, which may experience also prolonged period of reverse power flow at the HV/MV transformer.

During the operation of the network, the presence of DERs may cause the formation of unintentional islanding situations after the operation of a protective device of the network, interference with the network protection settings, rapid voltage changes, and power quality issues, which are of particular interest as most DSOs should comply with IEEE Std. 519 or IEC 61000. Other than limits on harmonic distortion, DERs connected through inverters – mostly PV units – are required to avoid direct current (DC) injections caused, in general, by improper operation of the control system.

National and international standards define connection rules in order to limit the impact of DG on the operation of the power distribution system, taking into account the methods and equipment currently adopted in distribution network for voltage regulation and fault protection. Furthermore, as PV applications are becoming more and more important, specific codes and standards are continuously updated [5]. For instances, recent standards in Germany (VDE-AR-N 4105) and Italy (CEI 0-16 and CEI 0-21), include requirements not only for power curtailment, but also for voltage support and low-voltage ride through (LVRT) capability. Voltage support is achieved by reactive power injection strategies, while the LVRT function avoids the loss of massive power production by PV units due to deep voltage sags and should be coordinated with the anti-islanding function [6].

Since DSOs are subject to performance requirements for quality of service (QoS) defined by energy regulatory bodies, with particular reference to continuity of supply and power quality defined by specific standards (eg, European Std. EN 50160), their approach to the operation of the distribution networks changes with the increased installations of DERs. Networks with a low penetration of DG, are usually operated making use the so-called "fit and forget" approach, which aims to foresee and solve all the potential issues at the planning stage without control over the customers installations during the operation. On the other hand, when the DG penetration increases, DSOs tend to apply automatic procedures that solve the grid problems once they occur, in general by forced restriction of production or consumption curtailments. The next step is expected to be the so-called active approach of operation of distribution network, in which available ICTs are more widely used in order to better exploit the ancillary services and the real-time flexibilities which may be provided by both DERs and responsive loads.

Moreover, while phasor measurement unit (PMU) applications for transmission system operation and control could be considered mature [7], there is a growing interest to also develop PMU-based applications for distribution networks and PMUs are predicted to be more commonly installed in future distribution equipment (see eg, [8] and references therein).

In order to provide an example of the analysis of the effects of the characteristics of the communication network on the performances of an active approach in the operation of the MV distribution network, the remainder of this chapter focuses on VVC, defined as the online coordination of reactive power resources and transformers equipped with OLTCs, in order to achieve an efficient and feasible operating condition of the power feeder.

2.2.1 Volt–VAR control

The following three possible approaches may be identified in order to address the issue of coordinating the outputs of DERs with the action of available control means, such as transformers equipped with OLTCs, mechanical switched, shunt capacitors, and static VAR compensators (SVC):

- Local approach that consists of local regulators at each distributed generator and control mean that use only local measurements;

- Centralized approach that consists of active network management (ANM) functions implemented in the distribution management system (DMS) coupled with the supervisory control and data acquisition (SCADA) master hardware and software located at the control center;
- Distributed approach that consists of a networked MAS [9] composed of numerous localized controllers with the ability to communicate with each other.

Both the centralized ANM scheme, in which the SCADA system communicates with the remote terminal units, and the MAS approach are based on the exchange of information using communication networks that are now largely internet protocol (IP)-based. DSOs typically use shared communication networks that are characterized by more stringent limitations than the communication links adopted for the operation of high voltage transmission networks [10,11]. As analyzed in Ref. [12], the implementation of all ANM functions in a central DMS is expected to require significant reinforcements of the communication infrastructures currently adopted by DSOs. A reduction of the communication requirements can be obtained by using MAS approaches.

When implemented as an ANM function, VVC can be formalized as a single optimization problem (see eg, [13] and [14], which also review previous contributions on the subject). Generally, in VVC, real power outputs of DERs are assumed to be defined by the availability of the energy resources and by market conditions. However, several characteristics of VVC are in common with more general optimal power flow problems that may include also the optimization of active power production and demand side control on the basis of costs and required reserves (see, eg, [15–19] and the references therein).

A general review of the applications of MAS in power systems, not limited to distributed control purposes, has been the object of a specific IEEE Power and Energy Society Working Group [20]. Control theory aspects, specifically consensus and cooperation topics, have been recently reviewed in Ref. [21]. These approaches avoid the concentration of all the information and decisions in a specific node of the communication network connected to a centralized processor with demanding computational tasks. Specific techniques of control over communication network are needed in order to minimize the effects of finite bandwidth, transmission delays, and packet loss, as well as to limit the risk of cyber attacks. These constraints due to the use of a shared communication network are common to all networked control systems approaches, as recently reviewed in [22] and [23].

Various MAS approaches have been proposed in the literature to solve the problem of VVC. A decomposition method of the inverse of the Jacobian of the power flow problem was proposed in Ref. [24] in order to decompose VVC in smaller size optimization problems that could be implemented in a

MAS. There is also an increasing literature relevant to the application of distributed optimization procedures to VVC, with particular reference to power networks with radial structure [25–28].

In some MAS schemes, special coordinating roles are assigned to specific agents. For example, in the scheme proposed in Ref. [29] a moderator collects all the sensitivity factors from the agents and sends backward the contracts to them stating that the amount of reactive power support is needed from each controlled DG. In Ref. [30], a similar coordinating role is attributed to a top feeder relay which has the additional role to provide the coordination with the central energy management system of the bulk transmission network. In other approaches, the sequence of agent actions should follow a specific order, based, as in Ref. [31], on the location of the controlled DG in the feeder. Moreover, leaderless MAS approaches have been recently proposed in Refs. [32] and [33]. These approaches are based on gossip-like algorithms that use the measurement of bus voltage synchrophasors and on the exchange of the information between two randomly chosen neighboring agents at a time. A related strategy has been presented in Ref. [34]. In Ref. [35] an algorithm is proposed in which each agent sets a target voltage value on the basis of the voltage measurements collected from all the neighbors. The action of each agent is cyclically activated by a token ring control strategy so to ensure that the output of only one DG or of few DGs at a time are adjusted depending on the number of circulating tokens. In Ref. [36] the solution of a fuzzy-based algorithm is achieved by means of an average consensus procedure between agent state vectors. The results show that a large number of iterations and therefore a large number of exchanged messages are needed to achieve a consensus on the mean value of the bus voltages that is used by the fuzzy algorithm.

2.2.2 Gossip-like VVC MAS procedure

We focus here on a leaderless asynchronous gossip like approach given by subsequent repetition of the execution of simple rules between different couple of agents. The procedure, based on the one proposed in Ref. [32–34], is enriched by several countermeasures against communication latency and packet loss and it also incorporates additional heuristic rules that improve the coordination with OLTCs [37].

Consensus algorithms in MAS can be described as a number of rules that periodically update the column vector of the agent states $\mathbf{x}(t + \Delta t)$ at time step $t + \Delta t$ by an exchange of information between the agents relevant to their present state:

$$\mathbf{x}(t + \Delta t) = \mathbf{P}(t)\, \mathbf{x}(t) \tag{2.1}$$

Element P_{ij} of matrix **P** represents the influence of the present state of agent j on the future state of agent i at each time step. Therefore, matrix **P** incorporates the available communication links between different agents.

In the described VVC application, the state of agent i represents reactive power, Q_i, injected in a bus of the feeder by the reactive power compensator controlled by agent i (eg, a DER interfaced with the grid through a power electronic converter). As the objective of VVC is the achievement of a feasible and efficient operating condition, the updated value of the reactive power depends on the state of the electrical network represented by the vector of bus voltage phasors **V**. Therefore, the consensus mechanism may be described by

$$\begin{aligned} \mathbf{Q}(t+\Delta t) &= \mathbf{Q}(t) + \Delta\mathbf{Q}(\mathbf{P}(t)\mathbf{V}) \\ \mathbf{V}(t) &= f(\mathbf{Q}(t),\mathbf{u}(t)) \end{aligned} \tag{2.2}$$

where ΔQ_i is the adjustment function of reactive power output of the compensator associated with agent i. It is a nonlinear function of the voltage phasors that is communicated to agent i by the available communication links represented by the nonzero elements of matrix $\mathbf{P}(t)$. Nonlinear function, f, represents the nonlinear relationship between the bus voltage phasors and the reactive power output of the compensators. It incorporates the power network equations, the voltage dependence of loads and generators as well as the effects of disturbances (switching, sudden change of load and generation, etc.) indicated by vector $\mathbf{u}(t)$.

We assume that N reactive power compensators are connected to different buses of the power distribution feeder with the capability to inject a controllable value of reactive power between the minimum limit, Q_{min}, and maximum limit, Q_{max}. The reactive power injection level is adjusted by an agent of the MAS connected to a node of the communication network. Each agent is equipped with a bus voltage sensor that incorporates the PMU function and with a memory buffer where it cyclically stores both the measured phasors and the corresponding measurement times for a predefined time interval equal to t_{wait}.

The algorithm corresponds to the repeated execution of the following steps (indicated as a compensation cycle).

Measurement, information exchange, and calculation of ΔQ:

1. An agent, which we denote as agent h, is assumed to be activated by another agent that also provides its updated priority index p_h as explained in the last steps.
2. Agent h randomly chooses a neighboring agent in the communication network, identified as agent k (by avoiding the agent that has activated him, if there is another one available).

3. Agent h sends to agent k the most updated values (present in the memory buffer) of both $|V_h|$ and θ_h of the positive sequence voltage phasor of the bus to which its compensator is connected and the indication of corresponding measurement time t_{meas}. Moreover it sends the identifier of bus h, the value p_h equal to its priority index, and the values of the margins between the current reactive output Q_h of its compensator and the relevant minimum and maximum limits, that is, $\delta_h^{max} = Q_{h,max} - Q_h$ and $\delta_h^{min} = Q_{h,min} - Q_h$.
4. When agent k receives the information from agent h, it accepts the assignment only if p_h is not lower than its priority index, otherwise it denies the assignment by sending the relevant message to agent h that concludes the compensation process with priority p_h.
5. If agent k accepts the assignment, if necessary, it updates its priority index and singles out the bus voltage phasor V_k measured at t_{meas} stored in his memory buffer. It calculates the values Q'_{hk} and Q''_{hk} that represent the reactive power transfer from the nodes of the networks in which agent h and agent k are connected

$$Q'_{hk} = \frac{X_{eff,hk}|V_h|^2}{|Z_{eff,hk}|^2} - \frac{X_{eff,hk}|V_h||V_k|}{|Z_{eff,hk}|^2}\cos(\theta_h - \theta_k) - \frac{R_{eff,hk}|V_h||V_k|}{|Z_{eff,hk}|^2}\sin(\theta_h - \theta_k) \tag{2.3}$$

$$Q''_{hk} = -\frac{X_{eff,hk}|V_k|^2}{|Z_{eff,hk}|^2} + \frac{X_{eff,hk}|V_h||V_k|}{|Z_{eff,hk}|^2}\cos(\theta_k - \theta_h) + \frac{R_{eff,hk}|V_h||V_k|}{|Z_{eff,hk}|^2}\sin(\theta_k - \theta_h) \tag{2.4}$$

where $|V_h|$, $|V_k|d$ are the root mean square (RMS) values and θ_h, θ_k are the phases of the positive-sequence voltage synchrophasors $\mathbf{V}_h$ and $\mathbf{V}_k$, respectively. Each agent is assumed to know the values of the effective impedances $Z_{eff,hk} = R_{eff,hk} + jX_{eff,hk}$ between the bus where its compensator is connected and the buses where the compensators of the neighboring agents are connected. In Ref. [32–34] it is shown that a sequence of repeated compensations of the mean value of (2.3) and (2.4) is globally convergent to the minimum network loss operating condition under some simplifying assumptions.
Moreover, analogous to agent h, it calculates reactive power margins δ_k^{max} and δ_k^{min} relevant to its compensator connected to bus k.

6. In order to define adjustment ΔQ of the compensator set point, agent k compares the value $\overline{Q_{hk}} = \min(|Q'_{hk}|,|Q''_{hk}|) \cdot \mathrm{sgn}(Q'_{hk})$ with the maximum allowed variations of the reactive output of both compensators, that is, $\delta_k^{\max}$, $\delta_k^{\min}$ and $\delta_h^{\max}$, $\delta_h^{\min}$:

$$\Delta Q = \max\left(\min\left(\overline{Q_{hk}}, \delta_k^{\max}, -\delta_h^{\min}\right), \delta_k^{\min}, -\delta_h^{\max}\right) \tag{2.5}$$

If Q'_{hk} and Q''_{hk} have different signs, then ΔQ is set equal to 0.

Implementation of ΔQ:

7. Agent k changes the output of its compensator by adding ΔQ only if at least one of the following two conditions is met:

$$\begin{aligned} &\Delta Q > 0 \text{ and } |V_k| < V_{\max} \\ &\Delta Q < 0 \text{ and } |V_k| > V_{\min} \end{aligned} \tag{2.6}$$

where $V_{\max}$ and $V_{\min}$ are two values a few percent higher and lower than bus voltage rated value, respectively, so to define the voltage interval of the normal operating state. In the simulations, $V_{\max}$ and $V_{\min}$ are chosen equal to 1.03 and 0.97, respectively. If none of (2.6) is met, the reactive power output is not changed.

8. Agent k sends back value ΔQ to agent h.
9. If agent h does not receive the message from agent k before delay t_{wait} after t_{meas}, it randomly selects another agent k (step 2). Priority index p_h remains unchanged.
10. If agent h receives the message from agent k, it changes the reactive output reference of its compensator by subtracting ΔQ only if at least one of the following two conditions is met:

$$\begin{aligned} &\Delta Q < 0 \text{ and } |V_h| < V_{\max} \\ &\Delta Q > 0 \text{ and } |V_h| > V_{\min} \end{aligned} \tag{2.7}$$

Selection of the new couple of agents

11. Agent h randomly choses another agent to be activated as new agent h.
12. When the chosen agent receives the relevant message from agent h with the priority p_h, it checks whether p_h is greater or lower than its priority index. If it is equal or greater, the receiving agent becomes the new agent h. If necessary, it updates its priority index to p_h and sends the relevant acknowledgment message back to the old agent h, which then returns in the idle state. If p_h is lower than the priority index of the receiving agent, it denies the assignment by sending the relevant message to agent h that concludes the compensation process with priority p_h.

13. The new agent h starts again the procedure from step 1, waiting at least t_{wait} after t_{meas} so to allow the stabilization of both compensators in the new operating conditions.
14. If the old agent h does not receive the acknowledgment message from the new one by t_{wait} after t_{meas}, it increments its priority index p_h and randomly selects another agent to be activated (step 11).

Conditions (2.6) and (2.7) exploit the fact that the connection to the transmission network through the substation transformer guarantees the reactive power balance in the feeder.

In order to guarantee the persistence of the procedure also in the presence of packet losses, the algorithm includes the possibility of concurrent multiple var compensation processes. In order to limit the possible negative effects of unwanted and unsynchronized multiple compensation process, the association of an increasing priority index (p_h) to each process is established. Concurrent compensation processes with priority indexes lower than the others are progressively stopped by the controllers. As a countermeasure against the complete failure of a critical communication links, a spontaneous activation is allowed after a predefined long time (eg, several minutes) in which a controller is never activated.

The countermeasure against communication latency is based on the availability of a memory buffer at each agent. The memory buffer stores the PMU–provided phasor data with the relevant time tag. This memory allows each couple of agents to estimate the reactive power flow by using synchronous values of voltage phasors. Another countermeasure against excessive communication latency is provided by the definition of a maximum delay t_{wait} after which the procedure carries on with the choice of a new active agent.

2.3 ROLE OF ICT IN THE INTEGRATION OF DISTRIBUTED ENERGY RESOURCES

In this section, an ICT power system cosimulation platform is used to assess the performances of the algorithm described in section 2.2.2 with a focus on the limitations due to the communication network, reproduced by means of a realistic model.

Cosimulation environments that integrate a simulator of the communication network with a power system simulator are very useful for the design and analysis of improved monitoring, control, and protection techniques in modern electric power systems, in particular when these functions rely on the exchange of information using a shared communication network.

In the literature, several approaches have been presented in order to develop ICT power system cosimulation platforms, as recently reviewed in Ref. [10,38]. The latter paper also presents an event-driven cosimulation environment implemented in Matlab/Simulink. Moreover, a review of cyber-physical system approaches in design and operation of power grids is available in Ref. [39].

One of the first platforms is the EPOCHS framework [40] that federates three off-the-shelf simulators: PSCAD/EMTDC for power system transients, Positive Sequence Load Flow for power system modeling, and Network Simulator 2 (ns-2) for communication network modeling. The same types of simulators are also included in the Global Event-driven CO-simulation (GECO) platform presented in Ref. [41] that uses a global event-driven mechanism in order to improve the synchronization. In Ref. [42] a cosimulation platform that integrates ns-2 with the utility power distribution system simulator OpenDss is used to analyze a compensation scheme of PV arrays outputs by means of distributed storage units controlled through a wireless communication network. An OpenDss/ns-2 integrated tool is used also in Ref. [43] to evaluate the impact of WiMAX communication system characteristics (with particular reference to rain fade) on DMS advanced functions. In Ref. [44] various cosimulation architectures are described and applied to the analysis of a DC power distribution system in a ship-board application. One of these architectures, based on the link between Opnet Modeler and the dynamic model of power electronic devices developed by using the Virtual Test Bed software environment, is described in detail in Ref. [45]. In Ref. [46] a hybrid simulation design based on high level architecture, IEC 61850, Object Linking and Embedding for Process Control and the Common Information Model is proposed with a focus on the evaluation of the real-time performance of wide-area monitoring, protection and control applications. In Ref. [47] a co-simulation tool based on the interface between the eMEGAsim real-time digital simulator and Opnet Modeler is proposed for the development of PMUs applications. In Ref. [48] a cosimulation environment built using OMNeT + + and OpenDSS is presented and in Ref. [49] this was used to test an integrated vehicle-to-grid, grid-to-vehicle, and RESs coordination algorithm. Examples of more general cosimulation tools for cyber-physical systems applied to power networks are the ADEVS/ns-2 integrated tool presented in Ref. [50] and the Modelica/ns-2 integrated tool presented in Refs. [51,52].

The developed platform is based on the interface between the communication simulator Riverbed Modeler (previously known as Opnet) and the electromagnetic transient program EMTP-rv, as described in Refs. [53,54]. As shown in Fig. 2.2, both the Riverbed Modeler and the EMTP-rv communicate with the outside environment through dynamic link libraries (DLLs) specifically

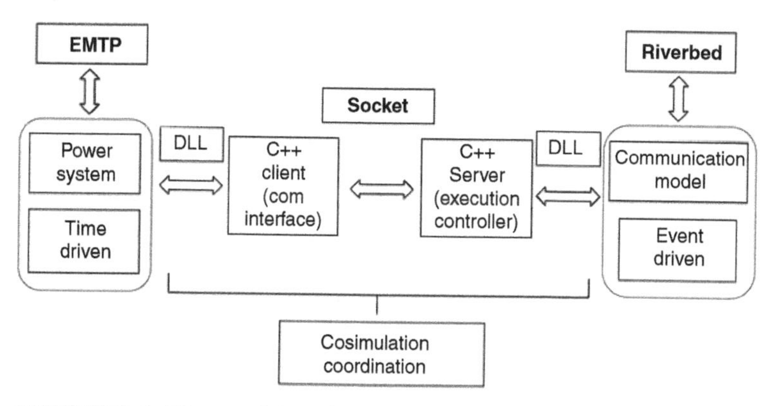

FIGURE 2.2 Architecture of the cosimulation platform.

developed for this cosimulation platform. The DLLs communicate with each other through socket application programming interfaces (APIs). The socket API allows the developed DLL to control and use the network sockets that are the endpoints of the interprocess communication flow.

In the socket communication, the Riverbed Modeler controller works as a server (execution controller), while the EMTP-rv controller acts as a client. At the simulation start-up, Riverbed enables the communication in the execution controller, opens a socket channel, sets the parameters, and starts to listening/waiting for a possible connection from the external environment. The cosimulation begins when the EMTP-rv sends the connection request as a client to the specific port and IP address provided by the server.

The synchronization mechanism between the two simulators is based on the typical waiting order of a communication through sockets. Simulation interval Δt is defined by the integration time-step adopted in EMTP-rv to solve the system of differential algebraic equations (for this paper $\Delta t = 1$ ms). Time step Δt is communicated to Riverbed Modeler that, in turn, executes the simulation until the subsequent sampling time $t + \Delta t$. As Δt is very small with respect to the analyzed transients, it is negligible the inaccuracy due the time shift of all Riverbed events that happen within a Δt interval to the end of the same interval.

2.3.1 Models of the communication networks

A client–server communication model has been implemented for the interface between each agent that regulates a reactive power compensator in the EMTP-rv model and the relevant node of the communication network in the Riverbed model.

As described in Refs. [55,56], the interface through DLLs is defined by five main components: external system (Esys) module and the corresponding process model, external system definition/domain (ESD) model, simulation description (SD) file; Esys API package, and the external simulation access (ESA) API package. The Riverbed node models are extended by an Esys module that enables the management and the delivery of the communication packets between the agents. ESD model is an attribute of the Esys module that defines an Esys interface for each agent. The Esys module uses the information contained in the SD file for the link to the DLL that includes the specific C/C++ functions defined by the Esys API Package for the initialization and the flow control of each interface. The main header of the Esys API package is the ESA API package that contains the initialization of the sockets for the communication with EMTP-rv.

A message generated by the EMTP-rv model of the agents, implemented by using a specific DLL, is first transferred to the socket communication and then to the relevant Esys interface of the client in the Riverbed agent model. The message is built into the datagram and sent to the server of the destination agent through the communication network. The destination server processes the received datagram, extrapolates the information from the payload, and returns the message to the associated EMTP-rv interface.

The simulator implements both the model of a wired communication network and the model of a cellular network.

2.3.1.1 *Wired communication network model*

As shown in Fig. 2.3, both transmission control protocol (TCP) and user datagram protocol (UDP) are represented using Riverbed models. The TCP model establishes a connection–oriented point-to-point communication link and includes connection set-up, data exchange, acknowledgment, retransmission, and connection termination functionalities. In the UDP model, the communication is connectionless, that is, a message is sent from one endpoint to another without prior arrangement or control.

In our application, the dimension of TCP packets is 408 bits for data exchange and 376 bits for set-up, acknowledgment and connection termination. The dimension of UDP packets is 312 bits.

In the Riverbed model, each agent node is connected to a router and to a background data traffic generator. The routers are connected to each other by a communication network with 64 kilobits per second (kbps) serial twisted-pair links and characterized by topology that follows the same tree configuration of the power distribution feeder. Each agent node is connected to the own router via a 10BaseT Ethernet link.

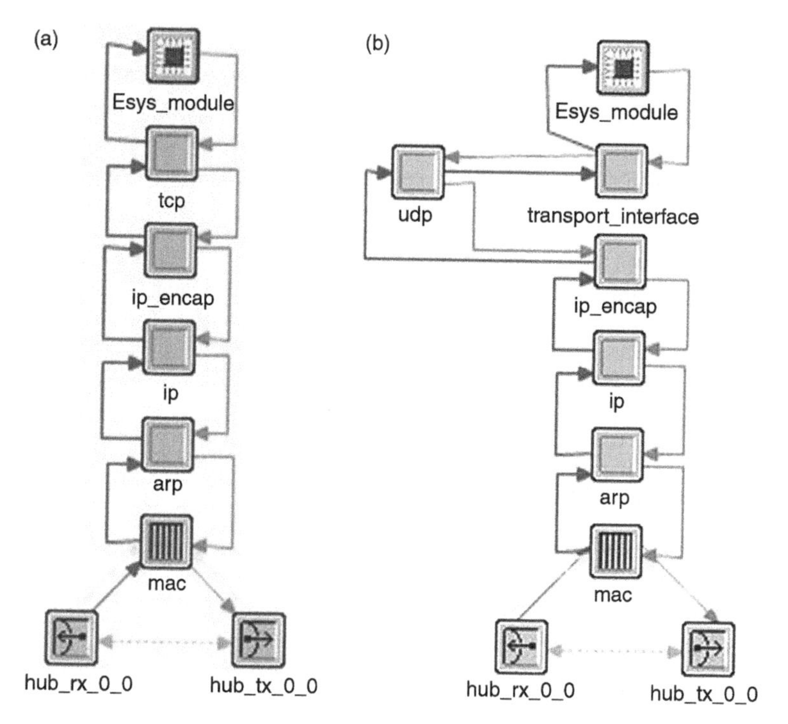

■ **FIGURE 2.3** (a) TCP and (b) UDP node models. hub_rx and hub_tx, physical layer; MAC (media access control) and arp (address resolution protocol), link layer; IP and ip_encap (which encapsulates packets into IP datagrams), internet layer; tcp or udp and transport_interface, transport layer.

The background traffic (BT) in the communication links is represented by an IP layer traffic flow from each node towards the router located at the substation. Each communication link is also characterized by a packet discard ratio (PDR), representing the probability of a packet to be lost in the link.

2.3.1.2 Cellular communication network model

As described in [57], the simulator includes the model of a third generation mobile cellular network, namely a Universal Mobile Telecommunications System (UMTS). The Riverbed model represents the three basic components of the UMTS network: the user equipment (UE), that is, the UMTS module of each agent, the UMTS terrestrial radio access network (UTRAN), and the Core Network (CN). UTRAN includes the Node B and the radio network controller (RNC), which manages the Node B logical resources and also the UE-Node B interface resources. Each Node B controls the radio transmission and reception of a cell and performs the packet relay between UE and the corresponding RNC. CN contains the serving GPRS support node (SGSN) and the gateway GPRS support node (GGSN). SGSN maintains access controls, security functions and also keeps track of UE locations. The GGSN encapsulates the packets and routes them to the SGSN that are received from the external network or internet.

Fig. 2.4 shows the Riverbed model of the UMTS UE. The GMM module (GPRS mobility/session management) manages GPRS attachment, PDP context establishment, service requests, and radio access bearer activation and it handles the interface with the IP stack. The radio link control (RLC)_media access control (MAC) module performs RLC and MAC functionalities, both in UEs and router control planes. As described previously, the node model is extended by an Esys module that enables the Riverbed interface with the EMTP simulation environment.

The air interface between each UE and the Node B is based on the wideband code division multiple access access scheme and uses a direct spread with a chip rate of 3.84 MHz and nominal bandwidth of 5 MHz. The model supports frequency division duplexing duplex mode. The radio frame has a length of 10 ms and it is divided into 15 slots. Spreading factors vary from 256 to 4 for uplink and from 512 to 4 for downlink, which allow data rates of up to 2 Mbps. The model accounts for the block error rate (BLER) that is the percentage of transport blocks with errors over the total number of transport blocks. The communication channels between each Node B and RNC (and CN) are assumed wired with large data rate.

The Riverbed model supports the UMTS four main types of QoS: background, interactive, streaming and conversational. These attributes include the traffic class, maximum and guaranteed bit rates, delivery order, transfer delay, maximum size of the service data unit (SDU) and SDU error ratio. The UE RLC interface could operate in either unacknowledged mode or acknowledged mode (AM). Retransmission decreases the effects of the BLER but increases the communication delay.

In the simulations, the gossip-like procedure uses the interactive QoS class communication that has higher priority than background QoS although, as background, it does not guarantee a bit rate. We have chosen to operate in AM. The uplink and downlink dedicated signaling channels scheduling is adopted with the maximum bit rate of 64 kbps (as a default value for the interactive QoS traffic class).

In order to test the robustness of the gossip-like procedure with respect to delays generated in the communication network, some simulations includes

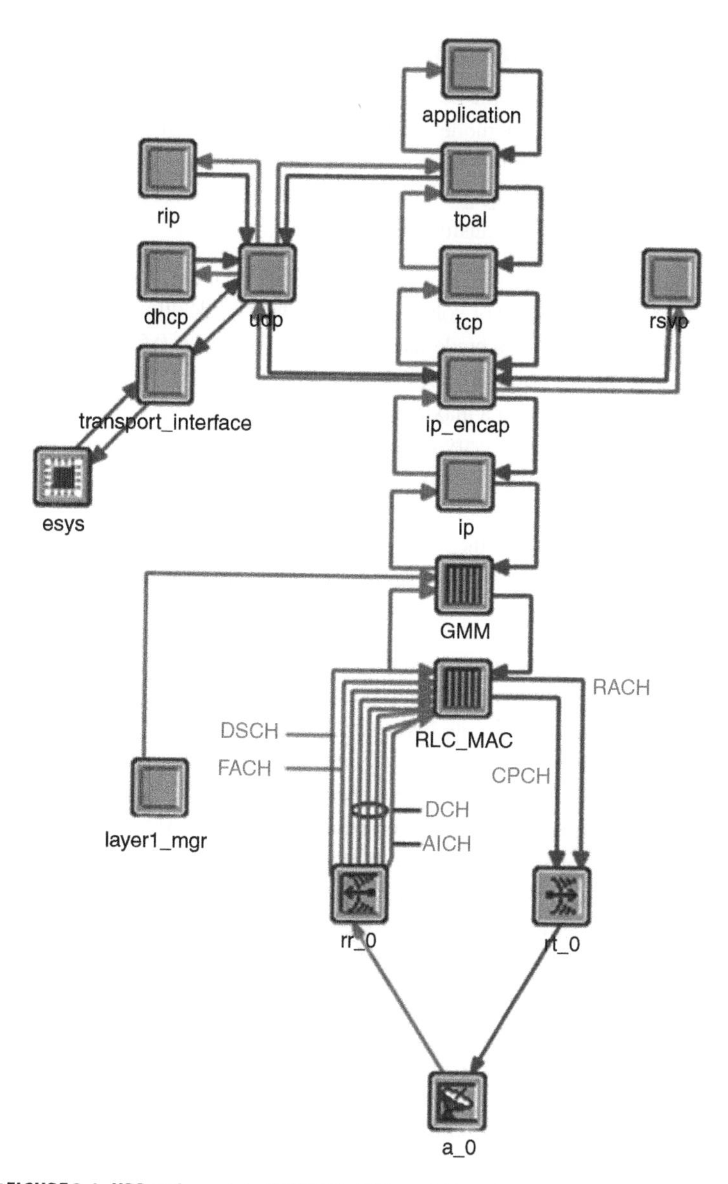

■ **FIGURE 2.4 UDP node model of UMTS UE.** AICH, acquisition indicator channel; CPCH, common packet channel; DCH, dedicated channel; DSCH, downlink shared channel; RACH/FACH, random access channel/forward access channel.

BT, due to the presence of additional UEs other than those associated with the agents. These new UEs and also the agents, generate the BT by using some default mobile user traffic profiles defined by Riverbed Modeler according to the 3rd Generation Partnership Project (3GPP) technical report TR 36.822. Also the mobile user traffic has been represented by using the interactive QoS.

2.3.2 Model of the power distribution feeder

The EMTP-rv model of the network is mainly composed by the three-phase constant-parameters PI models for the representation of the unbalanced lines, a three-phase transformer model at the substation fed by a positive sequence constant voltage generator, the models of reactive power compensator loads, and OLTC transformers.

The OLTC model is adapted from Ref. [58]. The OLTC regulator changes the tap when the RMS value of voltage at the secondary side differs from the reference value more than a predefined dead band for at least 0.5 s. The first tap change of each control action is postponed by a fixed delay, while subsequent changes are applied after a maximum delay time, fixed or with an inverse law. In order to avoid unnecessary operations and wear of the OLTCs of a series of cascaded transformers, the upstream transformer sends a message to downstream transformers in order to delay their actions if those actions are of the same type of the one that the upstream transformer is applying [59]. Once the upstream transformer terminates its action, after t_{wait} it sends another message to downstream transformers in order to release their actions.

The compensators are represented by components able to inject assigned and adjustable three-phase active and reactive powers. A first approximation of the quasi steady-state behavior of both synchronous generators and power electronic interfaced sources connected to an unbalanced network is provided by a three positive-sequence current sources in parallel with a 3×3 Y matrix, as described in, for example, Refs. [60,61]. The DG model implemented in EMTP-rv is composed by two positive-sequence triplets of current generators. The amplitude of one triplet is controlled by a feedback regulator in order to inject the requested value of three-phase active power, while the phase angle between current and bus voltage is regulated so to achieve a zero value of reactive power. The regulators of the second triplets have a reverse function, that is, amplitude is controlled in order to inject the requested reactive power and phase angle is controlled in order to cancel out active injection. Reference value, Q_i, of the reactive power injection of compensator, i, is dynamically changed by the associated agent taking into account the $Q_{\max}$ and $Q_{\min}$ limits. For the case of inverters, the values of $Q_{\max,i}$ and $Q_{\min,i}$ are dynamically updated on the basis of the active power value P_i, so that maximum inverter current $I_{\max,i}$ is met:

$$Q_{\max,i}, Q_{\min,i} = \pm\sqrt{\left(|V_i| I_{\max,i}\right)^2 - P_i^2} \tag{2.8}$$

A smooth transition between different power levels in a short time window of few hundreds of milliseconds is represented.

Moreover, each compensator i is equipped with a fast local regulator that adjusts Q_i, which is defined by the corresponding agent that participate to the gossip-like procedure, by a quantity $\Delta Q_{\text{local},i}$ if $|V_i|$ significantly differs from the rated value. Q_i is decreased if $|V_i| > V_{\max}$ and it is increased if $|V_i| < V_{\min}$:

$$
\begin{aligned}
\Delta Q_{\text{local},i} &= -\left(|V_i| - V'_{\max}\right)\frac{(Q_i - Q_{\min})}{\Delta V_{\max}} \quad \text{if } |V_i| > V'_{\max} \\
\Delta Q_{\text{local},i} &= \left(V'_{\min} - |V_i|\right)\frac{(Q_{\max} - Q_i)}{\Delta V_{\min}} \quad \text{if } |V_i| < V'_{\min}
\end{aligned}
\tag{2.9}
$$

where $V'_{\max}$ and $V'_{\min}$ are two values slightly lower and higher than $V_{\max}$ and $V_{\min}$, respectively, used in (2.6) and (2.7), while $\Delta V_{\max}$ and $\Delta V_{\min}$ indicate the voltage deviations with respect to $V_{\max}$ and $V_{\min}$, respectively, that cause a complete utilization of the available reactive power margins.

As EMTP-rv converts all the load models in RLC branches in time domain simulations, constant PQ and constant current three-phase unbalanced load models are represented by adopting the same two-triplets current generators structure used for the compensators, with the difference that a per-phase control of active and reactive power (negative) injections has been implemented.

The agent of each compensator also includes the model of a PMU that provides 10 estimates per second. The accuracy of PMUs is represented by the Normal distribution of the measurement errors of the PMU prototype described in Ref. [62], which addresses the issue of small phase shifts between different buses in MV distribution networks due to short line lengths and reduced power flows. In the simulations, the corresponding mean and standard deviation values are assumed equal to 10 μrad and 8.1 μrad for phase error and 120×10^{-6} pu and 9.3×10^{-6} pu for RMS error, respectively. The accuracy associated with a capacitive voltage divider is also included by means of a normal distribution with mean and standard deviation equal to −0.6 mrad and 7 μrad for the phase error and 2×10^{-3} and 58×10^{-6} pu for RMS error, respectively.

2.3.3 **Test results**

Numerical tests have been carried out for the two following test feeders (TFs) adapted from Ref. [63] with six additional three-phase reactive power compensators, indicated by Q1 – Q6:

TF1:	IEEE 37 node test feeder, with six reactive power compensators connected to buses 702, 712, 706, 703, 708, and 711, respectively, and with an OLTC transformer at the substation (the secondary side node is 701 indicated with a blue dot). The scheme with the wired communication network is shown in Fig. 2.5a) and the scheme with the UMTS communication network is shown in Fig. 2.5b), in which two Node Bs are needed to guarantee the coverage (the corresponding two cells include the UEs at nodes 701, 702, 712, and 706,703,708,711, respectively).
TF2:	IEEE 123 node test feeder, with six reactive power compensators connected to buses 13, 28, 47, 67, 87, and 108, respectively (Fig. 2.6) and with two OLTC transformers (with the secondary side connected to nodes 149 and 67, respectively).

The characteristics of unbalanced lines and loads have been defined as in Ref. [63]. In order to speed up the simulations, the constant power and constant current load models described in Section 2.3.2 have been applied only to the loads larger than 100 kW in TF1 and to the loads larger or equal than 40 kW in TF2, while the other loads are represented as constant impedances.

Substation transformers are equipped with OLTC with ±8 tap increments of 1.875%. The tap mechanical delay is 2 s, the time to first tap change is 20 s and the maximum delay time of the subsequent tap changes is 15 s with an inverse time law. An agent is associated to the bus at the secondary side of the substation transformer. It participates to the regulation cycles of the VVC procedure but it does not directly adjust the output of any compensator.

In TF2, the voltage regulator located between bus 160 and 67 has an OLTC with ±16 tap increments of 0.625%. The time to first change and the maximum time of subsequent changes inverse-law delay of are both 10 s and the mechanical delay is 2 s. The reactive power compensator connected to bus 67 participates to the MAS procedure only when the OLTC between bus 160 and 67 is not operating. When the OLTC is in operation, the agent set is divided into two independent groups: one relevant to the agents associated to nodes {149,13,28,47} and the other including the agents associated to nodes {67,87,108}. At each compensation cycle, both agent h and agent k must belong to the same group.

2.3.3.1 ***Results obtained with the wired communication network***

As shown by Fig. 2.5a and Fig. 2.6a, the wired communication network has eight nodes with a tree topology that follows the same configuration of the

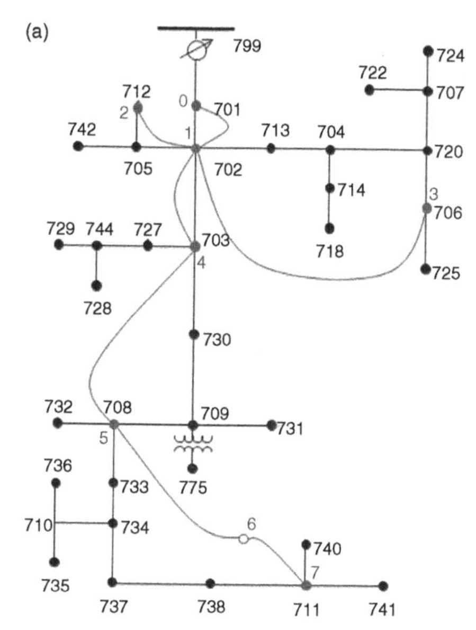

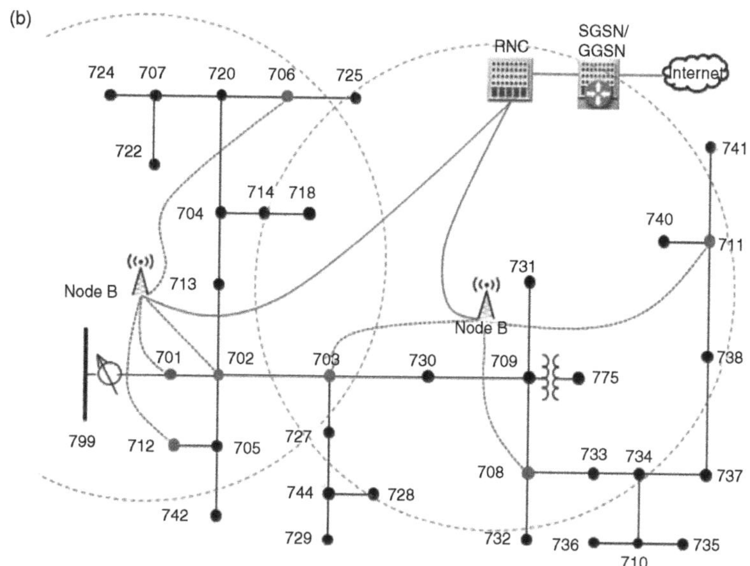

■ **FIGURE 2.5 TF1.** Power feeder in black and communication network in light gray (red in the web version): (a) wired communication, (b) cellular communication [solid light gray (red in the web version) lines represent wired channels, dotted light gray (red in the web version) lines represent UMTS channels]. Gray (red in the web version) dots indicate the agents associated to compensators while blue dots indicate an agent that does not directly adjust the output of any compensator. In subpart (b) the dark gray (green in the web version) circles indicate the estimated coverage areas of the Node B antennas.

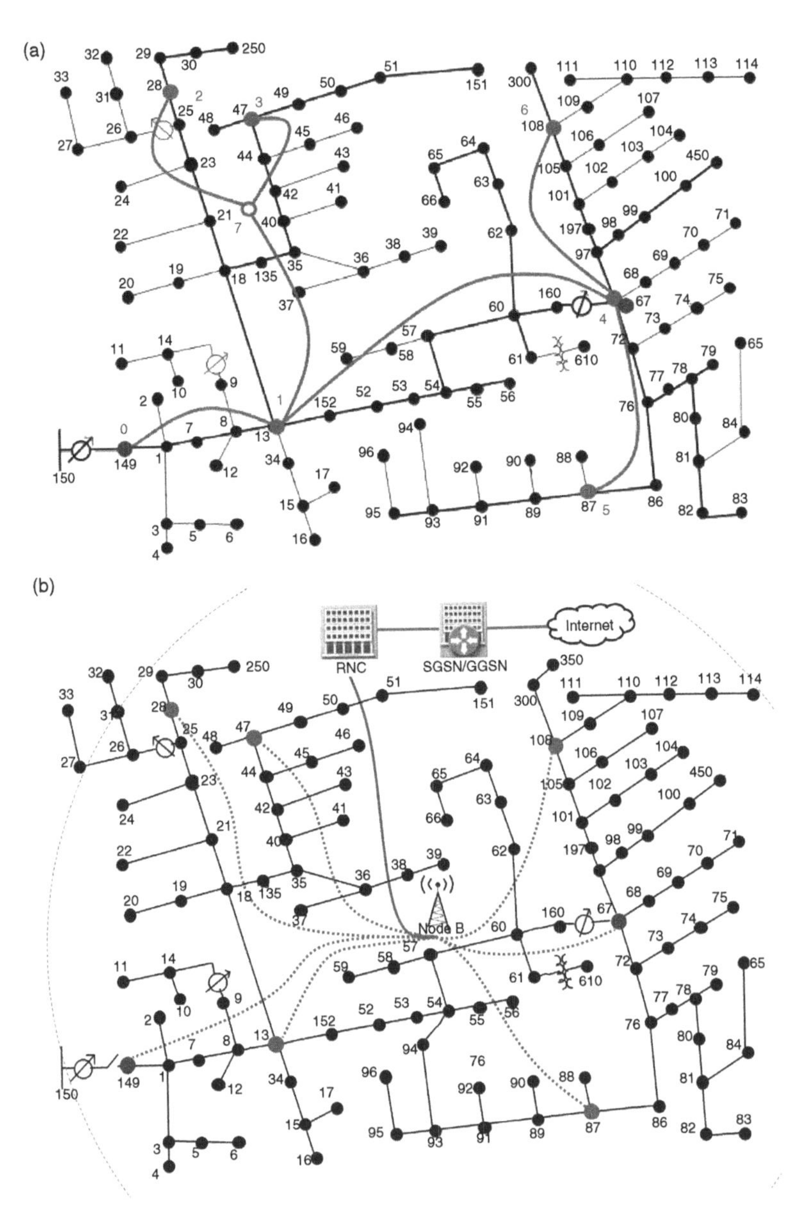

FIGURE 2.6 TF2. Power feeder in black and communication network in gray (red in the web version): (a) wired communication, (b) cellular communication [solid gray (red in the web version) lines represent wired channels, dotted gray (red in the web version) lines represent UMTS channels].

power feeder: node 0 is located at the feeder substation for both TF1 and TF2. For the case of TF1, nodes 1, 2, 3, 4, 5, and 7 connect the agents associated with the reactive power compensators and node 6 is located in the path between node 5 and node 7. For the case of TF2, nodes 1, 2, 3, 4, 5, and 6 connect the agents associated with the reactive power compensators and node 7 is located in the path between nodes 1, 2, and 3.

In the simulation relevant to the wired communication network, three different BT levels are analyzed, identified as BT0, BT1, and BT2, which correspond to 0, 4.75, and 9.5 kbps, respectively, that each BT generator sends towards node 0. Each BT level is analyzed both without PDR (case identified as PDR0) and by assuming a 5% PDR (case identified as PDR5) for each communication link. For BT levels 0 and 1 we assume t_{wait} =1 s. For BT level 2, we have compared the results obtained for two t_{wait} values, namely a) t_{wait} =1 s (BT2a) and b) t_{wait} =10 s (BT2b). For the simulations with t_{wait} = 1 s, the final time is t_f = 80 s, while when t_{wait} = 10 s, t_f is extended to 100 s. In order to compare the promptness of the procedure in the various scenarios characterized by different BT levels, we define settling time, t_{set}, as the time to enter and remain within a 500 W band for at least 30 s.

As a base case, limits for Q_{max} and Q_{min} are set equal to 500 and −500 var, respectively, for all the compensators. The active power injected by the compensators is null and initially the reactive power output is also null

For the case of TF1, Tables 2.1 and 2.2 compare the results obtained for the seven scenarios characterized by different values of BT and PDR by using TCP and UDP. Tables 2.3 and 2.4 compare the results for the same scenarios

Table 2.1 Mean and Standard Deviation Values of the Number of Compensation Cycles, Percentage of Incomplete Compensations, Power Loss Decrease, and Settling Time for TF1

BT PDR Level	No. of Compensation Cycles, Mean (Stdev)		% of Incomplete Compensations, Mean (Stdev)		Power Loss Decrease (kW), Mean (Stdev)		Settling Time (s), Mean (Stdev)	
	TCP	UDP	TCP	UDP	TCP	UDP	TCP	UDP
BT0 PDR0	70.5 (0.5)	75.8 (0.4)	0 (0)	0 (0)	13.4 (0.1)	13.5 (0.1)	26.0 (13.0)	28.3 (10.9)
BT1 PDR0	68.2 (0.5)	74.5 (0.5)	0 (0)	0 (0)	13.4 (0.2)	13.5 (0.1)	26.0 (10.6)	23.1 (13.5)
BT2a PDR0	59.2 (1.9)	63.4 (1.9)	0 (0)	0 (0)	8.2 (3.1)	7.9 (3.0)	32.8 (19.3)	31.6 (23.5)
BT2b PDR0	9.5 (0.6)	10.0 (0.1)	0 (0)	0 (0)	10.6 (2.5)	11.5 (1.4)	–	–
BT0 PDR5	30.6 (11.8)	67.9 (5.3)	32.3 (11.5)	6.1 (2.7)	12.1 (1.9)	13.0 (2.1)	33.5 (21.9)	28.1 (12.9)
BT2a PDR5	35.4 (5.1)	58.3 (3.8)	29.0 (10.3)	7.3 (3.8)	10.9 (3.3)	10.7 (3.1)	33.3 (15.3)	33.8 (19.3)
BT2b PDR5	7.4 (1.3)	9.2 (1.6)	11.8 (14.8)	5.0 (6.4)	9.8 (2.4)	10.7 (2.8)	–	–

Table 2.2 Mean and Standard Deviation Values of the Number of Packets, Percentage of Ignored and Lost Packets, Packet Delay and Number of Stopped Processes for TF1

BT (PDR) Level	No. of Packets, Mean (Stdev)		% of Packets Ignored or Lost, Mean (Stdev)		% of Packet Lost, Mean (Stdev)	Packet Delay (ms), Mean (Stdev)		No. of Stopped Processes, Mean (Stdev)	
	TCP	UDP	TCP	UDP	UDP	TCP	UDP	TCP	UDP
BT0 PDR0	282.0 (1.9)	303.2 (1.5)	0 (0)	0 (0)	0 (0)	49 (14)	11 (3)	0 (0)	0 (0)
BT1 PDR0	272.7 (1.7)	297.9 (1.9)	0 (0)	0 (0)	0 (0)	66 (33)	18 (16)	0 (0)	0 (0)
BT2a PDR0	245.9 (4.9)	264.4 (5.6)	4.2 (1.2)	4.2 (0.8)	0 (0)	76 (39)	22 (21)	0 (0)	0 (0)
BT2b PDR0	37.8 (2.5)	39.9 (0.5)	0 (0)	0 (0)	0 (0)	462 (918)	194 (523)	0 (0)	0 (0)
BT0 PDR5	188.6 (15.1)	276.1 (18.8)	47.2 (18.6)	5.7 (1.5)	5.7 (1.5)	52 (36)	12 (3)	2.6 (1.5)	1.8 (1.3)
BT2a PDR5	186.8 (7.3)	247.9 (10.6)	34.4 (6.6)	9.8 (1.5)	6.3 (1.4)	93 (88)	24 (33)	4.3 (1.8)	1.7 (1.3)
BT2b PDR5	30.7 (4.2)	37.7 (6.1)	11.2 (11.0)	4.5 (4.1)	4.5 (4.1)	890 (1509)	241 (561)	0.3 (0.6)	0.3 (0.6)

Table 2.3 Mean and Standard Deviation Values of the Number of Compensation Cycles, Percentage of Incomplete Compensations, Power Loss Decrease, and Settling Time for TF2

BT (PDR) Level	No of Compensations Cycles, Mean (Stdev)		% No. of Incomplete Compensations, Mean (Stdev)		Power Loss Decrease (kW), Mean (Stdev)		Settling Time (s), Mean (Stdev)	
	TCP	UDP	TCP	UDP	TCP	UDP	TCP	UDP
BT0 PDR0	70.7 (0.7)	77 (0.0)	0 (0)	0 (0)	11.4 (0.2)	11.4 (0.1)	24.7 (13.5)	21.0 (11.6)
BT1 PDR0	68.1 (0.8)	75.9 (0.4)	0 (0)	0 (0)	11.4 (0.2)	11.4 (0.2)	26.2 (12.6)	18.9 (12.9)
BT2a PDR0	59.1 (2.3)	65.5 (1.8)	0 (0)	0 (0)	8.9 (2.8)	8.9 (2.3)	30.3 (16.0)	34.8 (21.2)
BT2b PDR0	9.3 (0.7)	10.0 (0.2)	0 (0)	0 (0)	9.9 (2.2)	10.6 (1.4)	–	–
BT0 PDR5	32.6 (10.0)	65.7 (6.3)	34.8 (11.0)	7.2 (3.7)	10.8 (1.3)	11.3 (0.5)	31.7 (18.5)	26.4 (15.6)
BT2a PDR5	32.8 (5.1)	56.0 (10.4)	28.1 (8.3)	8.1 (4.1)	9.3 (2.2)	9.9 (2.0)	34.6 (20.0)	39.2
BT2b PDR5	7.3 (1.9)	8.6 (1.0)	11.4 (11.4)	5.1 (7.9)	8.1 (2.6)	9.3 (2.6)	–	–

Table 2.4 Mean and Standard Deviation Values of the Number of Packets, Percentage of Ignored and Lost Packets, Packet Delay, and Number of Stopped Processes for TF2

BT (PDR) Level	No. of Packets, Mean (Stdev)		% of Packets Ignored or Lost, Mean (Stdev)		% of Packet Lost, Mean (Stdev)	Packet Delay (ms), Mean (Stdev)		No. of Stopped Processes, Mean (Stdev)	
	TCP	UDP	TCP	UDP	UDP	TCP	UDP	TCP	UDP
BT0 PDR0	282.5 (2.5)	308.0 (0.1)	0 (0)	0 (0)	0 (0)	61 (19)	15 (5)	0 (0)	0 (0)
BT1 PDR0	271.7 (2.9)	303.5 (1.7)	0 (0)	0 (0)	0 (0)	83 (33)	22 (17)	0 (0)	0 (0)
BT2a PDR0	244.6 (7.3)	271.7 (5.4)	3.6 (1.1)	3.5 (0.7)	0 (0)	98 (43)	29 (24)	0 (0)	0 (0)
BT2b PDR0	36.8 (2.9)	39.9 (0.7)	0 (0)	0 (0)	0 (0)	629 (1078)	184 (513)	0 (0)	0 (0)
BT0 PDR5	188.0 (14.9)	273.8 (25.3)	44.5 (17.0)	6.8 (1.4)	6.8 (1.4)	63 (35)	15 (5)	2.8 (1.8)	2.9 (1.7)
BT2a PDR5	185.0 (10.9)	240.2 (43.8)	38.7 (8.0)	10.2 (1.4)	7.2 (1.9)	99 (66)	30 (36)	3.7 (1.6)	2.6 (1.3)
BT2b PDR5	30.6 (4.5)	36.1 (3.0)	13.7 (15.4)	6.4 (4.1)	6.4 (4.1)	814 (1355)	162 (413)	0.2 (0.4)	0.4 (0.7)

obtained for the case of TF2. These results are obtained with all OLTCs blocked in tap position 0 and the local voltage regulators not active.

Since the gossip-like procedure is based on the random choice of the active agents, the tables report the results of the statistical analysis carried out by performing 30 simulations for the same case. For each simulation, the pseudorandom number generator is initialized by different seed states associated with the computer system time. In Tables 2.1 and 2.3, the number of compensation cycles is the number of cycles in which at least one compensator changes its reactive power and the percentage of incomplete compensations indicates the percentage of cycles that do not complete regularly; the power loss decrease indicates the difference between the values of power losses at the starting time and at t_{set}. In Tables 2.2 and 2.4, the number of packets takes into account only those carrying compensation data sent by the agents; the percentage of ignored or lost packets refers to those that arrive at destination after t_{wait} or do not arrive at all (for UDP the specific percentage of these lost packets is also provided); packets delay indicates the traveling time of the packets that regularly arrive at destination before t_{wait}; and the number of stopped process is the number of stopping actions on the basis of the priority index value. The packed delay values indicated in Tables 2.3 and 2.4 are the mean values of the statistical parameters obtained for each of the 30 simulations.

For TF1 and scenario BT0-PDR0, the mean value (standard deviation) in kvar of the final reactive power outputs of the six compensators is: 146.5 (25.6), 63.1 (2.2), 148.2 (1.4), 144.7 (7.6), 169.8 (4.0), and 164.7 (1.6). The limited values of the standard deviations show that the reactive output scheduling at the end of different compensation cycles is almost the same. The significant value of the first standard deviation is due to the proximity of node 702, where the first compensator is located, to the slack bus. The low value of the standard deviation relevant to the loss decrease (fractions of a kW) shown in Table 2.1 indicates that an analogous power flow conditions is achieved at the end of different sequence of compensations. The procedure converges in less than 30 s to a mean value of the final power losses of about 76.3 kW. The reduction of nearly 13.5 kW is in agreement with the results shown in Ref. [34]. The results are also in reasonable agreement with those obtained by applying the three-phase version of the mixed integer linear programming (MILP) model proposed in Refs. [14] and [64] to the same distribution system, despite the significant differences between the two approaches. The solution of the MILP model provides the following reactive power scheduling (in kvar): 140, 75, 150, 140, 170, and 170 (the reactive power output of the compensators in the implemented MILP model can be changed in steps of 5 kvar). The minimum power loss calculated by the

MILP model is 76.27 kW, only some tens of watts lower than that achieved by the gossip procedure.

For TF2 and scenario BT0-PDR0, the mean value (standard deviation) in kvar of the final reactive power outputs of the six compensators is: 258.2 (67.7), 154.2 (2.3), 421.1 (2.3), 272.4 (56.2), -88.6 (17.0), and 132.7 (5.9). As for TF1, Table 2.3 shows that also for TF2 with efficient communication very small standard deviation values of loss decreases are obtained. Therefore analogous power flow conditions are achieved at the end of different sequence of compensations.

As shown in Tables 2.1 and 2.3 the results obtained for scenario BT1-PDR0 are similar to those of scenario BT0-PDR0. For the cases without packet loss (PDR0), as expected, the results show larger delays for increasing BT levels from 0 to 2. Even without PDR, in scenario BT2a due to the congestion of the communication links several packets do not reach the expected receiver within $t_{wait} = 1$ s and therefore they are ignored. Tables 2.2 and 2.4 show that in BT2b this problem is solved as t_{wait} is extended to 10 s. While for $t_{wait} = 1$ s the compensation procedure is fast enough so that t_{set} could be considered the time at which the procedure converges, this is no longer true for BT2b. Therefore for BT2b the power loss decrease is evaluated at $t_f = 100$ s, although the convergence is not yet reached due to the insufficient number of compensation cycles.

For the scenarios with PDR = 5%, the number of compensation cycles reduces in particular for TCP. The PDR causes an incomplete cycle when there is the loss of the packet that carries the information from agent k, which compensates first, to agent h. Moreover, the loss of the return packet from the new agent h and the old one causes the start of a process with increased priority index. The presence of concurrent processes is then eliminated by the stop of the process with lower priority index.

In Tables 2.2 and 2.4, we could distinguish between packets lost because of PDR5 from those ignored due to excessive delay only by using UDP, while this classification is not possible with the implemented TCP model because it closes the communication after t_{wait}.

The procedure has been also applied to both the case of a higher load level and a lower load level than that indicated in Ref. [63]. The former (high load) is obtained by multiplying both the original active and reactive power values at each load bus (normal load) by a different number obtained through a uniformed distribution between 1.3 and 1.7. The latter (low load) is obtained by using multipliers uniformly distributed between 0.3 and 0.7. The results have been obtained by using the same seed state for the random number generation, that is, by the same sequence of active agent pairs. The convergence of the

Table 2.5 Power Loss Reductions Obtained for TF1 Without OLTC for Different Load Levels

Load Level	Initial Power Loss (kW)	Power Loss Decrease (kW)			
		BT0-PDR0		BT2-PDR5	
		TCP	UDP	TCP	UDP
Normal load	89.8	13.5	13.5	11.8	12.9
Low load	25.2	3.9	3.9	3.4	3.8
High load	211.8	31.1	31.1	24.6	28.0

Table 2.6 Power Loss Reductions Obtained for TF2 Without OLTCs for Different Load Levels

Load Profile	Initial Power Loss (kW)	Power Loss Decrease (kW)			
		BT0-PDR0		BT2b-PDR 5	
		TCP	UDP	TCP	UDP
Normal load	125.8	11.4	11.4	11.2	11.3
Low load	33.0	1.05	1.06	1.03	1.03
High load	300.0	43.2	43.2	40.0	42.5

procedure is similar for all the load levels and the obtained reductions of power losses are shown in Tables 2.5 and 2.6 for TF1 and TF2, respectively, and both scenario BT0-PDR0 and scenario BT0-PDR5.

For TF1, the coordination between the VVC procedure and the action of the OLTC at the substation is illustrated in Figs. 2.7 and 2.8. The value of the desired regulated voltage at the secondary side is 1.02 pu with a dead band limit of ±0.015 pu. The results show that in BT2-PDR5 cases the OLTC starts in tap position 0, operates a first tap change at 22 s and, after other two changes, reaches a final tap equal to −3 at 40 s. In BT1-PDR0 cases, the OLTC operates the first tap change at the same time, and after one additional change it reaches a steady state condition at tap −2 at 30 s. The improved action by the agents with efficient communication results in a lower OLTC regulation. The increase of the power losses after each tap change is due to the increased consumption of voltage dependent loads.

For TF2, the sequence of compensations with both OLTCs in operation is illustrated by Figs. 2.9 and 2.10. The reference values of the secondary side voltage are: 1.01 pu for the substation transformer and 1 pu for OLTC transformer at bus 67, respectively, both with a dead band of ±0.015 pu. The substation transformer changes the tap to −1 position at 22 s and reaches the steady state condition. For BT1-PDR0 with both TCP and UDP, OLTC

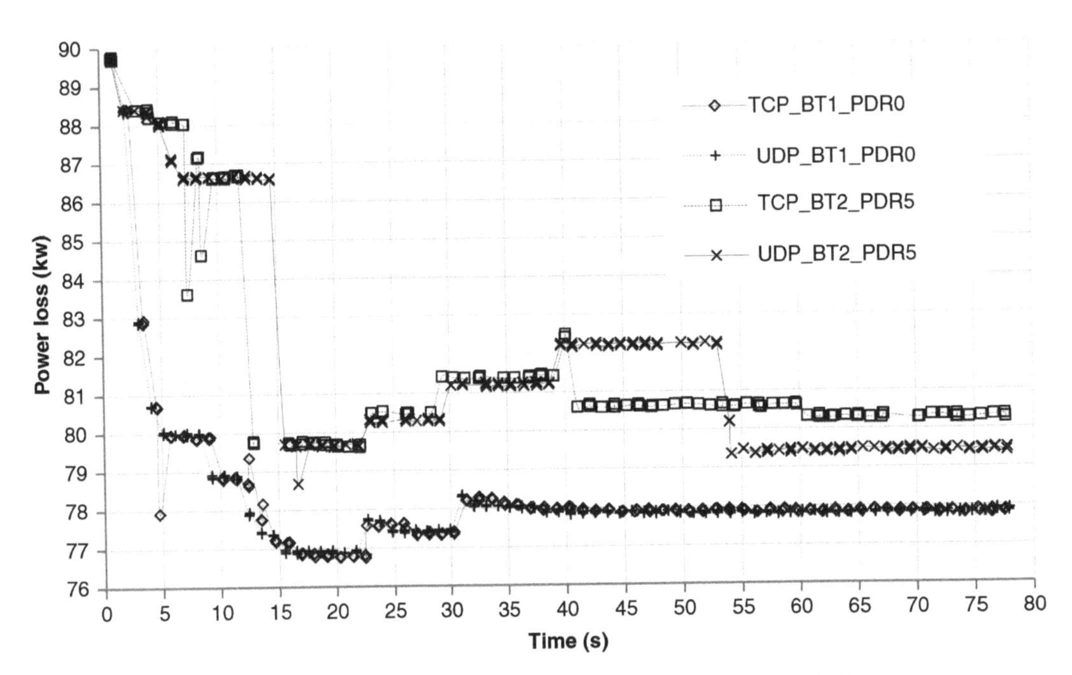

■ FIGURE 2.7 Power loss variation in TF1 with OLTC for different BT levels and PDR by using TCP and UDP.

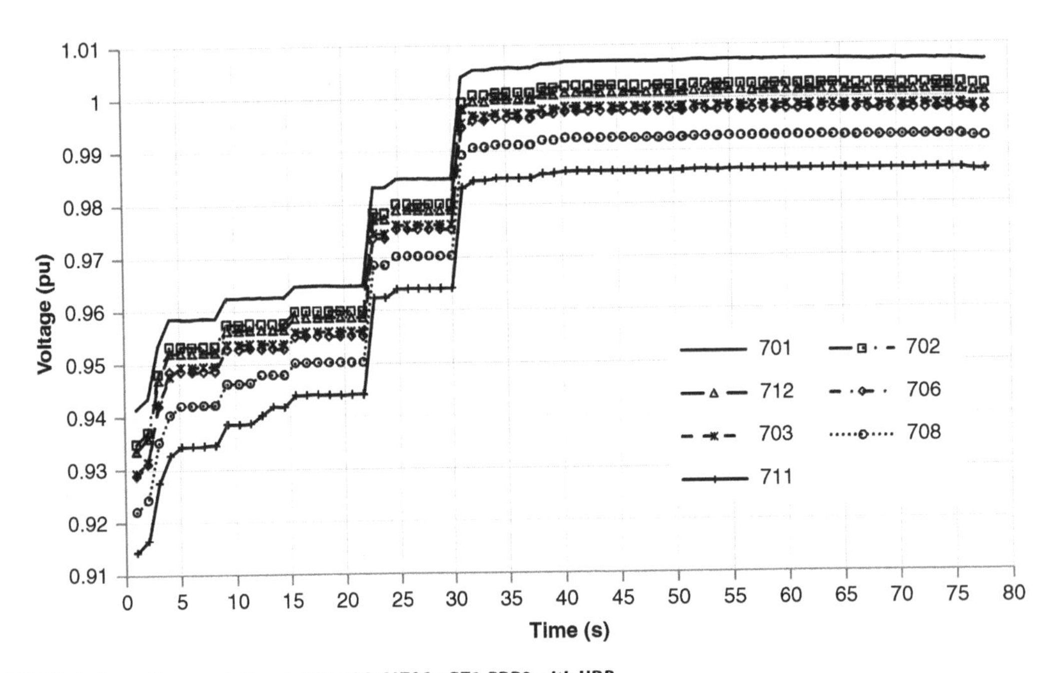

■ FIGURE 2.8 Bus voltage variations in TF1 with OLTC for BT1-PDR0 with UDP.

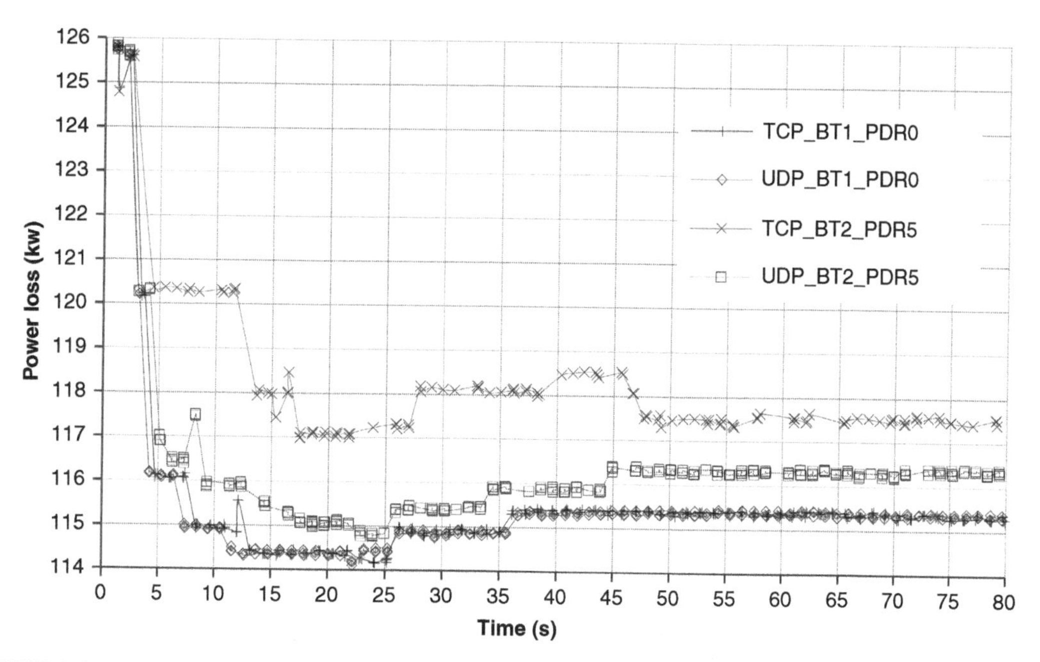

FIGURE 2.9 Power loss variation in TF2 with two OLTCs in operation for two different BT levels by using TCP and UDP.

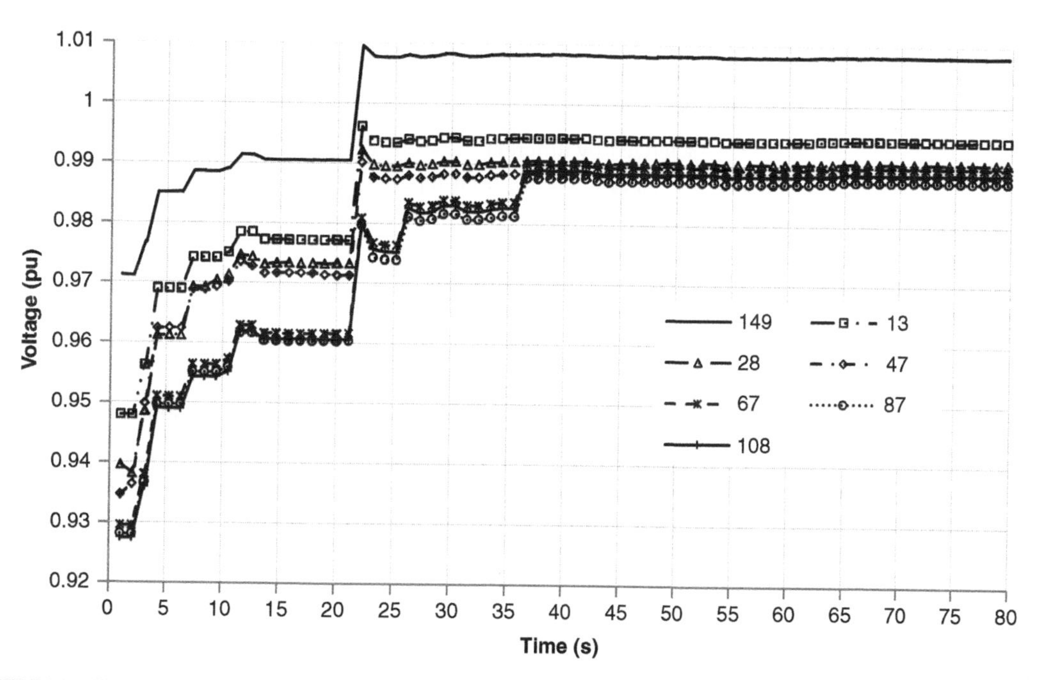

FIGURE 2.10 Voltage variations in TF2 with two OLTCs in operation for BT1 by using UDP.

transformer at bus 67 changes at first the tap to −1 at 25.6 s and then at about 35.6 s (UDP), reaching the final tap −2. For BT2-PDR5 case with UDP, this OLTC after the first tap variation at 25.6 s changes again two other times, reaching a steady state condition at tap −3 at 44.8 s. For BT2-PDR5 case with TCP, the high BT level hinders the communication between node 0 and node 4. Therefore, the OLTC transformer at bus 67 anticipates the first change to nearly 12 s and it reaches the steady state condition with tap −4 at time 39.1 after three other changes.

2.3.3.2 *Results obtained with the cellular communication network*

In the simulation relevant to the UMTS network the simulations are repeated for two different BLER values: $1E^{-5}$ (case indicated as BLER0) and 0.1 (BLER1), which is a typical reference performance [65]. Moreover the simulations are repeated with BT (case indicated as BT1) and without the BT (BT0). In order to obtain a significant value of the BT, the UEs associated to the agents and six additional UEs are forced to generate the BT towards two additional UEs receivers belonging to the same RNC subnetwork.

We assume that each of converters that act as a reactive power compensator also injects a constant active power equal to 400 kW, while the initial reactive outputs are null. The OLTCs start in tap position 0.

Tables 2.7 and 2.8 compare the results obtained for the eight scenarios characterized by different values of BLER, BT, and load levels. For each case the calculation has been repeated 20 times. In Table 2.8, the numbers of

Table 2.7 Number of Compensation Cycles, Percentage of Incomplete Compensations, Power Loss Decrease, and Settling Time for TF1 and TF2

BLER BT (Load Level)	No of Compensation Cycles, Mean (Stdev)		% of Incomplete Compensations, Mean (Stdev)		Power Loss Decrease (kW), Mean (Stdev)		Settling Time (s), Mean (Stdev)	
	TF1	TF2	TF1	TF2	TF1	TF2	TF1	TF2
BLER0 BT0 (normal)	154	154	0	0	15.8 (0.1)	5.5 (0.1)	62.4 (12.6)	31.5 (6.9)
BLER1 BT0 (normal)	117.5 (32.5)	117.2 (24.2)	15.5 (3.2)	16.6 (4.0)	15.3 (1.4)	5.5 (0.1)	63.0 (17.0)	40.4 (11.8)
BLER0 BT1 (normal)	119.5 (19.5)	122.9 (14.2)	16.0 (4.3)	11.6 (2.3)	15.3 (1.2)	5.4 (0.2)	71.0 (20.8)	43.6 (15.9)
BLER1 BT1 (normal)	102.0 (18.9)	107.0 (15.1)	29.9 (5.0)	26.6 (5.4)	15.5 (0.9)	5.4 (0.5)	77.5 (20.3)	42.6 (18.0)
BLER0 BT0 (high)	154	154	0	0	25.6 (0.05)	19.4 (0.5)	78.3 (11.5)	93.9 (36.4)
BLER1 BT0 (high)	129.3 (12.6)	115.9 (30.7)	17.6 (3.3)	16.3 (2.8)	25.5 (0.9)	18.0 (3.4)	102.2 (29.8)	91.2 (27.0)
BLER0 BT1 (high)	119.9 (18.8)	121.0 (23.1)	16.4 (3.9)	11.0 (3.3)	25.0 (2.3)	18.5 (3.3)	87.7 (16.3)	79.2 (17.5)
BLER1 BT1 (high)	101.4 (22.3)	108.3 (11.2)	32.4 (7.2)	25.6 (4.9)	24.3 (3.8)	17.9 (1.4)	105.2 (39.2)	122.2 (21.2)

Table 2.8 Number of Packets (Tx, Transmitted; Rx, Received), Percentage of Ignored and Lost Packets, Packet Delay, and Number of Stopped Process for TF1 and TF2

BLER BT (Load Level)	No. of Packets Mean (Stdev) Tx, Mean (Stdev) Rx		% of Packets Ignored, Mean (Stdev)		% of Packet Lost, Mean (Stdev)		Packet Delay (ms), Mean (Stdev)		No. of Stopped Processes, Mean (Stdev)	
	TF1	TF2	TF1	TF2	TF1	TF2	TF1	TF2	TF1	TF2
BLER0 BT0 (normal)	618.2(1.3) 618.2(1.3)	618.8(1.4) 618.8(1.4)	0	0	0	0	145.2 (10.2)	145.0 (10.1)	0	0
BLER1 BT0 (normal)	542.6(125.8) 447.3(104.8)	526.5(104.4) 434.6(87.0)	5.5 (1.2)	6.0 (1.0)	11.9 (5.7)	11.5 (1.5)	147.2 (32.7)	146.2 (29.6)	11.1 (3.4)	11.1 (3.6)
BLER0 BT1 (normal)	528.2(77.5) 443.9(68.7)	551.7(52.7) 478.2(50.6)	7.6 (1.3)	6.3 (1.1)	8.5 (1.3)	7.2 (1.0)	212.0 (154.1)	239.3 (173.5)	7.0 (2.6)	7.4 (2.7)
BLER1 BT1 (normal)	527(91.0) 371.6(62.6)	510.1(68.7) 378.4(55.3)	13.0 (1.4)	12.2 (1.8)	16.2 (4.2)	13.8 (1.7)	244.6 (197.8)	259.0 (202.3)	13.0 (5.3)	9.6 (3.0)
BLER0 BT0 (high)	618.5(1.3) 618.5(1.3)	618.4(1.5) 618.4(1.5)	0	0	0	0	145.1 (10.1)	145.4 (10.4)	0	0
BLER1 BT0 (high)	569.9(56.6) 474.5(30.5)	520.0(139.1) 433.1(116.1)	5.3 (0.7)	5.7 (1.0)	11.5 (1.7)	10.9 (1.3)	146.2 (26.4)	147.0 (31.3)	11.5 (3.4)	11.5 (4.0)
BLER0 BT1 (high)	548.6(49.1) 457.3(43.4)	543.5(102.9) 472.6(92.0)	7.2 (1.3)	6.1 (1.3)	9.3 (5.8)	7.3 (1.3)	213.0 (157.0)	239.0 (173.6)	6.9 (2.9)	7.9 (2.8)
BLER1 BT1 (high)	488.7(104.7) 347.5(75.0)	524.2(37.3) 389.6(36.5)	13.4 (2.3)	12.0 (1.4)	15.3 (1.9)	13.8 (1.9)	244.5 (195.7)	260.3 (205.3)	10.5 (4.0)	12.0 (3.7)

packets (transmitted and received, if the relevant numbers differ) take into account only those packets carrying data relevant to the gossip algorithm.

The final values of the bus voltages at the compensator buses are significantly improved with respect to the initial condition (eg, for TF1 in BLER0, BT0, and normal load the maximum deviation form 1 pu is equal to 1.1%). For BLER1, the number of compensation cycles reduces and the higher BLER causes both the interruption of incomplete processes and the start of concurrent processes (progressively stopped on the basis of their priority index) due to the loss of packets that carries critical information. Also for BT1, the additional transmissions cause a reduction of the number of compensation cycles and at the same time the increasing of the packets delay.

Additional simulation results are available in Refs. [37,57].

2.4 CONCLUSIONS

This chapter provides a review of the issues and most relevant typical countermeasures concerning the integration of DERs in distribution power systems. In general, modern solutions tend to increase the use of communication networks for the operation of power distribution systems. Moreover, improved equipment for both monitoring and control purposes are being installed. In particular, there is an increasing interest for PMUs, which are now commonly used for the operation of many transmission grids.

In order to provide an example, with a focus on the VVC of distribution feeders, this chapter analyses of the performances of a leaderless MAS scheme based on the information provided by PMUs. The analysis is carried out by means of a specifically developed ICT power cosimulation platform that accurately represents the transient phenomena in the power distribution system and the behavior of the communication network.

This application example illustrates the typical limitations induced by BT and loss of information in the communication channels, by means of the results obtained using both the model of a wired network and that of a cellular network. The specific countermeasures implemented in the MAS approach are described. In particular, a mechanism based on a priority index allows for the start of multiple concurrent compensations processes whenever packet carrying critical information is lost. The processes with low priority indexes are progressively stopped by the agents included in the scheme in order to guarantee the convergence performances of the algorithm. The results show that the performances of the adopted MAS approach appear reasonably good also for medium values of background data traffic and loss of information, taking also into account the expected accuracy of PMUs for MV distribution networks.

ACKNOWLEDGMENT

The application example shown in this chapter is based on results obtained in the framework of the research activities supported in part by University of Bologna (FARB FFBO122030) and by ENIACJU/CALL 2011-1/296131 E2SG "Energy to smart grid". These activities have been carried out in collaboration with several colleagues who are the coauthors of the papers in which the obtained results have been originally presented: Davood Babazadeh (Kungliga Tekniska Högskolan - KTH), Marina Barbiroli (University of Bologna), Riccardo Bottura (University of Bologna), Fabio Napolitano (University Bologna), Lars Nordstrom (KTH), and Kun Zhu (KTH). Riverbed modeler software has been used thanks to the Riverbed University Program. The EMTP-rv simulation environment has been used in the framework of the Powersys Company Education partnership.

REFERENCES

[1] Sioshansi FP. Distributed generation and its implications for the utility industry. San Diego: Elsevier Science; 2014.

[2] Zervos A, et al. Renewables 2014 – Global status report. Paris: REN21 Secretariat; 2014.

[3] Papathanassiou S, et al. Capacity of distribution feeders for hosting DER, Technical brochure 586, Paris: Cigré, June; 2014.

[4] DeBlasio R, Chalmers S, Basso TS, Ash B. IEEE Application Guide for IEEE Std 1547(TM), IEEE Standard for Interconnecting Distributed Resources with Electric Power Systems; 2009.

[5] Kouro S, Leon JI, Vinnikov D, Franquelo LG. Grid-connected photovoltaic systems: an overview of recent research and emerging PV converter technology. IEEE Ind Electron M 2015;9(1):47–61.

[6] Yang Y, Wang H, Blaabjerg F. Reactive power injection strategies for single-phase photovoltaic systems considering grid requirements. IEEE Trans Ind Appl 2014;50(6):4065–76.

[7] Phadke AG, Thorp JS. Synchronized Phasor Measurements and Their Applications (Power Electronics and Power Systems). New York: Springer; 2008.

[8] Sanchez-Ayala G, Aguerc JR, Elizondo D, Lelic M. Current trends on applications of PMUs in distribution systems, in 2013 IEEE PES Innovative Smart Grid Technologies Conference (ISGT); 2013, pp. 1–6.

[9] McArthur SDJ, Davidson EM, Catterson VM, Dimeas AL, Hatziargyriou ND, Ponci F, Funabashi T. Multiagent systems for power engineering applications – Part II: technologies, standards, and tools for building multiagent systems. IEEE Trans Power Syst 2007;22(4):1753–9.

[10] Mets K, Ojea JA, Develder C. Combining power and communication network simulation for cost-effective smart grid analysis. IEEE Commun Surv Tut 2014;16(3):1–26.

[11] Yan Y, Qian Y, Sharif H, Tipper D. A survey on smart grid communication infrastructures: motivations, requirements and challenges. IEEE Commun Surv Tutor 2013;15(1):5–20.

[12] Yang Q, Barria JA, Green TC. Communication infrastructures for distributed control of power distribution networks. IEEE Trans Industr Inform 2011;7(2):316–27.

[13] Deshmukh S, Natarajan B, Pahwa A. Voltage/VAR control in distribution networks via reactive power injection through distributed generators. IEEE Trans Smart Grid 2012;3(3):1226–34.
[14] Borghetti A. Using mixed integer programming for the volt/VAR optimization in distribution feeders. Electr Pow Syst Res 2013;98:39–50.
[15] Zhu Y, Tomsovic K. Optimal distribution power flow for systems with distributed energy resources. Int J Elec Power 2007;29(3):260–7.
[16] Borghetti A, Bosetti M, Grillo S, Massucco S, Nucci CA, Paolone M, Silvestro F. Short-term scheduling and control of active distribution systems with high penetration of renewable resources. IEEE Syst J 2010;4(3):313–22.
[17] Paudyal S, Cañizares CA, Bhattacharya K. Optimal operation of distribution feeders in smart grids. IEEE Trans Ind Electron 2011;58(10):4495–503.
[18] Bruno S, Lamonaca S, Rotondo G, Stecchi U, La Scala M. Unbalanced three-phase optimal power flow for smart grids. IEEE Trans Industr Inform 2011;58(10):4504–13.
[19] Pilo F, Pisano G, Soma GG. Optimal coordination of energy resources with a two-stage online active management. IEEE Trans Ind Electron 2011;58(10):4526–37.
[20] McArthur SDJ, Davidson EM, Catterson VM, Dimeas AL, Hatziargyriou ND, Ponci F, Funabashi T. Multiagent systems for power engineering applications – Part I: concepts, approaches, and technical challenges. IEEE Trans Power Syst 2007;22(4):1743–52.
[21] Chen Y, Lu J, Yu X, Hill DJ. Multi-agent systems with dynamical topologies: consensus and applications. IEEE Circuits Syst Mag 2013;13(3):21–34.
[22] Gupta RA. Networked control system: overview and research trends. IEEE Trans Ind Electron 2010;57(7):2527–35.
[23] Zhang L, Gao H, Kaynak O. Network-induced constraints in networked control systems – a survey. IEEE Trans Industr Inform 2013;9(1):403–16.
[24] Yu L, Czarkowski D, de Leon F. Optimal distributed voltage regulation for secondary networks with DGs. IEEE Trans Smart Grid 2012;3(2):959–67.
[25] Lam AYS, Zhang B, Tse DN. Distributed algorithms for optimal power flow problem. 2012 IEEE 51st IEEE Conference on Decision and Control (CDC). IEEE; 2012, pp. 430–437.
[26] Šulc P, Backhaus S, Chertkov M. Optimal distributed control of reactive power via the alternating direction method of multipliers. IEEE Transactions on Energy Conversion 2014;29(4):968–977.
[27] Zhang B, Lam AYS, Dominguez-Garcia A, Tse D. Optimal distributed voltage regulation in power distribution networks. arXiv:1204.5226v4 [math.OC]; 2015.
[28] Bolognani S, Carli R, Cavraro G, Zampieri S. A distributed control strategy for optimal reactive power flow with power constraints. 52nd, IEEE Conference on Decision, Control, IEEE; 2013, pp. 4644–4649.
[29] Baran ME, El-Markabi IM. A multiagent-based dispatching scheme for distributed generators for voltage support on distribution feeders. IEEE Trans Power Syst 2007;22(1):52–9.
[30] Aquino-Lugo AA, Klump R, Overbye TJ. A control framework for the smart grid for voltage support using agent-based technologies. IEEE Trans Smart Grid 2011;2(1):173–80.
[31] Farag HEZ, El-Saadany EF. A novel cooperative protocol for distributed voltage control in active distribution systems. IEEE Trans Power Syst 2012;28(2):1–11.
[32] Bolognani S, Zampieri S. A gossip-like distributed optimization algorithm for reactive power flow control. Proceedings of IFAC World Congress; 2011, pp. 1–14.

[33] Bolognani S, Zampieri S. A distributed control strategy for reactive power compensation in smart microgrids. IEEE Trans Automat Contr 2013;58(11):2818–33.

[34] Bolognani S, Carron A, Di Vittorio A, Romeres D, Schenato L, Zampieri S. Distributed multi-hop reactive power compensation in smart micro-grids subject to saturation constraints. 51st IEEE Conference on Decision and Control. Maui, Hawaii, USA; 2012, pp. 1118–1123.

[35] Tenti P, Costabeber A, Mattavelli P, Trombetti D. Distribution loss minimization by token ring control of power electronic interfaces in residential microgrids. IEEE Trans Ind Electron 2012;59(10):3817–26.

[36] Loia V, Vaccaro A, Vaisakh K. A self-organizing architecture based on cooperative fuzzy agents for smart grid voltage control. IEEE Trans Industr Inform 2013;9(3):1415–22.

[37] Bottura R, Borghetti A. Simulation of the volt/VAR control in distribution feeders by means of a networked multiagent system. IEEE Trans Industr Inform 2014;10(4):2340–53.

[38] Yang C, Zhabelova G, Yang C-W, Vyatkin V. Cosimulation environment for event-driven distributed controls of smart grid. IEEE Trans Industr Inform 2013;9(3): 1423–35.

[39] Khaitan SK, McCalley JD. Cyber physical system approach for design of power grids: a survey. 2013 IEEE Power & Energy Society General Meeting. IEEE; 2013, pp. 1–5.

[40] Hopkinson K, Wang X, Giovanini R, Thorp J, Birman K, Coury D. EPOCHS: a platform for agent-based electric power and communication simulation built from commercial off-the-shelf components. IEEE Trans Power Syst 2006;21(2):548–58.

[41] Lin H, Veda SS, Shukla SS, Mili L, Thorp J. GECO: global event-driven co-simulation framework for interconnected power system and communication network. IEEE Trans Smart Grid 2012;3(3):1444–56.

[42] Godfrey T, Mullen S, Griffith DW, Golmie N, Dugan RC, Rodine C. Modeling smart grid applications with co-simulation. 2010 First IEEE International Conference on Smart Grid Communications. IEEE; 2010, pp. 291–296.

[43] Celli G, Pegoraro PA, Pilo F, Pisano G, Sulis S. DMS cyber-physical simulation for assessing the impact of state estimation and communication media in smart grid operation. IEEE Trans Power Syst 2014;PP(99):1–11.

[44] Monti A, Colciago M, Conti P, Maglio M, Dougal R. A co-simulation approach for analysing the impact of the communication infrastructure in power system control. Proceedings of Conference on Grand Challenges in Modeling and Simulation (GCMS'09). Istanbul, Turkey: Society for Modeling & Simulation International; 2009, pp. 278–282.

[45] Li W, Monti A, Luo M, Dougal RA. VPNET: a co-simulation framework for analyzing communication channel effects on power systems. 2011 IEEE Electric Ship Technologies Symposium. IEEE; 2011, pp. 143–149.

[46] Muller SC, Georg H, Rehtanz C, Wietfeld C. Hybrid simulation of power systems and ICT for real-time applications. 2012 3rd IEEE PES Innovative Smart Grid Technologies Europe (ISGT Europe). IEEE; 2012, pp. 1–7.

[47] Babazadeh D, Chenine M, Zhu K, Nordström L, 2012. Real-time smart grid application testing using OPNET SITL. OPNETWORK2012.

[48] Lévesque M, Xu DQ, Joós G, Maier M. Communications and power distribution network co-simulation for multidisciplinary smart grid experimentations. SCS/ACM Spring Simul. Multi-Conf. Orlando, FL, USA: Society for Computer Simulation International; 2012, pp. 1–7.

[49] Xu DQ, Joos G, Levesque M, Maier M. Integrated V2G, G2V, and renewable energy sources coordination over a converged fiber-wireless broadband access network. IEEE Trans Smart Grid 2013;4(3):1381–90.

[50] Nutaro J, Kuruganti PT, Miller L, Mullen S, Shankar M. Integrated Hybrid-Simulation of Electric Power and Communications Systems. 2007 IEEE Power Engineering Society General Meeting. IEEE; 2007, pp. 1–8.

[51] Liberatore V, Al-Hammouri A. Smart grid communication and co-simulation. IEEE 2011 EnergyTech. IEEE; 2011, pp. 1–5.

[52] Al-Hammouri AT. A comprehensive cosimulation platform for cyber-physical systems. Comput Commun 2012;36(1):8–19.

[53] Bottura R, Borghetti A, Napolitano F, Nucci CA. ICT-power co-simulation platform for the analysis of communication-based volt/Var optimization in distribution feeders. 5th Innovative Smart Grid Technologies Conference ISGT. Washington, IEEE; 2014, pp. 1–5.

[54] Bottura R, Babazadeh D, Zhu K, Borghetti A, Nordstrom L, Nucci CA. SITL and HLA cosimulation platforms: tools for analysis of the integrated ICT and electric power system. EUROCON, 2013 IEEE; 2013, pp. 918–925.

[55] Molnár K, Hošek J, Růčka L. Mutual cooperation of external application and OPNET Modeler simulation environment. Proceedings of International Conference on Research in Telecommunication Technology (RTT 2009). Srby, Czech Republic; 2009, pp. 1–5.

[56] Lu DZ, Yang H. Unlocking the power of OPNET Modeler. Cambridge: Cambridge University Press; 2012.

[57] Bottura R, Borghetti A, Barbiroli M, Nucci CA. Reactive power control of photovoltaic units over wireless cellular networks. PowerTech, 2015 IEEE Eindhoven; 2015, pp. 1–7.

[58] Khodabakhchian B. Modeling on-load tap changers in EMTP-rv. EMTP-rv Newslett 2005;1(1):11–5.

[59] Larsson M. Coordination of cascaded tap changers using a fuzzy rule-based controller. Fuzzy Set Syst 1999;102(1):113–23.

[60] Kamh MZ, Iravani R. Unbalanced model and power-flow analysis of microgrids and active distribution systems. IEEE Trans Power Deliver 2010;25(4):2851–8.

[61] Chen T, Chen M, Inoue T, Kotas P, Chebli EA. Three-phase cogenerator and transformer models for distribution system analysis. IEEE Trans Power Deliver 1991;6(4):1671–81.

[62] Paolone M, Borghetti A, Nucci CA. A synchrophasor estimation algorithm for the monitoring of active distribution networks in steady state and transient conditions. 17th Power Systems Computation Conference (PSCC'11) Stockholm, Sweden; 2011, pp. 22–26.

[63] Fuller J, Solanki J, Kersting B, Dugan R, Carneiro SJ. IEEE PES distribution test feeders. Available from: http://www.ewh.ieee.org/soc/pes/dsacom/testfeeders/.

[64] Borghetti A, Napolitano F, Nucci CA. Volt/VAR optimization of unbalanced distribution feeders via mixed integer linear programming. Int J Elec Power 2015;72:40–7.

[65] Mishra AR, editor. Advanced cellular network planning and optimisation: 2G/2.5G/3G. Evolution to 4G. Chichester: John Wiley & Sons; 2007.

Chapter 3

Operational aspects of distribution systems with massive DER penetrations

Tomonobu Senjyu*, Abdul Motin Howlader**

*Department of Electrical and Electronics Engineering, University of the Ryukyus, Okinawa, Japan;

**Postdoctoral Fellow University of Hawaii, Manoa Honolulu, Hawaii

CHAPTER OUTLINE

3.1 Introduction 51

3.2 Control objectives 53

3.2.1 Importance of distributed generations 54

3.2.2 Challenges of distributed generations system 56

3.2.3 Overview of control system 57

3.3 Control method 58

3.3.1 The objective function and constraints 58

3.4 Particle swarm optimization 60

3.4.1 PV generator system 61

3.4.2 BESS at the interconnection point 62

3.4.3 Plug-in electric vehicle 63

3.5 Simulation results 63

3.5.1 Dynamic responses for the without optimization approach 66

3.5.2 Dynamic responses for the comparison method 66

3.5.3 Dynamic responses for the proposed method 70

3.6 Conclusions 73

References 73

3.1 INTRODUCTION

World primary energy consumption enlarged by 2.3% in 2013, an acceleration of over 2012 (+1.8%). Conventional fossil fuels such as coal, oil, and gas remain the largest sources of world energy. Oil, coal, and natural gas supply the 32.9, 30.1, and 23.7% of global energy consumption, respectively. Unreliable nuclear power provides 4.4% of global energy consumption, and hydroelectric contributes for 6.7% of global energy consumption. The green renewable energy sources deliver a record 2.7% of global energy consumption [1]. According to the BP Energy Outlook 2035, the world's population is expected to reach 8.7 billion which means an additional 1.6 billion people will require energy. Primary energy consumption of the world will be increased by 37% between 2013 and 2035, with averaging 1.4% per annum (p.a.) growth rate. Therefore, energy production will increase at 1.4% p.a. from 2013 to 2035, matching the expansion of energy consumption. Fossil fuels in cumulative lose share but remain the dominant form of energy in 2035 with a share of 81%, down from 86% in 2013. Renewable energies (eg, wind, solar, tide, biofuel) gain share rapidly, from around 3% today to 8% by 2035, overtaking nuclear in the early 2020s and hydro in the early 2030s. Total carbon emissions from energy consumption will rise by 25% between 2013 and 2035 (1% p.a.), with the rate of growth declining from 2.5% over the past decade to 0.7% [2].

Renewable energies are expected to play a vital role for the reduction of emission and to meet the future energy demand [3]. High penetration of renewable energy sources is required to integrate with the existing power grid. Therefore, the conventional power grid has to be reconfigured for the integration of renewable sources. The reconfiguration form of the existing power grid with renewables is known as the smart grid (SG). Distributed generations (DGs) will play an important role for the future SG [4,5]. DGs consist of wind generator, PV generator, fuel cell, biofuel, smart house, electric vehicle, battery energy storage system (BESS), and so on and are sources of energy which can transport a variety of benefits including improved reliability if they are operated properly by the SG [6]. Furthermore, DGs are energy cost saving [7]. In addition, DGs can improve power system efficiency as well as prosper the power quality of the distribution network [8]. Generation of most DGs mainly depends on the weather conditions. When a large number of DGs are integrated with the power system, there are adverse effects such as power fluctuation, voltage flicker inside the power grid, frequency deviation, energy losses, and so on. Sometimes these problems become severe which cannot be solved by only the tap changing of a load ratio control transformer (LRT) in the substation and step voltage regulators (SVRs) in the distribution line.

To solve that, some control methodologies, which mainly use reactive power controllers such as additional reactive power compensators [9–10] or the reactive power control of DGs [11–12], have been proposed in some literatures. However, these control methods are subject to the power factor control of DGs in spite of being a detriment to the DGs operators by causing a reduction in the active output power. On the other hand, interfaced inverters which are used for variable speed wind turbines and PV generators can control the reactive output power independently within their capacity [13]. Furthermore, a BESS is introduced to the power system to compensate the power quality loss, voltage, and frequency fluctuations [14–16]. In recent years, BESS technologies have been developed to achieve cost reduction, longer operating life, and higher efficiency [17,18]. Therefore, introducing a BESS into the power distribution systems may help to solve some problems caused by a large-scale integration of DGs, based on renewable energies and to achieve some economic operation goals. Centralized control methods and decentralized autonomous control methods have been proposed [19,20] which use the reactive power control function of PV generators. However, in these works, reverse power flows occur at daytime when the output power is large. In addition, interconnection power flow fluctuates considerably, which has a harmful effect for the operation of the modern SG systems.

In the next few years, due to the commercialization of PEVs, an expansion in high market shares is expected [21,22]. The overall fuel conversion efficiency of PEVs is approximately at 22.5–45%, while that of conventional vehicles is estimated at only about 20% [23]. PEVs are environmental friendly because they emit no tailpipe pollutants. With PEVs, electric motors provide quiet, smooth operation and stronger acceleration and require less maintenance than internal combustion engines [24]. Another beneficial feature of the PEVs is that they can perform charge and discharge operations while connected to the home. Therefore, it can be considered as a controllable load, and the use of the PEV as the controller of distribution systems is expected. Another way to reduce distribution losses is to reconfigure the distribution system using electromagnetic switches between specific buses. Much research has been conducted on the optimal network reconfiguration problem for the power loss reduction [25]. Some studies have focused on the loss reduction using only the reconfiguration of the system, some of them have been considered the existence of tap transformers, and others have been taken DGs into consideration. However, none of the studies has combined the usage of PVs, BESS, PEVs, and tap transformers together with reconfiguration for reducing distribution losses. Thus, this problem is addressed in this chapter. It is formulated as an optimization problem and solved to find the best feeder configuration and power references of PVs, PEVs, and tap transformer positions in order to reduce distribution losses of DGs for the SG applications.

3.2 CONTROL OBJECTIVES

Nowadays, electric power grids are slowly getting smarter. The use of DGs is increasing in the modern power grid. Though SG promoters tout the ability of a smarter grid to enable greater consumption of DGs, the benefits could flow in both directions [26]. A zero emission-based smart DG system

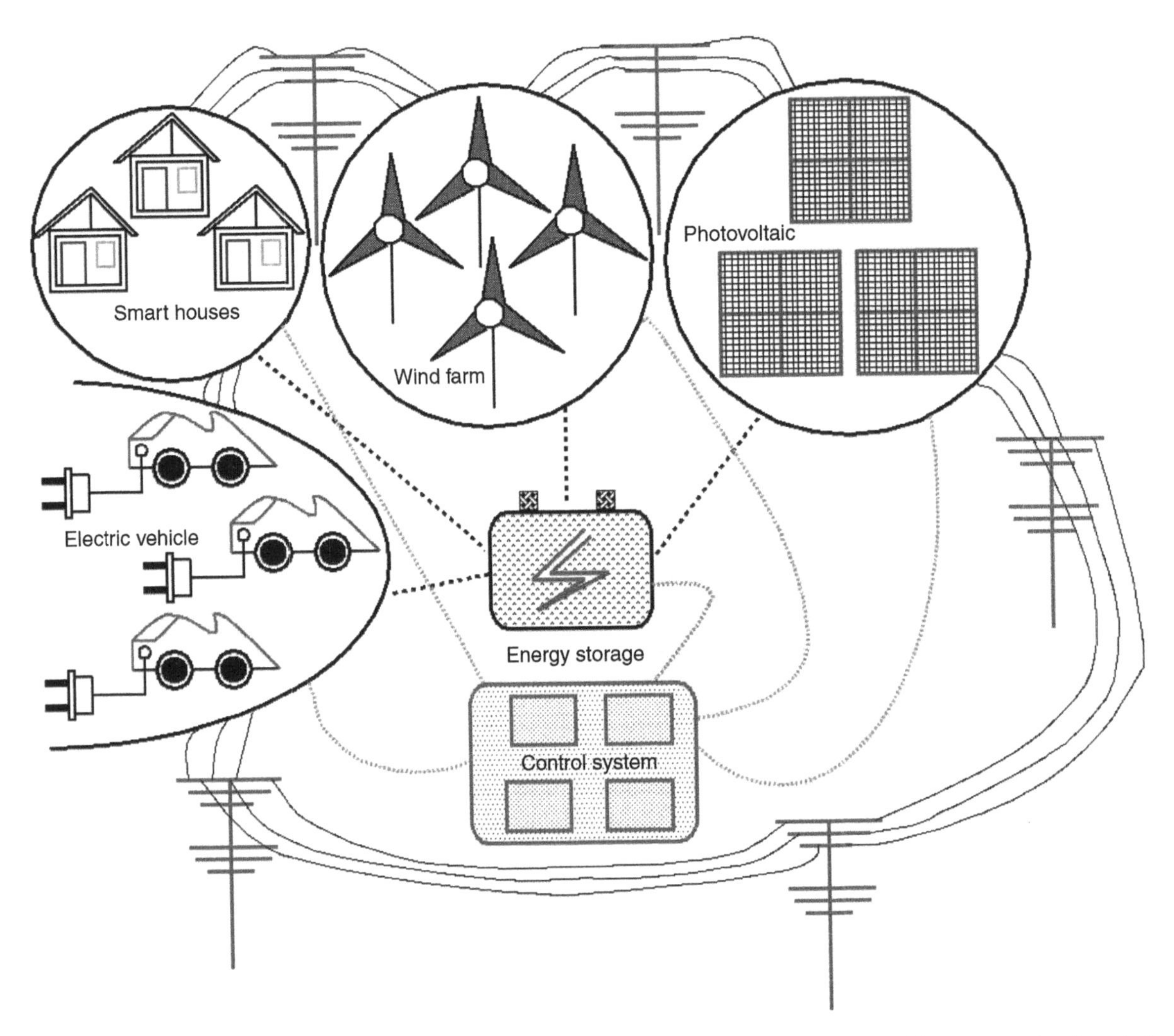

■ **FIGURE 3.1 A zero emission-based smart distributed system.**

is shown in Fig. 3.1. From this figure, all renewable sources, that is, wind farm, PV, smart houses, electric vehicles, and energy storage are connected in a power network which are controlled by an intelligence power controller.

3.2.1 Importance of distributed generations

The conventional power generation system depends on the centralized generation system but there are numerous benefits of DGs. The benefits of DGs are as follows [27,28]:

- *Reduction the transmission loss*. Usually, the DGs are located near the demand side. Therefore, they can avoid the expensive and inefficient long-distance transmission of power. A strong power grid is required to transmit power where it is consumed. Installation and maintenance costs of the power grid are expensive. On the other hand, long-distance energy transmission is inefficient due to the transmission loss. Generation and distribution in the locally can improve the power system efficiency.

- *Lessening peak load demand.* Due to the generation in the consumer-side, the DGs can decrease peak demand of a power network. It is an effective solution of the problem of high peak load shortages.
- *Impacts on local economics and communities.* DGs generate clean power locally, which creates job opportunities within the local community. Usually, DGs provide more job opportunities than developing traditional, centralized power stations. DGs can develop within underused spaces such as parking lots, unused fields, and rooftops. Therefore, they do not require the cost involved in building large power plants on undeveloped land. Also, DGs can play an important role to supply power for remote and inaccessible areas which develops the economy of remote communities. In addition, smart houses are gaining interest as a DG. The house receives power from the power grid when generation is lower than the load demand. On the other hand, the house can transmit power to the power grid when generation is more than the load demand. As a result, a house owner can be benefited to sale extra power to the power company.
- *Faster response to new power demands.* The centralized power plant requires a long installation time to deliver power for new demands. However, DGs are small scaled and often require lower installation period, it enables faster response when additional power demands are required.
- *Improvement of energy security.* DGs provide more energy security than the large centralized power plants in case of severe weather condition, natural disasters, human error, and acts of terrorism. If the centralized power plants collapsed, a huge blackout may occur in the large residential area and it becomes disaster for the residents. For example, following an earthquake, tsunami, and the failure of cooling systems at Fukushima I Nuclear Power Plant and issues concerning other nuclear facilities in Japan on March 11, 2011, a huge blackout occurred for a long period of time. A nuclear emergency had been declared in Japan and 140 thousand residents within 20 km (12 mi) of the plant were vacated [29,30].
- *Improved supply reliability and power management.* Each DG of the power system is controlled by the independence utility of grid system

which offers easy maintenance of power, voltage, and frequency. It also provides the possibility of combining energy storage and management systems, with reduced obstruction.

- *Enhancement of the power system.* An intelligent grid can support high levels of renewable energy, reducing our dependence on fossil fuels which also reduces emissions of greenhouse gases. A modern power grid incorporating with clean local energy that can fully support electric vehicles which can reduce the reliance on foreign oil.

3.2.2 Challenges of distributed generations system

Conventionally, power generation, transmission, distribution, and load demands have been managed as independent processes. Due to the large amount of DGs in the power network, the traditional approach of the power system management has gradually been shifting [31]. DGs are a vital part of the modern smart power system but integration of DGs in the conventional power network is challenging for the power system engineers. The following are issues regarding the DGs network:

- *High renewable energy penetration.* Renewable energy resources are considered as power generation sources of DGs. Because of high penetration of renewable energy resources in DGs network, active energy resources such as loads, storages, and electric vehicles will be increased. Integration of active power resources into the DGs increases the complexity of power system. Furthermore, generation of renewable sources depends on the weather condition. Power fluctuation from the renewable energy sources is serious concern for the DG networks because it may create frequency deviations, voltage flicker inside the power grid, and power system instability [32–35]. Research has been conducted in order to resolve this problem using energy storage devices [36–40]. However, installation and maintenance costs of energy storage devices are high. If power fluctuation of the DGs is increased, the capacity of the energy storage devices will be increased. Power fluctuations of some DG networks are very high and sometimes it is very difficult to install a large capacity of energy storage in power networks.
- *Power balancing.* DGs may offer operational challenges for a power system. The irregular natural power sources such as wind power and solar always provide a variable energy supply with both predictable (day–night and seasonal) fluctuations and unpredictable fluctuations driven by medium-term weather conditions and forecast errors. Such

power fluctuations will require complex power-balancing mechanisms between demand and supply. Hence, alternative power capacities such as conventional generation and energy storage are used to fill up demand/supply gaps when production from renewables is low and high.

- *Power losses.* DGs such as wind farms and PV may be located far from the demand side which requires a long transmission line to deliver the electric power. It may increase the system costs and transmission losses.
- *Impacts of real-time power market.* Many electrical sources have been taken over by independent power producers. This trend leads to the competition in power trades and power sectors [41,42]. Therefore, electricity producers should use conventional fossil fuels based power units efficiently so that they can reduce the operational cost and maximize the profit. Optimum operation of conventional units incorporated with DGs is an important issue for the SG system. Hence, unit commitment programs are introduced in order to deliver the optimum operation of the smart power system [43–47]. On the other hand, power-forecasting error of the DGs system is challenging for the real-time power market. It may increase the energy price of the power company. A reconfiguration or replanning approach of the power system can lessen the forecasting error and expansion the benefits of the power company.

3.2.3 Overview of control system

The objective of this chapter is to perform cooperative control between existing voltage control devices such as LRTs and SVRs, PVs, and BESSs introduced to interconnection points and at the part of a distribution system where it is assumed to have important loads. The proposed method aims to achieve voltage regulation within the acceptable range, smooth the power flow at the interconnection point, to ensure output power if a power outage occurs, and to reduce the total distribution losses.

According to the Electricity Law in Japan, the statutory range of voltage at the residential consumer side must be within 101 ± 6 V. In the distribution systems, between pole transformers, for the 6.6 kV high-voltage distribution system, and on the consumer side, for the 100 V system, when the load is heavy, it can cause up to the 6.5 V voltage drop and flow in reverse. Also, when the power flows to the system from distributed generators (DGs), it can cause up to 2 V voltage rise [48]. Thus, the range of voltage at the residential consumer side is from 101.5 V (0.967 pu) to 105 V (1.0 pu). If all pole transformers are setup at 6.6 kV:105 V, the system should maintain the voltage range within 6.380 kV (0.967 pu) to 6.6 kV (1.0 pu), that is, the high-voltage distribution system. This is the defined acceptable voltage range in this chapter.

In order to smooth interconnection point power flow and to increase the flexibility in the distribution system operation, a bandwidth is defined. In this research, the bandwidth of the interconnection point active power flow is defined as ±0.1 pu (500 kW) from the average of a daily load curve, and the interconnection point reactive power flow range is defined to keep the power factor between 0.8 and 1.0. The interconnection point power flow is controlled by the BESS introduced at that point.

Therefore, the purpose of this research is to maintain the node voltages within the acceptable range which is defined previously, reduce total distribution losses, and interconnect point power flow smoothing.

3.3 CONTROL METHOD

The following section describes the decision technique for the tap changing of existing voltage control devices, and the reactive power control method, using inverters interfaced with PVs, BESSs, and PEVs.

3.3.1 The objective function and constraints

The objective function is used to minimize distribution losses in terms of the node voltages, the tap positions and the reactive output power of the inverters interfaced with PV and BESS. It can be formulated as follows:

3.3.1.1 *Objective function*

$$\min : \mathrm{F}(\mathrm{P_B}, \mathrm{Q_B}, \mathrm{Q_{PV}}, \mathrm{P_{EV}}, \mathrm{Q_{EV}}, \mathrm{T_k}) = \sum_{t}^{24} \sum_{i}^{x} P_{\mathrm{L}i} \tag{3.1}$$

3.3.1.2 *Constraints*

$$V_{i\_\min} \le V_i(t) \le V_{i\_\max} \tag{3.2}$$

$$P_{\mathrm{f}}^{\min} \le P_{\mathrm{f}}(t) \le P_{\mathrm{f}}^{\max} \tag{3.3}$$

$$Q_{\mathrm{f}}^{\min} \le Q_{\mathrm{f}}(t) \le Q_{\mathrm{f}}^{\max} \tag{3.4}$$

These constraints are the voltage constraints in the distribution system, active power flow constraints, and reactive power flow constraints, respectively. The bandwidth of voltage and active power flow are already mentioned. The reactive power flow bandwidth is as follows.

$$Q_{\mathrm{f}}^{\min} = -P_{\mathrm{f}}(t) * \tan\left[\cos^{-1}(0.8)\right] = -P_{f}(t) * 0.75 \tag{3.5}$$

$$Q_{\mathrm{f}}^{\max} = P_{\mathrm{f}}(t) * \tan\left[\cos^{-1}(0.8)\right] = P_{f}(t) * 0.75 \tag{3.6}$$

$$\sqrt{P_{\mathrm{B}}^{2} + Q_{\mathrm{B}}^{2}(t)} \leq S_{\mathrm{B}} \tag{3.7}$$

$$\zeta_{\mathrm{B}}(t+1) = \begin{cases} \zeta_{\mathrm{B}}(t) - \dfrac{P_{\mathrm{B}}(t)/\eta}{C_{\mathrm{B}}(t)} & (P_{\mathrm{B}}(t) \geq 0) \\ \zeta_{\mathrm{B}}(t) - \dfrac{P_{\mathrm{B}}(t) * \eta}{C_{\mathrm{B}}(t)} & (P_{\mathrm{B}}(t) < 0) \end{cases} \tag{3.8}$$

$$20 \leq \zeta_{\mathrm{B}}(t) \leq 80 \tag{3.9}$$

These constraints are the large capacity BESS inverter constraints, considering the loss in charging and discharging of the BESS for each power constraint, and state of charge (SOC) constraints for the purpose of rapid degradation control of the BESS, respectively.

$$\sqrt{P_{\mathrm{PV}}^{2}(t) + Q_{\mathrm{PV}}^{2}(t)} \leq S_{\mathrm{PV}} \tag{3.10}$$

$$\sqrt{P_{\mathrm{EV}}^{2}(t) + Q_{\mathrm{EV}}^{2}(t)} \leq S_{\mathrm{EV}}(t) \tag{3.11}$$

$$P_{\mathrm{EV}}(t) \leq S_{\mathrm{EV}} \tag{3.12}$$

$$\zeta_{\mathrm{EV}}(t+1) = \begin{cases} \zeta_{\mathrm{EV}}(t) - \dfrac{P_{\mathrm{EV}}(t)/\eta}{C_{\mathrm{EV}}(t)} & (P_{\mathrm{EV}}(t) \geq 0) \\ \zeta_{\mathrm{EV}}(t) - \dfrac{P_{\mathrm{EV}}(t) * \eta}{C_{\mathrm{EV}}(t)} & (P_{\mathrm{EV}}(t) < 0) \end{cases} \tag{3.13}$$

$$20 \leq \zeta_{\mathrm{EV}}(t) \leq 100 \tag{3.14}$$

$$T_{k}^{\min} \leq T_{k}(t) \leq T_{k}^{\max} \tag{1.15}$$

These constraints are the PV inverter capacity constraints, PEV inverter constraints, SOC of the PEV constraints, and tap transformer's tap position constraints, respectively. The symbols of the aforementioned equations indicate as follows:

P_{B}	Active power of the battery
Q_{B}	Reactive power of the battery
P_{PV}	Active power of the PV
Q_{PV}	Reactive power of the PV
P_{EV}	Active power of the PEV
Q_{EV}	Reactive power of the PEV
T_{K}	Tap position
T_K^{min}	Minimum tap position
T_K^{min}	Maximum tap position
P_{L}	Total power flow to the distribution network
V_i	Node voltage
$V_{\mathrm{i\_min}}$	Minimum node voltage
$V_{\mathrm{i\_max}}$	Maximum node voltage
P_{f}	Active power flow to the interconnection point
$P_{\mathrm{f}}^{\mathrm{min}}$	Minimum active power flow to the interconnection point
$P_{\mathrm{f}}^{\mathrm{mix}}$	Maximum active power flow to the interconnection point
Q_{f}	Reactive power flow to the interconnection point
$Q_{\mathrm{f}}^{\mathrm{min}}$	Minimum reactive power flow to the interconnection point
$Q_{\mathrm{f}}^{\mathrm{mix}}$	Maximum reactive power flow to the interconnection point
S_{B}	Marginal capacity of the inverter for BESS
ζ_{B}	SOC of the battery
ζ_{EV}	SOC of the PEV
S_{PV}	Marginal capacity of the inverter for PV
S_{EV}	Marginal capacity of the inverter for PEV
η	Efficiency of the system

3.4 PARTICLE SWARM OPTIMIZATION

There are many methods to solve the aforementioned optimization problem. In this work, particle swarm optimization (PSO) [49,50] was chosen. PSO is an optimization method which uses the general idea that a flock of birds can find the path to food by cooperation. This is modeled by the particle swarm which has the search position and velocity information in multidimensional space. The PSO algorithm is follows:

Step 1	Generate an initial searching point for each swarm.
Step 2	Evaluate the objective function using each swarm's searching point.
Step 3	Finish searching if stopping conditions are satisfied. If not, go to Step4.
Step 4	Search the next point considering the best of the current swarm's searching point and every swarm's best searching point. Go to Step2.

The searching algorithm communicates the best positions information to all swarms, and each continues updating their own positions and velocities until finished searching. The updating of velocity and search position is decided by following equation:

$$V_{k+1}(i) = w \cdot V_k(i) + c_1 \cdot \text{rand} \cdot (\text{pbest}(i) - S_k(i)) + c_2 \cdot \text{rand} \cdot (\text{gbest} - S_k(i)) \quad (3.16)$$

$$s_{k+1}(i) = s_k(i) + V_{k+1}(i) \quad (3.17)$$

where,

$V_{k+1}(i)$	ith particle velocity in k + 1th search
$rand_1$	Uniform random numbers from 0 to 1
S_{k+1}	Search position of ith particle in kth search
w	Weighting of inertia
c_1	Weighting for best position of self particle
c_2	Weighting for best position of particle swarm
pbest	Best position of self particle
gbest	Best position of particle swarm.

3.4.1 PV generator system

In this chapter, the DGs are considered based on PV generators which are becoming more popular in smart houses. The reactive output power from the inverters interfaced to the DGs is used to control the distribution system voltage. The reactive power is maintained within the capacity of the DG

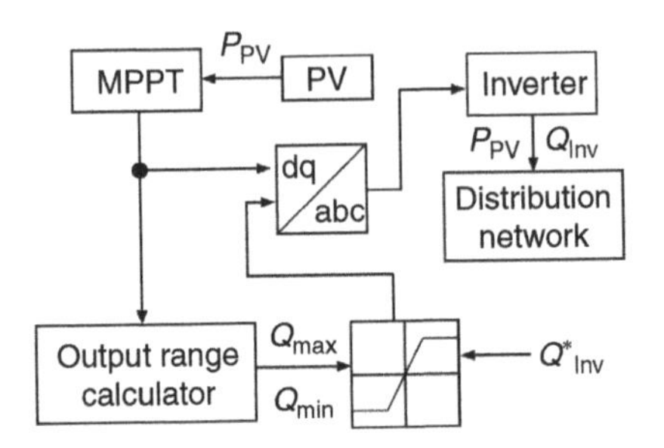

■ **FIGURE 3.2 Reactive power control system for the PV.**

inverters while maximizing the PV active power. The reactive power control system scheme is shown in Fig. 3.2. The active power, P_{PV}, from the PV feeds to the maximum power point tracking (MPPT) control system to generate the maximum output power [51]. Then, the output power range is generated from the active power of the PV generator, P_{PV}. It determines the margin of the inverter capacity, S_{PV}. The optimal scheduling reactive power reference, Q^*_{PV} adapts the range of the reactive power for the PV generator. The active and reactive powers of the PV generator are estimated the dq to abc conversion [52]. The three-phase voltage reference (abc) is the input of a voltage source inverter, and the inverter delivers the active P_{PV} and reactive power Q_{PV} to the distribution networks.

3.4.2 BESS at the interconnection point

To suppress large variations of power flow at the interconnection point, the active and reactive powers are controlled using the BESS at the interconnection point. Fig. 3.3 shows the active and reactive powers control systems

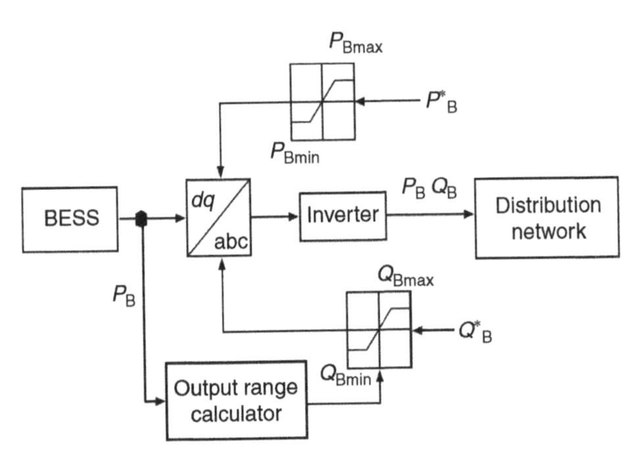

■ **FIGURE 3.3 BESS control system.**

Table 3.1 Parameters of the Distribution System

Parameter	Comparison Method	Proposed Method
Line impedance of each section	$0.04 + j0.04$ pu	
Rated capacity of PV node	0.08 pu (400 kW)	
Rated capacity interfaced inverter of PV	0.08 pu (400 kW)	
Large BESS capacity	5.0 pu (25 MWh)	2.0 pu (10 MWh)
Rated capacity interfaced inverter of BESS	0.4 pu (2 MW)	0.03 pu (1.5 MW)
PEV capacity		0.24 pu (1200 kW)
Rated capacity interfaced inverter of PEV		0.06 pu (300 kW)

of the BESS. The control strategy of the BESS system is similar as the PV system.

In this control system, active power reference signal P^*_{B} and the reactive power reference signal Q^*_{B} are used for the optimal schedule algorithm. Based on the control references, the PSO-based optimization method considering the forecast information of the BESS is controlled to satisfy the power flow bandwidth constraints at the interconnection point. In this chapter, the BESS is assumed to have a sodium–sulfur battery, and the efficiency of the BESS is 80% without considering self-discharging of the BESS.

3.4.3 Plug-in electric vehicle

The PEV assumed in this chapter is based on commercial PEVs, and the parameters are described in Table 3.1. When the PEV is connected to the grid, we can control the active and reactive powers of the grid. In addition, this chapter considers two different demand areas. Moreover, when the PEV is connected to the grid, the SOC of the PEV corresponding to each area has been taken into consideration. The PEV's SOC are assumed with a percentage of ownership in a residential area of 80%, and office area of 30%. To cope with the reverse power flow, PEVs are charging in daytime and discharging at nighttime.

3.5 SIMULATION RESULTS

In this chapter, in order to show the effectiveness of the proposed cooperative control method based on optimal reference control, simulations are performed based on the distribution system model which is shown in Fig. 3.4.

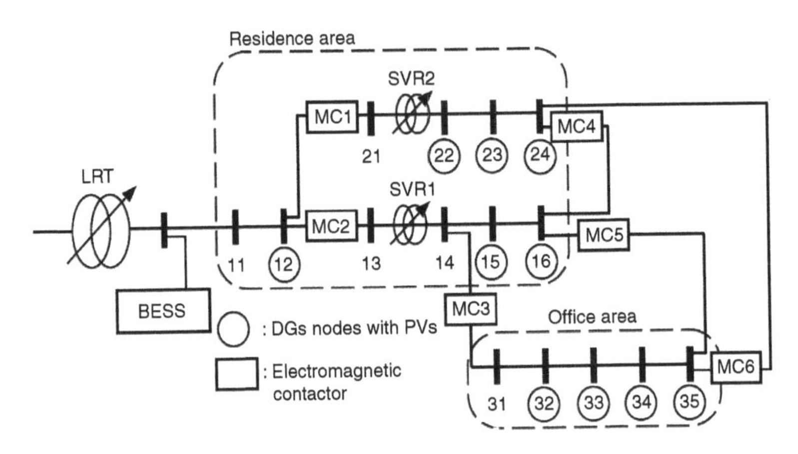

■ **FIGURE 3.4 Model of distribution system.** MC, Magnetic connector.

The system consists of total 15 DGs nodes at two different areas (ie, residential and office areas). In DGs, all nodes are connected to the PEVs and 10 nodes are connected to the PVs (circles indicate of these nodes). Due to the electromagnetic contactor in Fig. 3.4, the distribution systems can be considered as the reconfigurable distribution systems. Line impedances, power factor of the load demand, rated capacity of the inverters interfaced PVs, BESSs, and rated capacity of PEVs for DGs are listed in Table 3.1, where the nominal capacity of the DG and the nominal voltage are 5 MVA and 6.6 kV, respectively.

This research assumes to get the high accuracy forecast information of the PV output power and load demand for the next day. The actual PV output power profile for an hour average value is shown in Fig. 3.5a. This power profile has been used for the optimum scheduling method. The load demand profile of a day is shown in Fig. 3.5b, and it is assumed that nodes 11–24 are considered for residential area, while nodes 31–35 are considered for the office area.

The proposed system (Fig. 3.4) is a reconfigurable distribution system, thus, the distributions can consider for eight cases. The configuration and optimized distribution loss for the 24-h of each case are listed in Table 3.2. It is confirmed from the Table 3.2, that case 2 contains the smallest distribution loss (1469 kWh) among the all cases. Therefore, case 2 is selected for the comparison method. The comparison method considers the optimize approach where the distribution loss is the smallest. Finally, the proposed method considers the 1-h optimization for all cases. After comparing the all

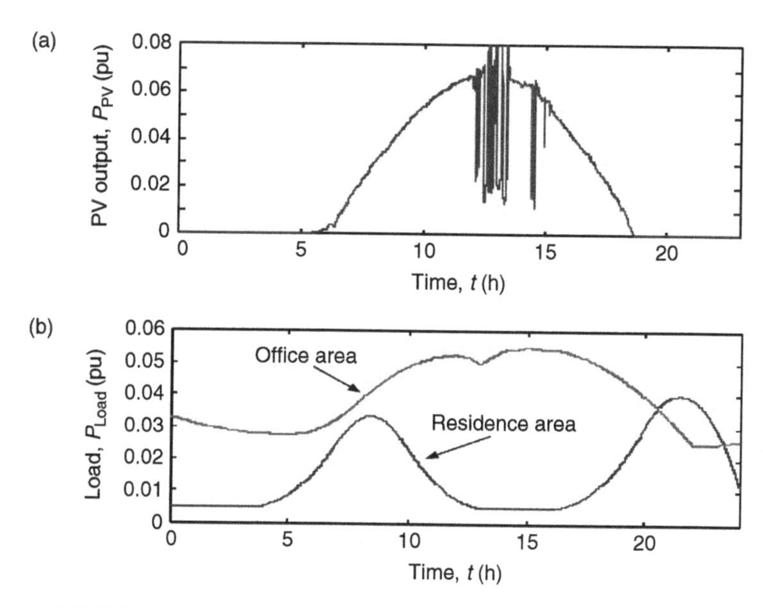

FIGURE 3.5 (a) PV output power, (b) load demand.

Table 3.2 Distribution Loss of Each Case

Case	Close EC	Distribution Loss (kWh)	Case	Close EC	Distribution Loss (kWh)
1	1,2,3	1541	5	1,2,5	1910
2	2,3,4	1469	6	1,3,4	2920
3	2,3,6	1451	7	1,4,5	2422
4	1,2,6	1757	8	1,3,6	2242

cases results, the proposed method selects the best case where the distribution loss is minimal. This action repeats for 24 h.

Table 3.3 shows the selected case for every hour of the proposed method. From this table, case 1 and case 3 are selected relatively more than other cases. However, there is a very important matter to select case 8 at time 10–18 h. The reason behind this is that all DGs are connected in series for case 8. Usually, in that time, the output power of the PV is maximum, and it is difficult to reduce the distribution loss of the system. Since, the DGs are connected in series for case 8, it is more flexible for the voltage control using the reactive power control from these DGs. From Table 3.3, the distribution loss of the proposed method (1129 kWh) is smaller than the comparison

Table 3.3 Results of the Proposed Method

Time	1	2	3	4	5	6	7	8	9	10	11	12
Case	3	3	1	1	1	2	1	1	4	3	7	8
Time	13	14	15	16	17	18	19	20	21	22	23	24
Case	8	8	8	3	3	3	3	3	1	3	1	1
Distribution loss									1129 kWh			

method (1469 kWh). The simulations have been conducted in three patterns as follows: without optimization method (ie, without any control devices), comparison method, and proposed method. The dynamic responses of each method are described in the following subsection.

3.5.1 Dynamic responses for the without optimization approach

Simulation results for the without optimization method are shown in Fig. 3.6. Fig. 3.6a shows the node voltages of the distribution system, where the dashed lines indicate the acceptable voltage range or constraint of the node voltages. From this figure, the acceptable range of the node voltages is very narrow and fluctuates over the acceptable range. Therefore, the without optimization approach cannot control the node voltages and the system may fall into an imbalance situation. From Fig. 3.6b–c, active and reactive powers of the system are also fluctuated over the acceptable ranges. Therefore, the without optimization method cannot control the system appropriately.

3.5.2 Dynamic responses for the comparison method

Simulation results of the comparison method are shown in Fig. 3.7. The dashed lines of these figures mean the acceptable range or constraint of the system. Fig. 3.7a shows the node voltages of the comparison method. From this figure, the node voltages satisfy the constraint of the system, and due to the active output powers of the PV generators, the node voltages are high at the time 10–18 h. If the lower-node voltages are required at this time, the distribution must be controlled by the active powers of the PV generators. The active and reactive powers at the interconnection point are shown in Fig. 3.7b–c, respectively. Both figures also fulfill the constraints. Again, from Fig. 3.7b, the active power flow is always in the lower acceptable range. It means that the active power flow satisfies the constraint even if the acceptable range is lower. The dashed lines in Fig. 3.7c are always a constant value because the active power flow is almost constant at every time.

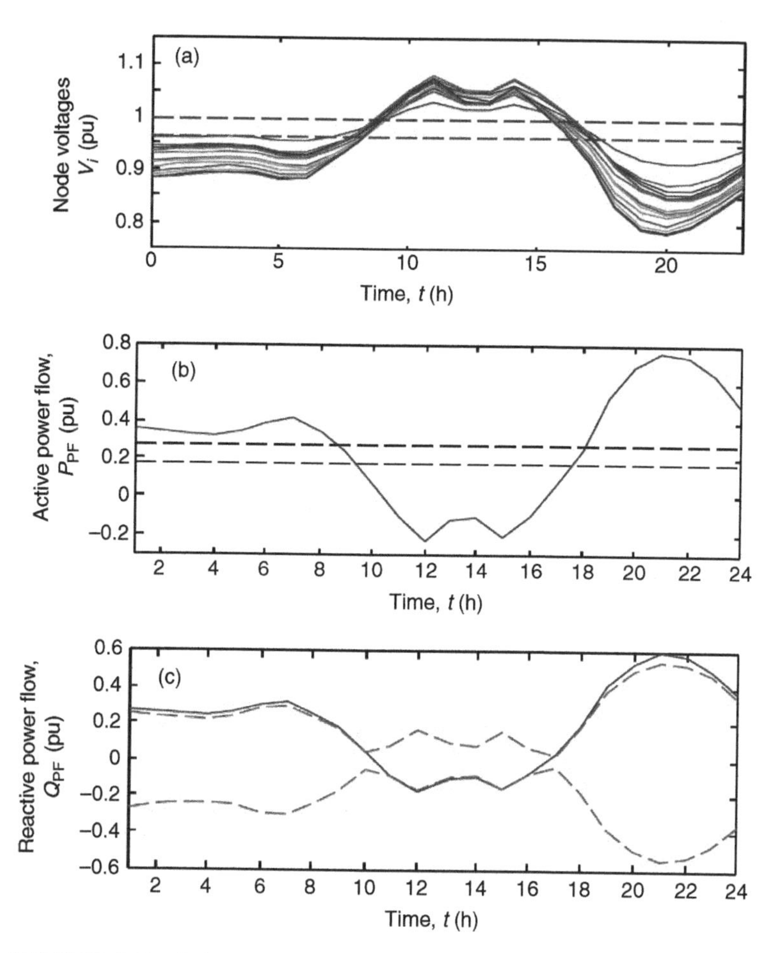

■ **FIGURE 3.6 Simulation results of without optimization control.** (a) Node voltages, (b) active power flow at interconnection point, and (c) reactive power flow at interconnection point.

The active and reactive powers of the BESS are depicted in Fig. 3.7d. Most of the time, the active power is larger than the reactive power of the battery. Therefore, it can confirm that the main work of the BESS is the active power control at the interconnection point. At the time 10–18 h, the active power of the battery is negative (which means battery is charging at this period) because the generation of the PV is high in this period. The SOC of the BESS is indicated in Fig. 3.7e. From this figure, the SOC satisfies the constraint of the BESS. Therefore, it is possible to construct a stable charge and discharge actions of the BESS. Figs. 3.7f–g show the active and reactive powers from the PEV at residence and office areas, respectively. It is

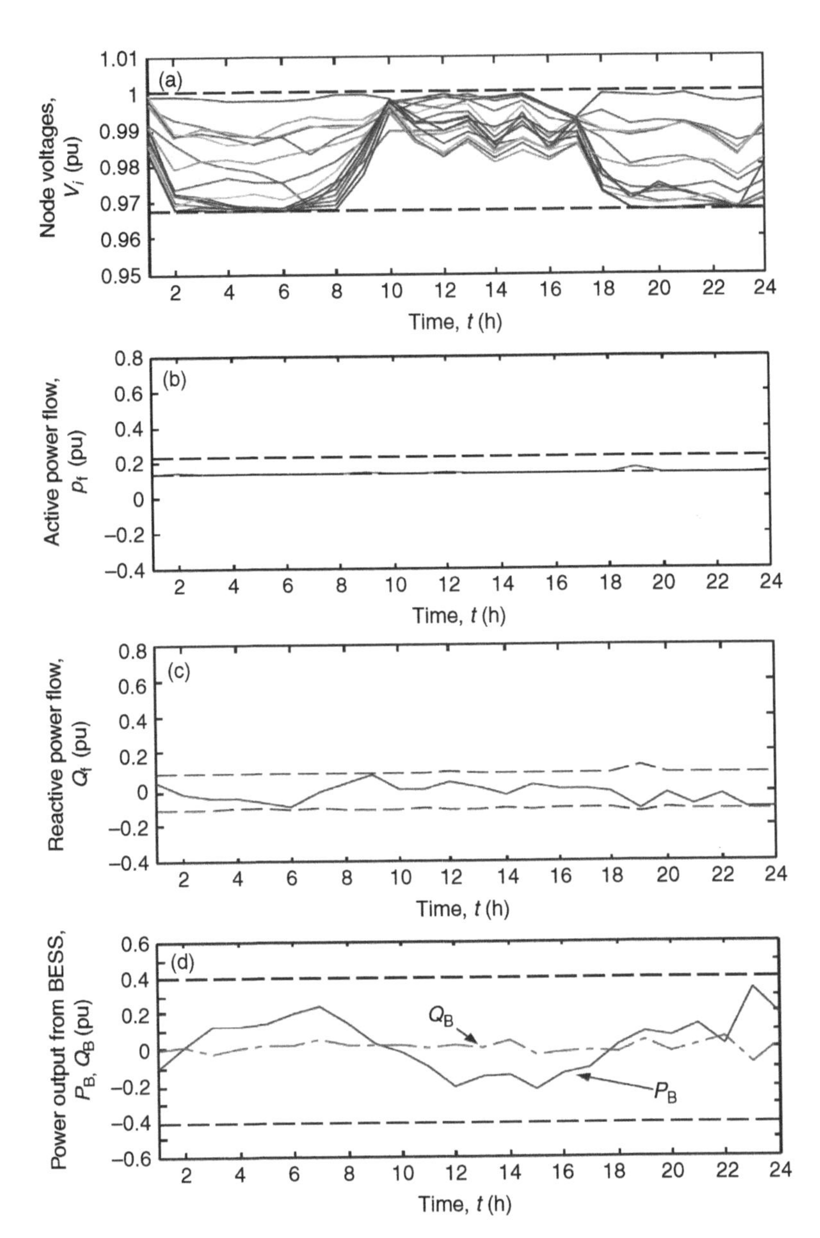

■ **FIGURE 3.7 Simulation results of the comparison method.** (a) Node voltages, (b) active power flow at interconnection point, (c) reactive power flow at interconnection point, (d) active and reactive output powers from BESS,

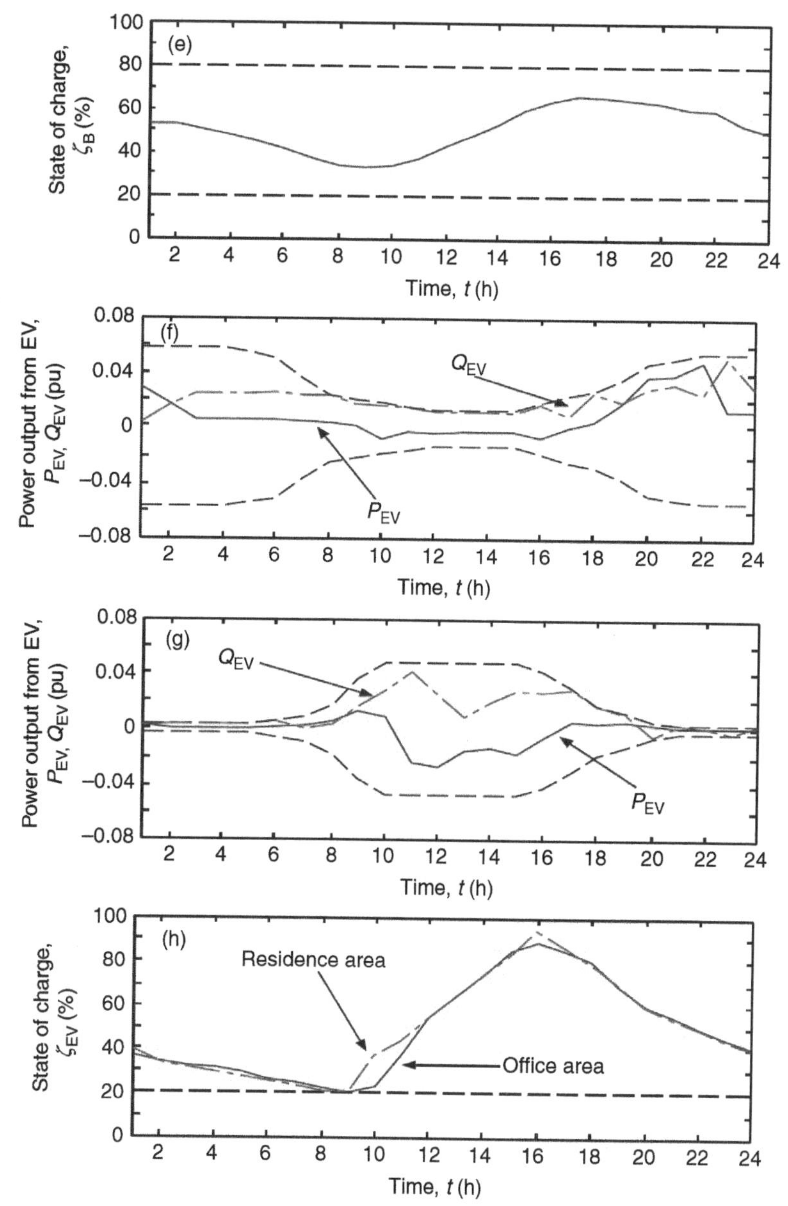

■ **FIGURE 3.7** (*Continued*) (e) state of charge of the BESS, (f) active and reactive output powers of PEV at residence area, (g) active and reactive output powers of PEV at office area, (h) state of charge each area of PEV,

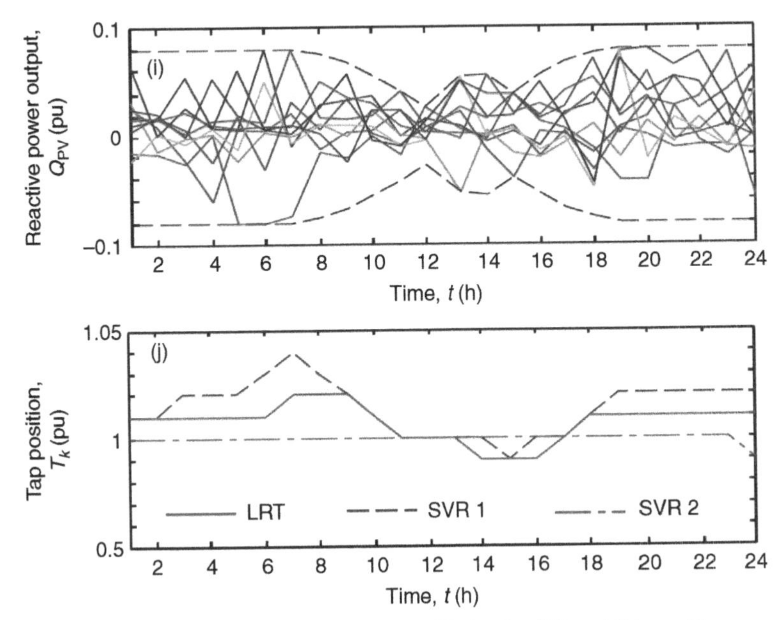

FIGURE 3.7 (*Continued*) (i) reactive output power of the interfaced DGs, and (j) tap positions.

confirmed that the reactive power is higher than the active power in both areas. It means the voltage control by the reactive power is more effective than the active power control to improve the distribution qualities.

The SOC of each PEV is shown in Fig. 3.7h. It can be seen from this figure, the SOC of each area is almost same. Therefore, it reflects that the SOC of the PEV is in balance at each area. Fig. 3.7i shows the reactive output power from the PV generator. The reactive output power is positive for the voltage enlargement. The tap positions of LRT and SVRs are depicted in Fig. 3.7j. The limit of the tap position is defined as $0.90 \leq T_k \leq 1.10$. From Fig. 3.7j, the tap position can be controlled within constraints. From the comparison method, the voltage control is accomplished by the reactive power from PV and PEV, and power flow control is achieved by the active power control of the BESS.

3.5.3 Dynamic responses for the proposed method

Fig. 3.8 shows the simulation results for the proposed method. The constraints of the proposed method are followed as similar as the comparison method (Fig. 3.7). In spite of the same SOC of the PEV (Figs. 3.8h and 3.7h) in both methods, the SOC of the BESS for the proposed method in Fig. 3.8e is around 10% higher than the comparison method (Fig. 3.7e). Therefore, it is confirmed that the proposed method can reserve the energy of the BESS.

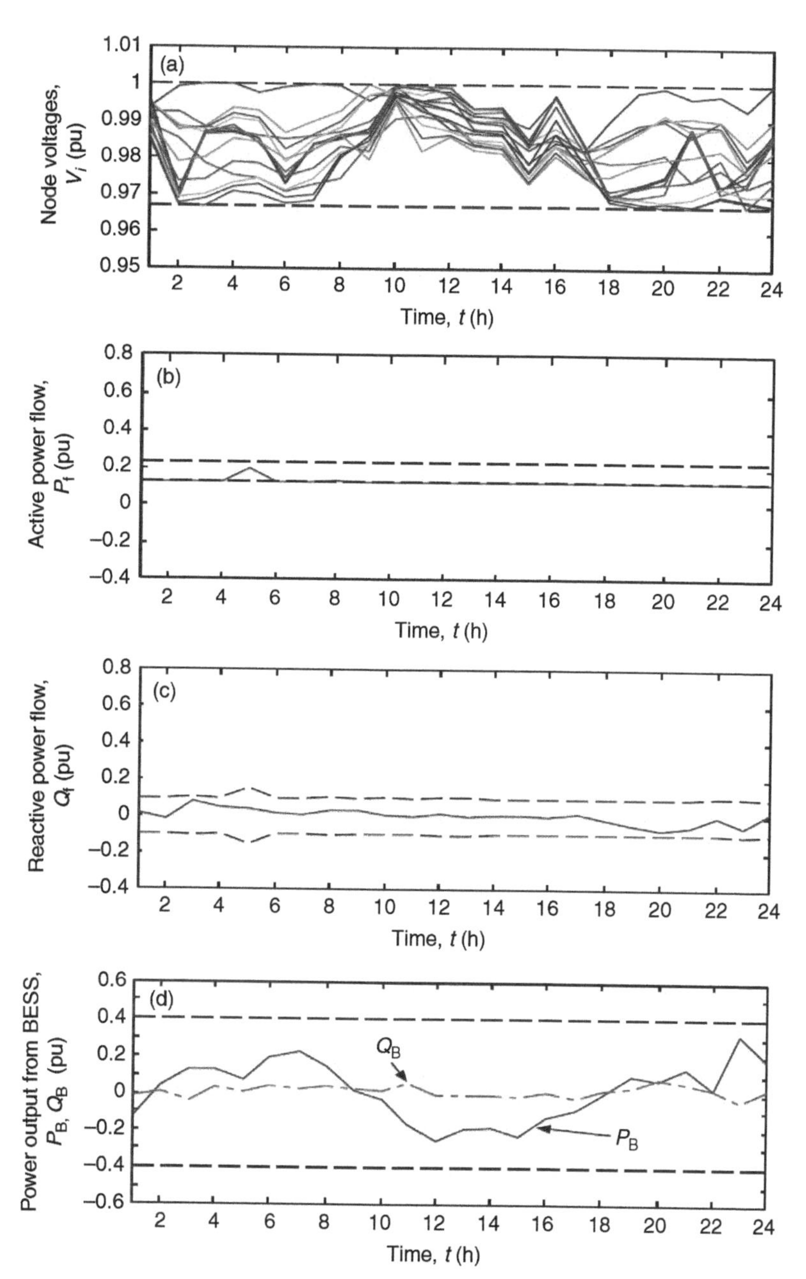

■ **FIGURE 3.8 Simulation results of the proposed method.** (a) Node voltages, (b) active power flow at interconnection point, (c) reactive power flow at interconnection point, (d) active and reactive output powers from BESS,

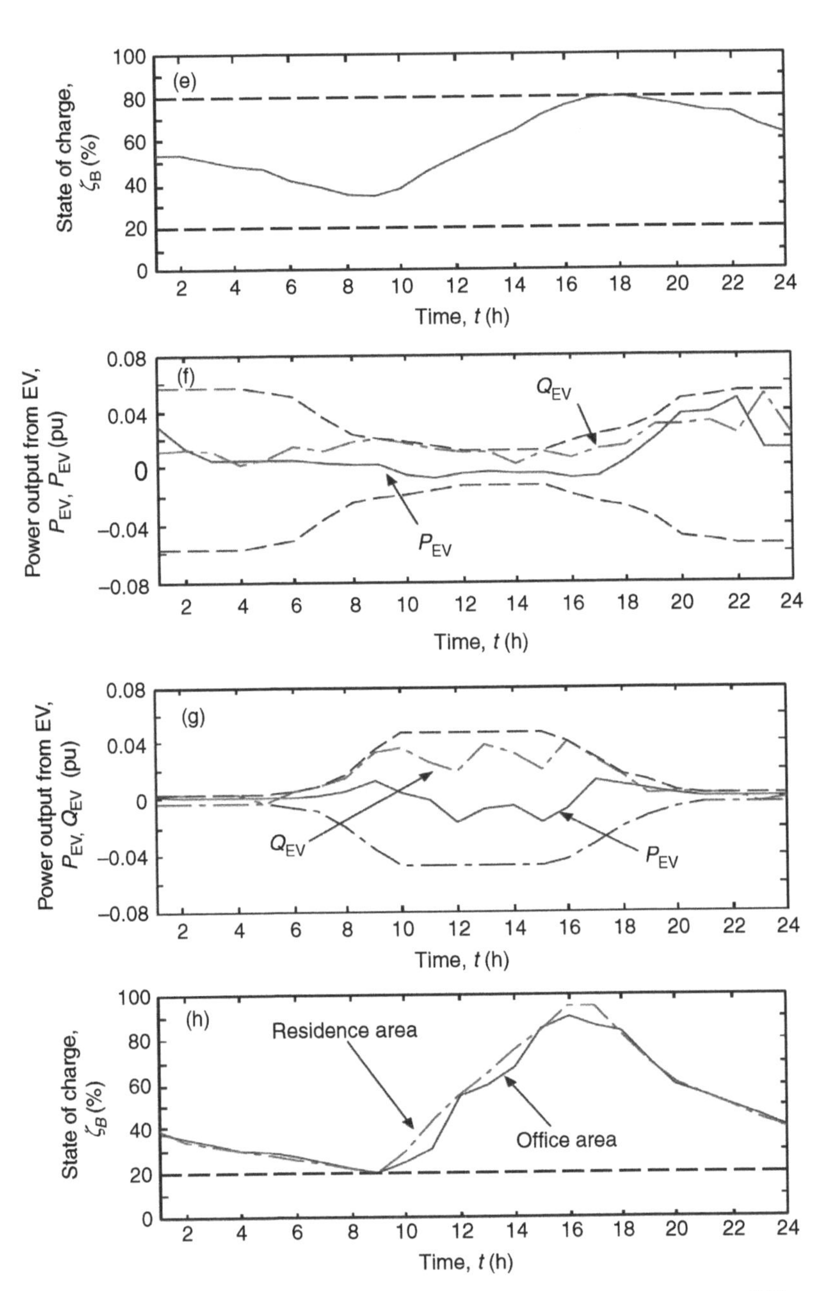

FIGURE 3.8 *(Continued)* (e) state of charge of the BESS, (f) active and reactive output powers of PEV at residence area, (g) active and reactive output powers of PEV at office area, (h) state of charge each area of PEV,

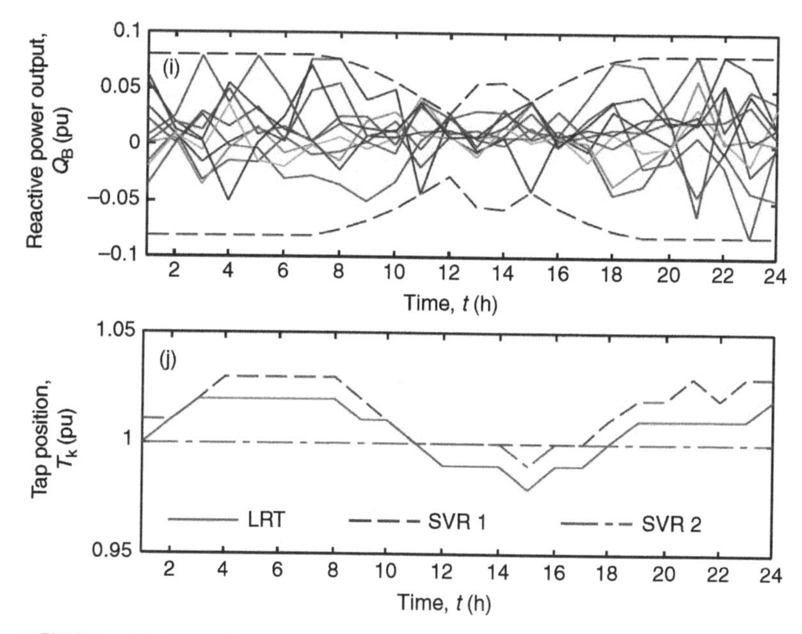

FIGURE 3.8 (*Continued*) (i) reactive output power of the interfaced DGs, and (j) tap positions.

3.6 CONCLUSIONS

This chapter analyses an optimal control method of PV interfaced inverters, LRTs and SVRs, house BESSs, large BESSs, and PEVs in reconfigurable distribution systems. This chapter also introduces distributed energy sources, energy storages, and SG application. From the simulations, it is confirmed that the proposed method can reduce the distribution losses as compared with the comparison method. The reference schedule is optimized based on constraints of these control devices. Comparisons of the results are analyses in different cases. The distributed energy sources are an important element of the SG systems. Therefore, analyses of this chapter are effective to reduce the overall losses for the SG systems.

REFERENCES

[1] Statistical Review of World Energy, http://www.bp.com/en/global/corporate/about-bp/energy-economics/statistical-review-of-world-energy.html; 2014.

[2] Energy Outlook 2035, http://www.bp.com/en/global/corporate/about-bp/energy-economics/energy-outlook.html.

[3] Howlader AM, Urasaki N, Yona A, Senjyu T, Saber AY, et al. Control strategies for wind farm based smart grid system. IEEE 10th International Conference on the Power Electronics and Drive Systems, KitaKyushu, Japan; 2013, pp. 463–467. doi: 10.1109/PEDS. 2013.6527063.

[4] Ferrari ML, Traverso A, Pascenti M, Massardo AF, et al. Plant management tools tested with a small-scale distributed generation laboratory. Energ Convers Manage 2014;78:105–13.

[5] Howlader HOR, Matayoushi H, Senjyu T, et al. Distributed generation incorporated with the thermal generation for optimum operation of a smart grid considering forecast error. Energ Convers Manage 2015;96:303–14.

[6] Choudar A, Boukhetala D, Barkat S, Brucker JM, et al. A local energy management of a hybrid PV-storage based distributed generation for microgrids. Energ Convers Manage 2015;90:21–33.

[7] Picciariello A, Reneses J, Frias P, Soder L, et al. Distributed generation and distribution pricing: why do we need new tariff design methodologies? Electr Pow Syst Res 2015;119:370–6.

[8] Shayeghi H, Sobhani B, et al. Zero NDZ assessment for anti-islanding protection using wavelet analysis and neuro-fuzzy system in inverter based distributed generation. Energ Convers Manage 2014;79:716–25.

[9] Thatte AA, Ilic MD, et al. An assessment of reactive power/voltage control devices in distribution networks. Power Engineering Society General Meeting, vol. 1; 2006, pp. 8–15.

[10] Radman G, Raje RS, et al. Dynamic model for power systems with multiple FACTS controllers. Electr Pow Syst Res 2008;78:361–71.

[11] Viawan FA, Karlsson D, et al. Voltage and reactive power control in systems with synchronous machine-based distributed generation. IEEE Trans Power Deliv 2008;23(2):1079–87.

[12] Atmaca E, et al. An ordinal optimization based method for power distribution system control. Electr Pow Syst Res 2008;78:694–702.

[13] Liang R, Wang Y, et al. Fuzzy-based reactive power and voltage control in a distribution system. IEEE Trans Power Deliver 2003;18(2):610–8.

[14] Wasiak I, Thoma MC, Foote ET, Mienski R, Pawelek R, Gburczyk P, Burt GM, et al. A power-quality management algorithm for low-voltage grids with distributed resources. IEEE Trans Power Deliver 2008;23(2):1055–62.

[15] Howlader AM, Izumi Y, Uehara A, Urasaki N, Senjyu T, Yona A, Saber AY, et al. A minimal order observer based frequency control strategy for an integrated wind-battery-diesel power system. Energy 2012;46(1):168–78.

[16] Howlader AM, Izumi Y, Uehara A, Urasaki N, Senjyu T, Saber AY, et al. A robust H_∞ controller based frequency control approach using the wind-battery coordination strategy in a small power system. Int J Elec Power 2014;58:190–8.

[17] Diva KC, Ostergaard J, et al. Battery energy storage technology for power systems – an overview. Electr Pow Syst Res 2009;79(4):511–20.

[18] Nair NC, Garimella, et al. Battery energy storage systems: assessment for small-scale renewable energy integration. Energ Buildings 2010;42(11):2124–30.

[19] Oshiro M, Tanaka K, Uehara A, Senjyu T, Miyazato Y, Yona A, Funabashi T, et al. Optimal voltage control in distribution systems with coordination of distribution installations. Int J Elec Power 2010;32:1125–34.

[20] Tanaka K, Oshiro M, Toma S, Yona A, Senjyu T, Funabashi T, Kim CH, et al. Decentralized control of voltage in distribution systems by distributed generators. IET Gener Transm Dis 2010;4:1251–60.

[21] Hadley SW, Tsvetkova A, et al. Potential impacts of plug-in hybrid electric vehicles on regional power generation. Electr J 2009;22(10):56–68.

[22] Santini D, Duvall M. Global prospects of plug-in hybrids. Presented at the EVS-22, Yokohama, Japan; October 2006.

[23] Shao S, Pipattanasomporn M, Rahman S, et al. Demand response as a load shaping tool in an intelligent grid with electric vehicles. IEEE Trans Smart Grid 2011;2(4):624–31.

[24] US Department of Energy. All-Electric Vehicles, http://www.fueleconomy.gov/feg/evtech.shtml

[25] Oliveiraa LW, Carneiro S Jr, Oliveirab EJ, Pereirab JLR, Silva IC Jr, Costa JS, et al. Optimal reconfiguration and capacitor allocation in radial distribution systems for energy losses minimization. Int J Elec Power Energy Syst 2010;32(8): 840–8.

[26] Power, Business and Technology for the Global Generation Industry. http://www.powermag.com/

[27] Clean-Coalition. http://www.clean-coalition.org/

[28] India Smart Grid Knowledge Portal. http://indiasmartgrid.org/

[29] Howlader AM, Urasaki N, Yona A, Senjyu T, Saber AY, et al. Review of output power smoothing methods for wind energy conversion. Renew Sust Energ Rev 2013;26:135–46.

[30] Wiesenthal J., et al. Japan declares nuclear emergency, as cooling system fails at power plant, Business Insider Report; March 11, 2011.

[31] Pertti J, Sami R, Antti R, Jarmo P, et al. Smart grid power system control in distributed generation environment. Annu Rev Control 2010;34:277–86.

[32] Howlader AM, Izumi Y, Uehara A, Urasaki N, Senjyu T, Yona A, Saber AY, et al. A minimal order observer based frequency control strategy for an integrated wind–battery–diesel powers system. Energy 2012;46:168–78.

[33] Howlader AM, Urasaki N, Yona A, Senjyu T, Saber AY, et al. Design and implement a digital H_∞ robust controller for a MW-class PMSG based grid-interactive wind energy conversion system. Energies 2013;6:2084–109.

[34] Kamel RA, Chaouachi A, Nagasaka K, et al. Wind power smoothing using fuzzy logic pitch controller and energy capacitor system for improvement micro-grid performance in islanding mode. Energy 2010;35:2119–29.

[35] Howlader AM, Urasaki N, Pratap A, Senjyu T, Saber AY, et al. A fuzzy control strategy for power smoothing and grid dynamic response enrichment of a grid-connected wind energy conversion system. Wind Energy 2013;17:1347–63.

[36] Abbey C, Joos G, et al. Super capacitor energy storage for wind energy applications. IEEE T Ind Appl 2007;43:769–76.

[37] Teleke S, Baran ME, Bhattacharya S, Huang A, et al. Validation of battery energy storage control for wind farm dispatching. Power Engineering Society General Meeting; 2010, pp. 1–7.

[38] Sheikh MRI, Muyeen SM, Takahashi R, Tamura J, et al. Smoothing control of wind generator output fluctuations by PWM voltage source converter and chopper controlled SMES. Eur Trans Electr Power 2011;21:680–97.

[39] Liu C, Hu C, Li X, Chen Y, Chen M, Xu D, et al. Applying SMES to smooth short-term power fluctuations in wind farms. 34th annual conference of IECON; 2008, pp. 3352–3357.

[40] Cardenas R, Pena R, Asher G, Clare J, et al. Power smoothing in wind generation systems using a sensorless vector controlled induction machine driving a flywheel. IEEE Transactions on Energy Conversion, 2004 2004;19:206–16.

[41] Kyoho R, Senjyu T, Yona A, Funabashi T, et al. Optimal control of thermal units commitment for renewable energy generators considering forecast error. The 5th International Conference on Advanced Power System Automation and Protection (APAP2013), Jeju, Korea; October 28–31, 2013.
[42] Higa S, Mengyan W, Yona A, Senjyu T, Funabashi T, et al. Optimal operation method considering uncertainly of renewable energy and load demand in micro-grids. The 5th International Conference on Advanced Power System Automation and Protection (APAP2013), Jeju, Korea; October 28–31, 201.
[43] Kyoho R, Senjyu T, Yona A, Funabashi T, Kim CH, et al. Optimal operation of thermal units commitment with demand response considering uncertainties. Seventeenth International Conference on Intelligent System Applications to Power Systems (ISAP2013), 824; Tokyo, Japan, 1–4, 2013.
[44] Yoza A, Kyoho R, Uchida K, Yona A, Senjyu T, et al. Optimal scheduling method of controllable loads in smart house with uncertainties. Seventeenth International Conference on Intelligent System Applications to Power Systems (ISAP2013), 908; Tokyo, Japan, 1–4, 2013.
[45] Goya T, Senjyu T, Yona A, Urasaki N, Funabashi T, et al. Optimal operation of thermal unit in smart grid considering transmission constraint. Int J Elec Power 2012;40:21–8.
[46] Chakraborty S, Senjyu T, Saber AY, Yona A, Funabashi T, et al. A fuzzy binary clustered particle swarm optimization strategy for thermal unit commitment problem with wind power integration. IEEJ T Electr Electr 2012;7:478–86.
[47] Wang J, Shahidehpour M, Li Z, et al. Security-constrained unit commitment with volatile wind power generation. IEEE T Power Syst 2008;23:1319–27.
[48] Matsuda K, Uemura S, et al. Study on voltage control method for distribution systems with distributed generation, Central Research Institute of Electric Power Industry Report, No. T02016; 2003 (in Japanese).
[49] Chakraborty S, Senjyu T, Saber AY, Yona A, Funabashi T, et al. A novel particle swarm optimization method based on quantum mechanics computation for thermal economic load dispatch problem. IEEJ T Electr Electr 2012;7(5):461–70.
[50] Chakraborty S, Ito T, Senjyu T, Saber AY, et al. Unit commitment strategy of thermal generators by using advanced fuzzy controlled binary particle swarm optimization algorithm. Int J Elec Power 2012;43(1):1072–80.
[51] Oshiro Y, Ono H, Urasaki N, et al. A MPPT control method for stand-alone photovoltaic system in consideration of partial shadow. IEEE 9th International Conference on the Power Electronics and Drive Systems, Singapore; 2011. doi: 10.1109/PEDS.2011.6147382.
[52] Datta M, Senjyu T, et al. Fuzzy control of distributed PV inverters/energy storage systems/electric vehicles for frequency regulation in a large power system. IEEE Trans Smart Grid 2013;4(1):479–88.

Chapter 4

Prediction of photovoltaic power generation output and network operation

Takeyoshi Kato
Institute of Materials and Systems for Sustainability (IMaSS), Nagoya University, Nagoya, Japan

CHAPTER OUTLINE

4.1 Needs for forecasting photovoltaic (PV) power output in electric power systems 78
4.2 Power output fluctuation characteristics 79
4.2.1 Fluctuation characteristics of irradiance at single point 81
4.2.2 Fluctuation characteristics of spatial average irradiance in utility service area 84
4.3 Forecasting methods 86
4.3.1 Overview 86
4.3.2 Accuracy measures 88
4.3.3 NWP models 89
4.3.4 Satellite cloud motion vector approach 94
4.3.5 All Sky images 96
4.3.6 Statistical models 96
4.4 Examples of forecasted results 98
4.5 Smoothing effect on forecast accuracy 100
4.6 Power system operation considering PV power output fluctuations 101
4.7 Energy management examples of smart house with PV 104
4.7.1 United States/Japan demonstration smart grid project in Los Alamos 104
References 106

4.1 NEEDS FOR FORECASTING PHOTOVOLTAIC (PV) POWER OUTPUT IN ELECTRIC POWER SYSTEMS

The installation of photovoltaic power generation system (PV system) is increasing rapidly in many countries, because the electricity production cost of PV system is becoming competitive to the other power generation resources. As shown in Fig. 4.1, the world cumulative installed capacity of PV system was 140 GW in 2014 [1]. In some European countries such as Germany and Italy, the annual power supply by PV system has already reached 7% of overall electricity supply. In the last few years, the newly installation was large in Asian countries such as China and Japan. The installed capacity in these two countries accounted for 60% of the world installation of 40 GW in 2014.

The high penetration of PV systems would negatively affect the stability of electric power system due to the intermittent and uncertain nature of PV power outputs. Then, power system operators are facing increased levels of variability and uncertainty of residual electricity load, which is the electricity demand minus the aggregated PV power outputs. When the power supply is larger than the electricity demand, the power system frequency rises. Therefore, the power supply and demand balancing capability must be increased in order to avoid the unstable operation of electric power system due to the high penetration PV system. The balancing capability on supply side includes conventional thermal power plants and natural gas combined cycle power plants. However, the start-up times of these power plants takes

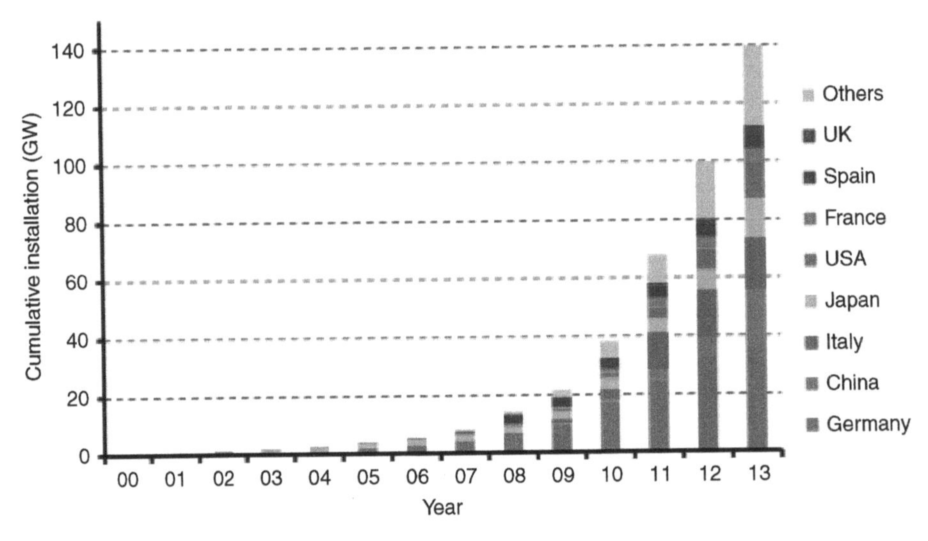

FIGURE 4.1 World cumulative installed capacity of PV system [1].

a few hours depending on fuel types. Therefore, the needs of balancing capability of quicker response such as pumped hydro power plants and storage batteries will be increased as the installation of PV systems increases.

For the best use of balancing capability, highly accurate and reliable forecasting methods play a very important role in various time horizons from several hours ahead to several days ahead. A day-ahead forecast is the most important and influential for the power-demand balancing. On the basis of day-ahead forecast of electricity demand, power system operators usually determine unit commitment scheduling of required generation resources to meet the electricity demand for each hour of the next day. The scheduling is determined at around the noon of the day before the operating day. Depending on the timing of the unit commitment scheduling or electricity markets, at least 36 h-ahead forecasting must be utilized. An hour-ahead and several hours ahead forecasting of the electricity demand are also important for the economic load dispatching control of the power system. The passage of clouds through a project area can cause sudden increases or decreases in irradiance, which are called ramp events. In order to mitigate the impact of ramp events, several-hours-ahead forecasting of the ramp rate and width is important.

An accurate and reliable forecasting method should be employed not only in power system but also in individual energy management system in a microgrid, industrial facilities, commercial buildings, and residential buildings. Based on the forecasted PV system power output as well as forecasted electricity demand, the operation scheduling of conventional generators and storage batteries is optimized so that PV power output is effectively utilized and the fuel consumption of conventional generators is minimized.

According to the needs of forecasting in power system operations or individual system operations, various forecasting methods and resources are utilized as shown in Fig. 4.2. Available resources are numerical weather prediction (NWP) models, satellite images, all-sky images, and measured PV power output data. The usefulness of these resources depends on the forecast time-horizon as described in the Section 4.3.

4.2 POWER OUTPUT FLUCTUATION CHARACTERISTICS

To improve the forecast accuracy, proper understanding of the power output fluctuation characteristics of PV system is essential. This section briefly describes the fluctuation characteristics of single point power output and aggregated power output of high penetration PV systems. The fluctuation

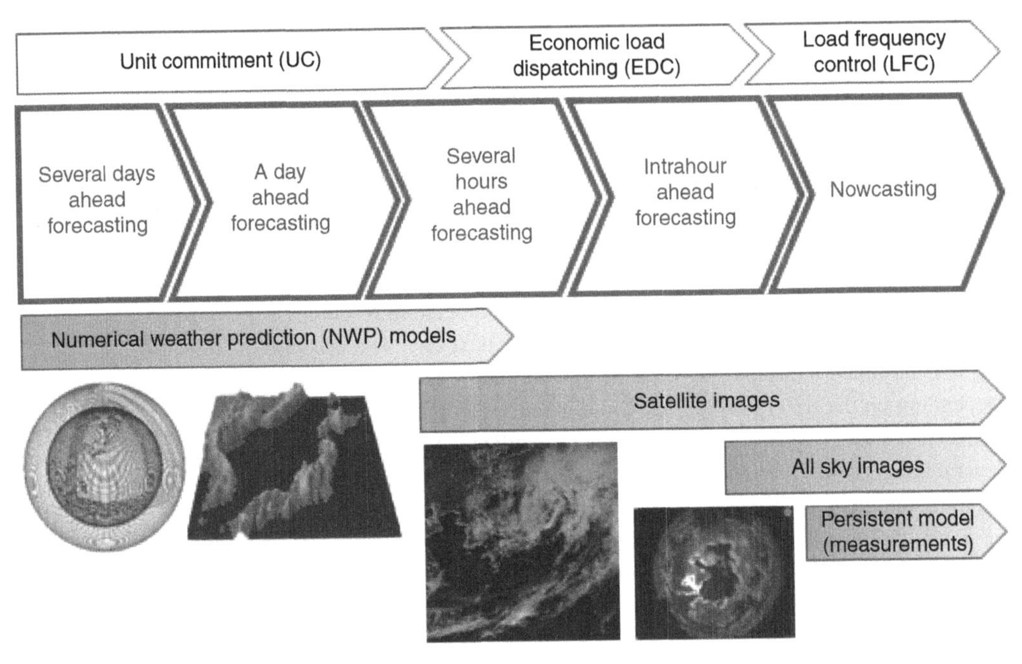

■ **FIGURE 4.2 Various forecasting methods according to needs in power system operations.**

characteristics of global horizontal irradiance (GHI) are sometimes discussed instead of PV power outputs, because the irradiance is primary factor which determines PV power outputs. In fact, the performance ratio (PR), which is a metric commonly used to measure how effectively PV system converts the irradiance into the alternating current (AC) electricity relative to what would be expected from the panel nameplate rating, is defined based on the irradiance in IEC 61724 as follows [2].

$$PR = \frac{\sum_i EN_{AC\_i}}{\sum_i P_{STC}\left(\frac{G_{POA\_i}}{G_{STC}}\right)} \tag{4.1}$$

Where EN_{AC} is the measured AC electrical generation (kW), P_{STC} is the summation of installed modules' power rating from flash test data (kW), G_{POA} is the measured plane of array irradiance (kW/m^2), G_{STC} is the irradiance at standard test conditions (=1000 W/m^2), and is i: a given point in time. Note that the performance ratio given in Eq. (4.1) is the traditional expression. The current expression is corrected so that the temperature effect is taken into account.

4.2.1 Fluctuation characteristics of irradiance at single point

The power outputs of PV systems have a 24-h fluctuation cycle given by the entirely predictable sun motion. Therefore, PV power outputs essentially increase in the morning hours and decrease in the afternoon. In addition to a 24-h fluctuation cycle, PV power outputs have shorter fluctuation cycles due to the clouds motion over PV systems. The fluctuation cycle depends on the types of cloud passing over PV systems. The clouds are typically classified into 10 types as described in Table 4.1.

Table 4.1 Classification of Cloud Types

Clouds	Type
High	Cirrus Cirrostratus Cirrocumulus
Mid	Altostratus Altocumulus Nimbostratus
Low	Cumulus Stratus Cumulonimbus Stratocumulus

Fig. 4.3 shows an example of GHI change in a day at single point. Fig. 4.4 shows the visible satellite image on the same day in Fig. 4.3. As shown in the magnified image on the left, the Chubu area is covered by a number of broken clouds. Therefore, a single point GHI largely fluctuates on second-to-second basis because of the clouds movement over GHI observation point.

Temperature is also a key factor which affects PV power outputs. The PR described above quantifies the overall effect of losses due to various factors such as inverter inefficiency, module temperature, reflection from the module front surface, shading, etc. Some of these factors, especially module temperature, are weather-dependent. The strong dependence of PR on temperature causes a large seasonal variation in PR, which can be as large as ±10%. PV power outputs can be different in along seasons even if the

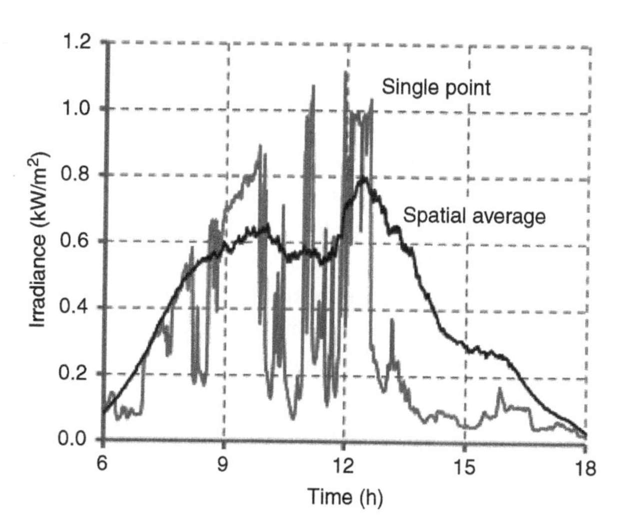

FIGURE 4.3 An example of observed global horizontal irradiance at single point (4/28/2011).

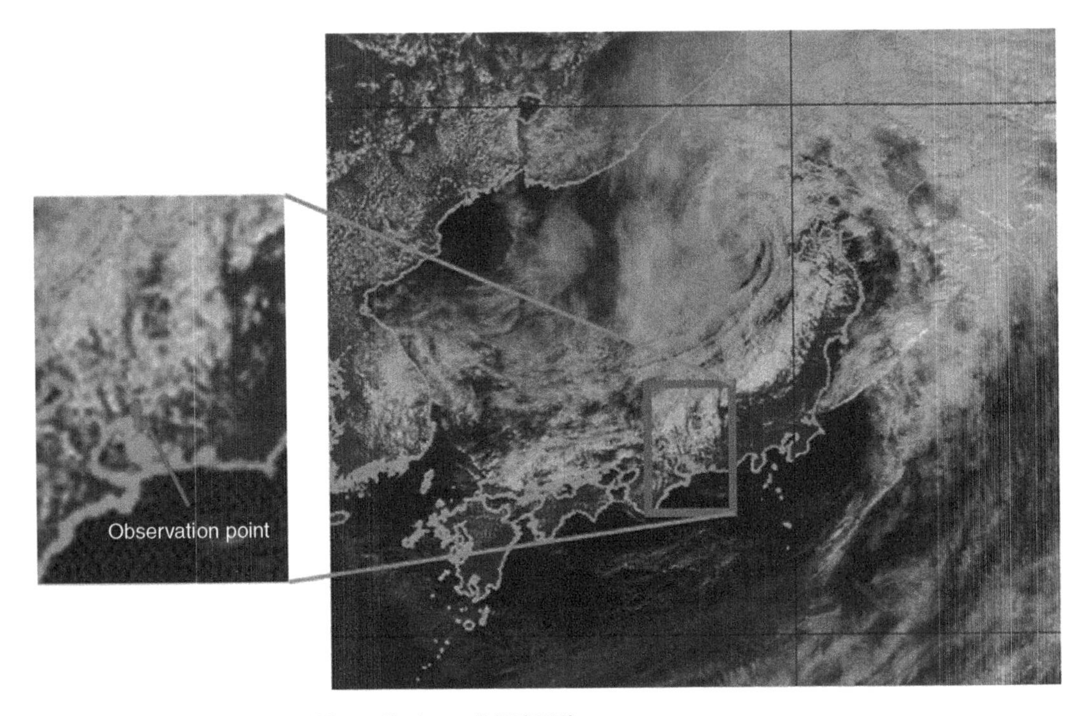

■ FIGURE 4.4 An example of visible satellite image (4/28/2011).

irradiance is the same. Therefore, the performance ratio shown in Eq. (4.1) is corrected so that the temperature effect is taken into account [2].

4.2.1.1 Smoothing effect

When the high penetration of PV systems is realized, a number of PV systems are widely dispersed in electric power utility service area. Because a cloud moves over different PV systems at a different time, the power output fluctuation can be different among PV systems. In such a situation, depending on the distribution of PV systems and the characteristics of clouds, the aggregated PV power outputs can be relatively small compared with those expected for individual PV system. This is referred to as spatial smoothing effect. The proper evaluation of aggregated power output characteristics by properly taking the smoothing effect into account is essential for preparing the practical and economically feasible measures against the negative impacts of high penetration PV systems.

In Fig. 4.3, the ensemble average of GHI at 61 points in the Chubu region in Japan is also shown. A second-to-second basis fluctuation of GHI at single point is significantly mitigated by the smoothing effect. Therefore, for estimating the aggregated power fluctuation of high penetration PV systems dispersed in large coverage area, multipoint observations of PV system power output or irradiance is necessary. Besides, in the real situation, a number of PV systems can be installed even between GHI observation points. As a result, because of larger smoothing effect, the aggregated power output of PV systems can be further smoothed.

From a practical point of view, however, the available number of observation points is limited because the service area of electric power utility is so large. Therefore, upscaling techniques should be employed so that the aggregated power output of all PV systems installed in a given area can be calculated by using the data from a subset representative of those systems. Upscaling techniques have been extensively used in forecasts of wind power output, because the forecasts for each wind turbine in a given area can be a time-consuming task.

In the case of aggregated power output of PV systems or spatial average irradiance, different upscaling techniques from wind power output should be employed because of the difference in fluctuation characteristics. Based on a "Transfer Hypothesis" to describe spatial smoothing effect, a method to calculate fluctuation spectrum of spatial average irradiance $S_{ave}(f)$ in a certain area is proposed [3]. The principle of the proposed method is as follows. In Fig. 4.5, $S_{mea.15}$ shows the fluctuation profiles of ensemble average GHI of 15 points in the Hokuriku region in Japan [4]. Because very long-cycle

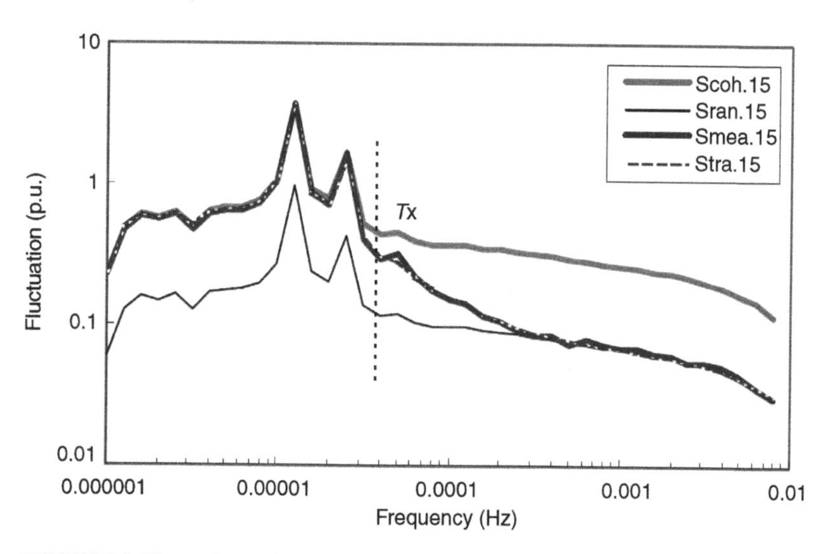

FIGURE 4.5 Fluctuation profiles of ensemble average irradiance of 15 sites [4].

fluctuations can be coherent among different locations, the spectrum is the same as the simple sums of individual spectrum of 15 points as indicated by $S_{coh.15}$ in Fig. 4.5. On the other hand, because very short-cycle fluctuations can be random each other, the spectrum is the same as the Pythagoras sums of individual spectrum of 15 points as indicated by $S_{ran.15}$ in Fig. 4.5.

Because very long-cycle fluctuations of irradiance can be coherent among different locations, the very long-cycle spectrum of spatial average irradiance can be the same as the spectrum $S_1(f)$ at representative single point in a certain area. On the other hand, because very short-cycle fluctuations can be random among different locations, the very short-cycle spectrum of spatial average irradiance can be given by $S_1(f)/\sqrt{N}$. As a result, based on a Transfer Hypothesis, $S_{ave}(f)$ is given by the following equation:

$$S_{ave}(f) = \left| \frac{S_1(f) + j \cdot T_X \cdot f \cdot \frac{S_1(f)}{\sqrt{N}}}{1 + j \cdot T_X \cdot f} \right| \tag{4.2}$$

T_x is the cycle below which the long cycle fluctuations among different points are not independent any more, and is determined so that the calculated spectrum fits the observed spectrum of spatial average irradiance fluctuations. N is the number of dummy observation points, and is determined to be as large as possible in a certain area while the short-cycle fluctuations can be seen independent between neighboring two points. In other words, the distance between neighboring two points is set as long as possible while the short-cycle fluctuations can be seen independent. Depending on the fluctuation cycles which should be dealt as independent, the minimum length between neighboring two points can be different. For example, the fluctuation cycle of around 30 min should be dealt as independent, the length of two points should be around 5 km. Because T_x and N is determined as a function of the size of area concerned, Eq. (4.1) works as low-path filter to calculate $S_{ave}(f)$ based on $S_1(f)$.

4.2.2 Fluctuation characteristics of spatial average irradiance in utility service area

As an example, the fluctuation characteristics of spatial average irradiance in the Chubu region in Japan are shown in this section. The geographical size of the Chubu region is about 200 × 250 km, which corresponds to average size of electric power utility service area in Japan. The spatial average irradiance is calculated as a weighted ensemble

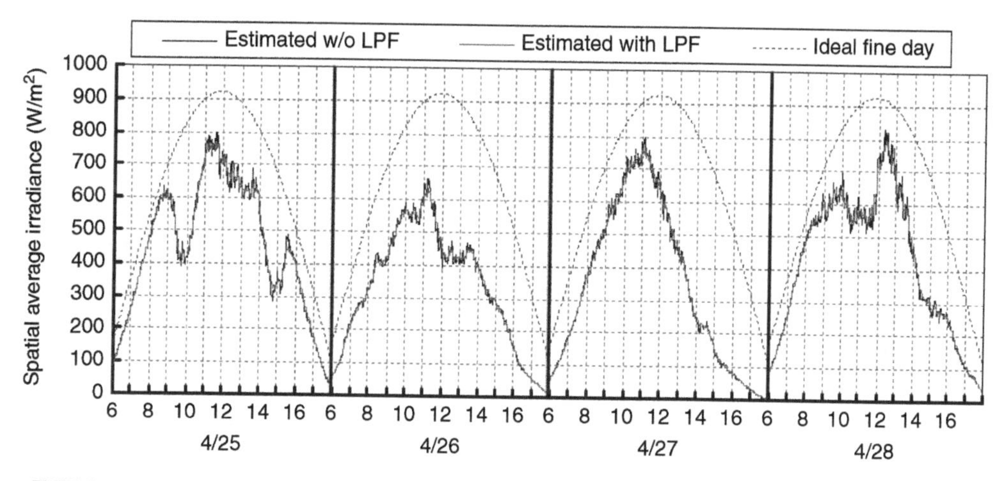

FIGURE 4.6 Examples of change in low-path filter applied spatial average GHI [5].

average of filtered irradiance observed at 61 points in the Chubu region, in which the weight value is determined based on the distribution of detached houses over the area.

Fig. 4.6 shows examples of change in the observed insolation and the low-path filter applied insolation [5]. In these examples, the fluctuation of spatial average insolation shorter than about 30 min is estimated to be only 20% of observed insolation by applying the low-path filter. On the other hand, the GHI shown in Figure 4.6 includes the large change even though it is the change in spatial average GHI in the electric power utility service area. Such a change is called as ramp events. Ramp events occur when the sudden and large change in GHI almost simultaneously occurs in a certain area by the passage of clouds over the area.

Ramp events are one of the major issues to be addressed for the stable operation of electric power system. In the case of the spatial average irradiance in the Chubu region, where ramp events occur 42 days (12 %) per year, the ramp event is defined as follows: the ramp rate is larger than 160 W/m^2 per h, the ramp duration is longer than 1 h, and such a change is caused by clouds movement. Fig. 4.7 shows the relation between duration time and fluctuation width of ramp events. The duration of ramp event varies mainly between 30–120 min. The ramp width increases linearly as the duration increases. In the case of downward ramp of relatively large width, the short-cycle fluctuations before and during the ramp event are as large as that in quasifine day. On the other hand, the

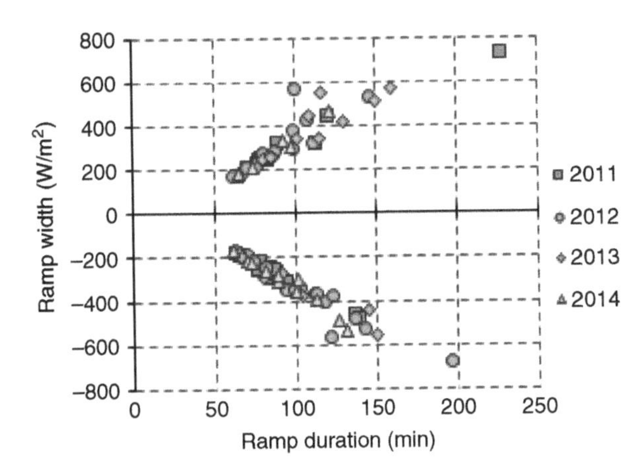

FIGURE 4.7 Relation between duration time and fluctuation width of ramp events [6].

short-cycle fluctuations after the downward ramp event are small enough as in a cloudy day [6].

4.3 FORECASTING METHODS

4.3.1 Overview

Forecasts may apply to the power output of single PV system or the aggregated power output of a number of PV systems dispersed over an extended geographic area. According to the applications of forecasting in power system operations or individual system operations and the required forecasting time-horizon, various forecasting methods are utilized as shown in Fig. 4.2. Forecasting methods are characterized by methodologies and available information, and useful methods depend on the time-horizon of forecasting. Fig. 4.8 shows an example of comparison of annual root mean square errors (RMSEs) over a year as a function of time horizon among different forecasting methods [7]. The satellite reference means the satellite nowcasting, that is, the satellite forecast for the time when the satellite image was taken. For longer time-horizons (from several hours ahead to a few days ahead), NWP models are an essential. For the time-horizon up to several hours, satellite image analysis methods perform better than NWP models. For very short time-horizon or nowcasting, the persistence forecast model based on the online data of PV system power outputs or irradiance is preferable. The all-sky images (not included in Fig. 4.8) can be used for short time-horizon up to a few tens of minutes ahead.

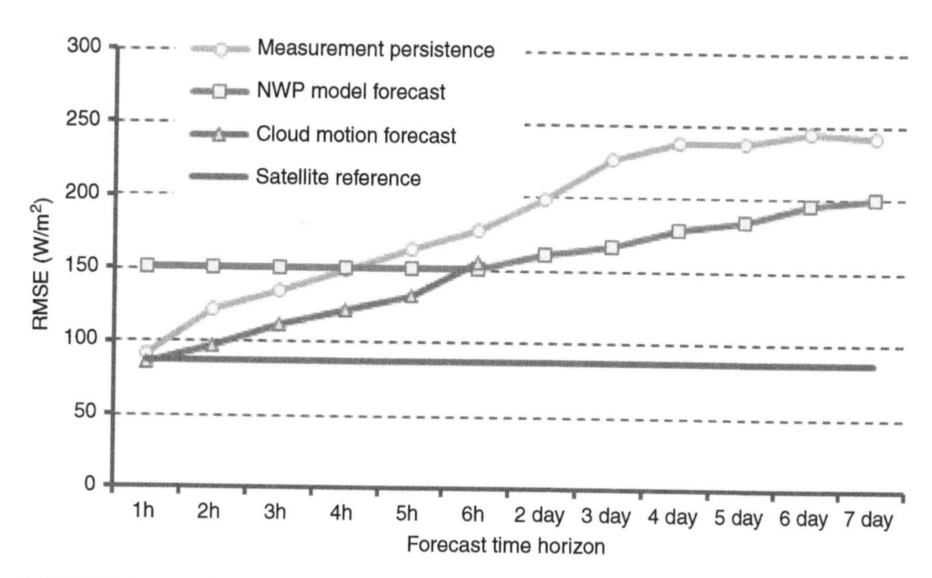

FIGURE 4.8 Comparison of annual RMSEs over a year as a function of time horizon among different forecasting methods [7].

Forecasting methods can be broadly characterized as physical or statistical. The forecasting based on NWP models is physical approach, which forecasts the irradiance based on the computation of an atmospheric motion in space and time and translates it to PV system power outputs as the end product as a function of relevant weather variables and PV system characteristics (manufacturer specifications).

The forecasting based on the persistence models is the statistical approach which relies primarily on past data to "train" models, with little or no reliance on physical models. In that sense, a statistical approach is closely related to and often overlaps with machine learning techniques. Various training techniques such as conventional regression models, pattern recognition models can be applied. Statistical approaches are also useful as a post processing to improve the accuracy of physical approach or to directly forecast the PV power output based on various weather elements forecasted in NWP models. Such approaches are called model output statistics (MOS). Fig. 4.9 shows basic steps of a typical physical approach. The primary influential variables to PV output power are the irradiance in the PV array plane and the temperature at the back of the PV modules.

The field of irradiance forecasting is rapidly evolving. This section describes a review of various forecasting methods for various time horizons ranging from a few minutes ahead to a day ahead.

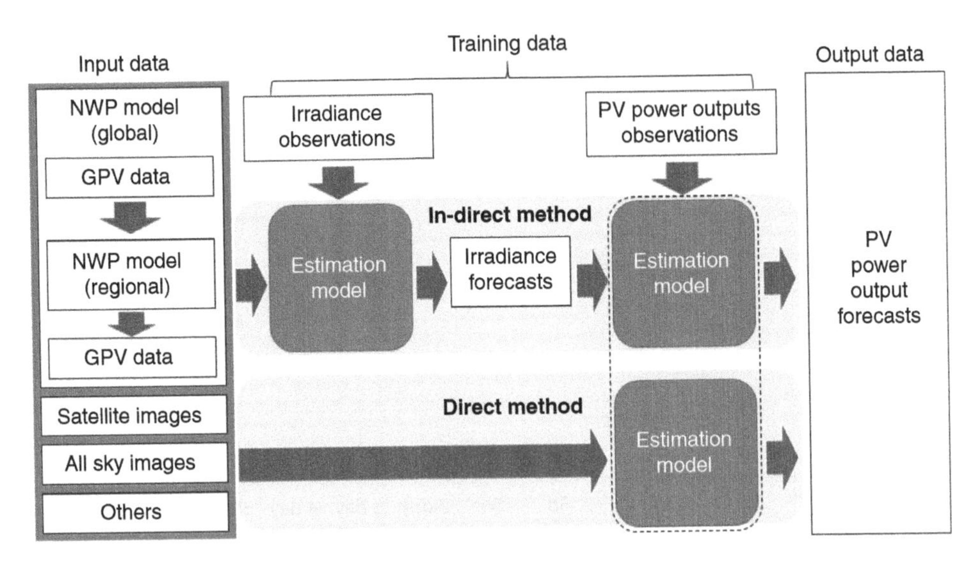

■ FIGURE 4.9 Basic steps of a typical physical approach of PV power output forecasting.

4.3.2 Accuracy measures

Common metrics to evaluate the forecast accuracy include mean bias error (MBE, or bias), mean absolute error (MAE), and RMSE. These are defined as follows:

$$MBE = \frac{1}{N}\sum_{i=1}^{N}(x_{f,i} - x_{o,i}) \quad (4.3)$$

$$MAE = \frac{1}{N}\sum_{i=1}^{N}\left|x_{f,i} - x_{o,i}\right| \quad (4.4)$$

$$RMSE = \sqrt{\frac{1}{N}\sum_{i=1}^{N}\left(x_{f,i} - x_{o,i}\right)^2} \quad (4.5)$$

where $x_{f,i}$ and $x_{o,i}$ are the ith forecast and observation, respectively.

The bias or MBE is the average forecast error representing the systematic error of a forecast model to under or overforecast. As described below, a post-processing of model output is useful to significantly reduce the bias. MAE gives the average magnitude of forecast errors, while RMSE (and MSE) give more weight to the largest errors. RMSE without systematic error (SD) captures the part of the RMSE that is not due to systematic error. The relation between MBE, RMSE, and SD is as follows:

$$SD = \sqrt{RMSE^2 - MBE^2} \quad (4.6)$$

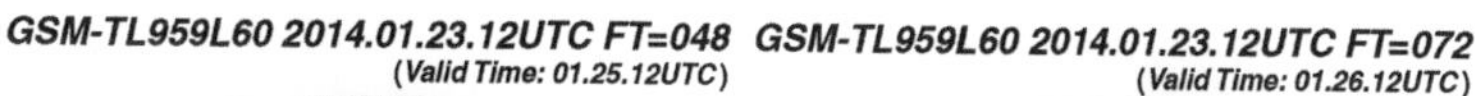

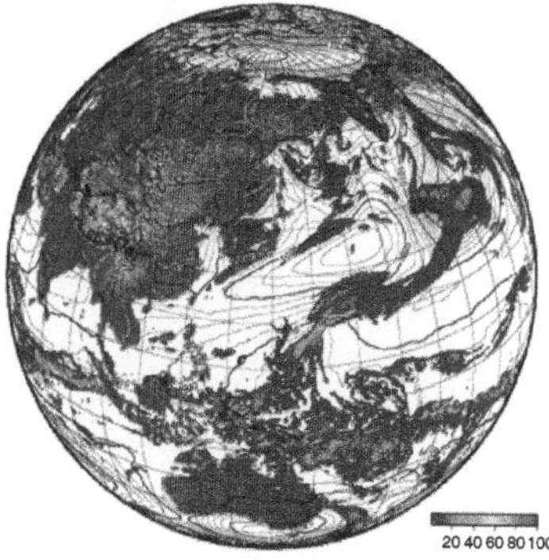

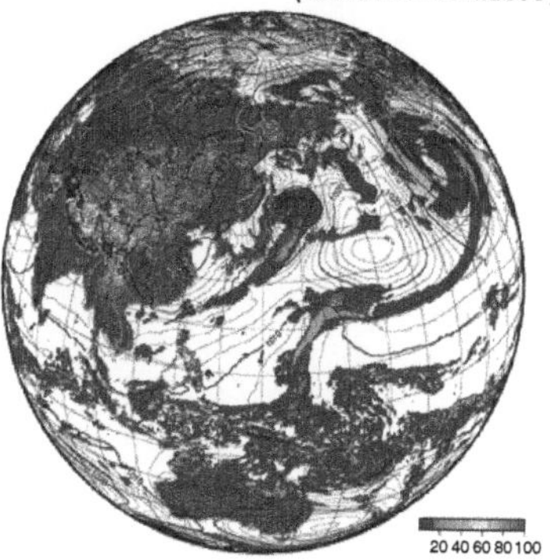

FIGURE 4.10 An example of forecast result of global NWP model by the JMA.

Therefore, SD provides an indication of the RMSE that can be achieved when MBE is zero.

The normalized value by dividing with a reference value may be used for all these metrics to facilitate comparisons. In the case of PV system power output, the normalized value by the rated capacity is commonly used. Meanwhile, in the case of irradiance, the normalized value by the average irradiance over the certain period is often used.

4.3.3 NWP models

NWP models are the computer program that simulates an atmospheric motion in space and time. Fig. 4.10 shows an example of forecast results by the global NWP model of the Japan Meteorological Agency (JMA) [8]. NWP models are essentially required to the forecasting of longer time-horizon from several hours ahead to a few days ahead. NWP models can also be applied to several hours ahead forecasting.

In NWP models, the atmosphere is assumed to be composed of a number of lumps in which corner points are called as the grid points. Simulation using NWP model generates the future state of the model atmosphere at all grid points from its initial state. Fig. 4.11 shows an example of a latitude–longitude grid used in the global NWP model operated by JMA [8]. The model equations and inputs are discretized on a three-dimensional (3D) grid extending vertically from the surface of the Earth. The higher the number of lumps indicates the more elaborate simulation.

A variety of weather phenomena can be analyzed and predicted by the different types of NWP models. Global models are fundamental NWP models covering the whole Earth. The initial conditions of global models are derived from satellite, radar, radiosonde, and ground station measurements that are

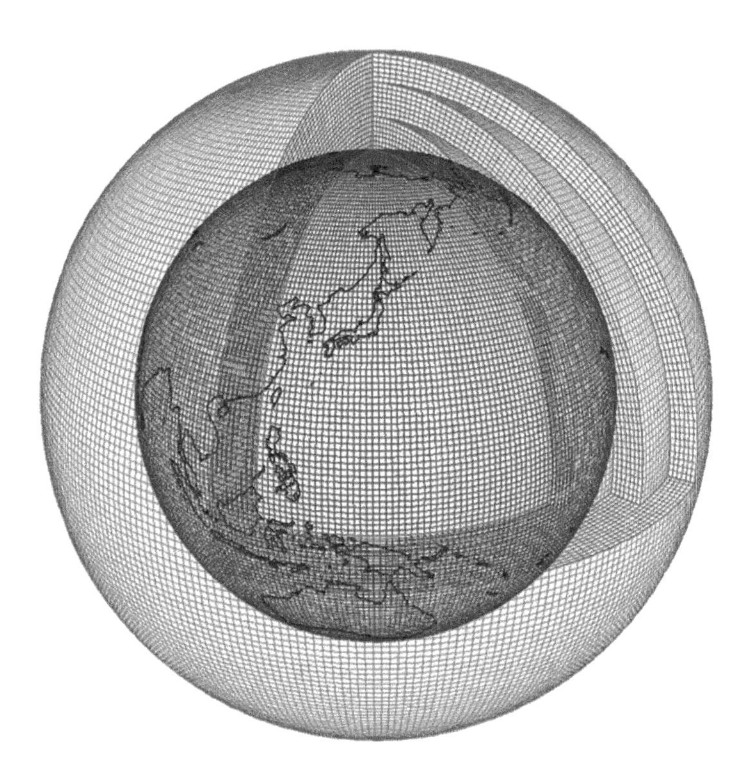

■ **FIGURE 4.11 An example of latitude–longitude grid used in global NWP model by the JMA.**

processed and interpolated to the 3D grid. Global models are used to provide the boundary conditions of a mesoscale NWP model is described below.

Various global NWP models are run in different countries. The European Center for Medium range Weather Forecasting (ECMWF) forecast model is one of the most famous and accurate global NWP model. Global Forecast System (GFS) produced by the National Centers for Environmental Prediction (NCEP) of the US National Oceanic and Atmospheric Administration (NOAA) is also famous. Because GFS data is freely available at the GFS homepage, GFS is mostly used by research institutes and private companies. NCEP runs GFS four times per day at 0, 6, 12, and 18 Universal Time Coordinated (UTC). The horizontal resolution of GFS is 13 km for the first 10 days and 27 km from 240–384 h (16 days).

Mesoscale or limited area models are NWP models that cover a limited geographical area with higher resolution, and that attempt to account for local terrain and weather phenomena in more detail than global models. Initial conditions for these models are extracted from the global models. The

Weather Research and Forecasting (WRF) Model developed at the National Center for Atmospheric Research is used extensively for research and real-time forecasting throughout the world [9]. WRF is a next-generation meso-scale numerical weather prediction system designed for both atmospheric research and operational forecasting needs. In the United States, WRF is currently in operational use at NCEP, Air Force Weather Agency, and other centers. On the other hand, because WRF is an open source model, it has a large community of more than 30,000 registered users in more than 150 countries. WRF serves a wide range of meteorological applications across scales from tens of meters to thousands of kilometers. The setup of Advanced Research WRF (Version 3.5.1) consists of a main grid with horizontal grid spacing of 30 km and one nested domain with 10-km grid spacing. Because WRF is an open source model, the calculation conditions such as initial condition, lateral boundary condition, horizontal resolution, etc. can be adjusted flexibly according to the purpose.

In Japan, the JMA currently operates several NWP models to cover various types of prediction, including very-short-range forecasts, short/medium-range forecasts, typhoon track forecasts, and aviation forecasts. Table 4.2 shows the specifications of NWP models utilized by the JMA [8]. Because the time horizon of the MSM model is 39 h, the MSM model is useful for a day-ahead forecasting. Since the outputs of JMA NWP models do not include the irradiance forecast outputs as of July 2015, postprocessing techniques are applied to generate irradiance forecasts. Possible inputs to generate irradiance are cloud cover, temperature, probability of precipitation, relative humidity, wind speed, and direction, etc.

Because of deficiencies and nonlinearities of NWP models, the forecast results may deviate from their true trajectories as the time-horizon increases, resulting in the large error in the end. Therefore, several techniques described below are applied to improve the forecast results of NWP models.

Table 4.2 Specifications of the JMA's NWP models

Specification	Global Spectral Model (GSM)	Mesoscale Model (MSM)	Local Forecast Model (LFM)
Forecast range	84 h (00, 06, 18 UTC), 264 h (12 UTC)	39 h (00, 03, —, 18, 21 UTC)	9 h (hourly)
Number of horizontal grid points and/or grid spacing (no. of truncation wave)	0.1875° [TL959]	817 × 661 (5 km at 60°N and 30°N)	1581 × 1301 (2 km at 60°N and 30°N)
Model domain	Globe	Japan and its surrounding areas	
Vertical levels	100 levels up to 0.01 hPa	50 levels up to 21.8 km	60 levels up to 20.2 km

4.3.3.1 Ensemble forecast of NWP models

One way to improve the forecast accuracy and reliability of NWP models is an ensemble forecast. Most NWP models and most forecast results reflect a deterministic approach. In the actual situation, however, depending on uncertainties in the model, initial conditions, or atmospheric conditions, forecast results also have the uncertainties. In an ensemble forecast, initial conditions or physical parameterizations are varied within a single NWP model. The ensemble mean is more accurate on average than any individual forecast results. In addition, the distribution level of forecast results by the ensemble forecast means the confidence level of forecasting. If the distribution is small, the forecast results do not depend on the initial conditions and can be seen to be very reliable. On the other hand, if the distribution is large, the forecast results depend highly on the initial conditions and are not so reliable. Because the information on uncertainty of forecast results can be practically very useful, the uncertainty is also provided as a confidence interval of forecast results.

4.3.3.2 Spatiotemporal interpolation and smoothing

NWP models provide the forecast results at discrete grid points. Therefore, spatiotemporal interpolation is practically important when NWP models are utilized for forecasting irradiance at a specific single point. The simplest method is to use the forecast results at the nearest grid point to the location of interest, though the spatial resolution is 5 km even in mesoscale NWP model. Other approaches involve interpolating forecasts from grid points surrounding the point of interest. In addition to spatial interpolation, temporal interpolation must be used when available NWP model outputs have a lower temporal resolution than desired.

4.3.3.3 Postprocessing by statistical model

Forecast results of NWP models may include systematic errors or bias errors. Therefore, if measured irradiance data is available, postprocessing contributes to improve the accuracy of forecast results of NWP models. The simplest postprocessing model is a linear function given as follows:

$$I_c = aIo + b$$

where I_c is the corrected forecast results and Io is the original forecast results of NWP model. Coefficients a and b are estimated by using measured irradiance data and original irradiance forecasts of NWP model during the training period in the past. The simplest way to estimate the coefficients is the least squares method. The Kalman filter is practically useful method to obtain suitable coefficients day by day. Fig. 4.12 shows an example of time series of coefficients obtained from the Kalman filter [10].

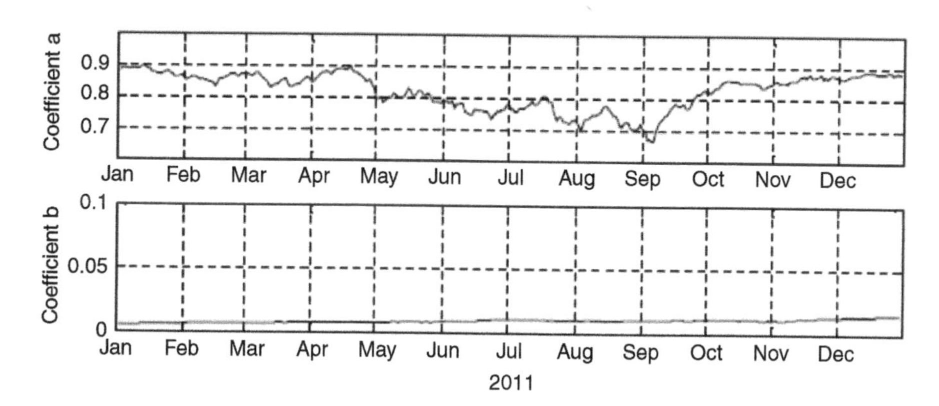

FIGURE 4.12 An example of time series of coefficients obtained from the Kalman filter [10].

Because forecast errors often depend on the time of the day and of the year, on sky conditions, etc., the training should be done separately over individual time in a day, different conditions or regimes. In order to reduce the training process, the sky clearness index is often used together with the information on the position of the sun (cosine of the solar zenith angle).

MOS is applied to correct the original irradiance forecasts of NWP models as a postprocessing technique, when the ground observations of irradiance are available. MOS relates observed weather elements to appropriate variables (predictors) by a statistical approach such as multiple regression. For example, in Ref. [11], a genetic algorithm and artificial neural networks are applied for regression and the National Digital Forecast Database (NDFD) by the US National Weather Service and used as training data.

As of July 2015, because direct outputs of the JMA NWP models did not include the irradiance forecast outputs, the irradiance was forecast by MOS approach using the forecasts of the other weather elements. Possible inputs to generate irradiance are cloud cover, temperature, probability of precipitation, relative humidity, wind speed, and direction, etc. For instance, in Ref. [12], the special average irradiance is forecasted by using a simple linear function of cloud cover forecasts at three levels of GPV (MSM) data and extraterrestrial irradiance. In Ref. [13], the support vector machine is applied.

4.3.3.4 Combination of different forecast models

The combination of different forecast models is a practically feasible way to improve the forecasting accuracy. Various combination is available, that is, the combination of different NWP models such WRF and MSM/JMA, the combination of different boundary condition data for mesoscale NWP

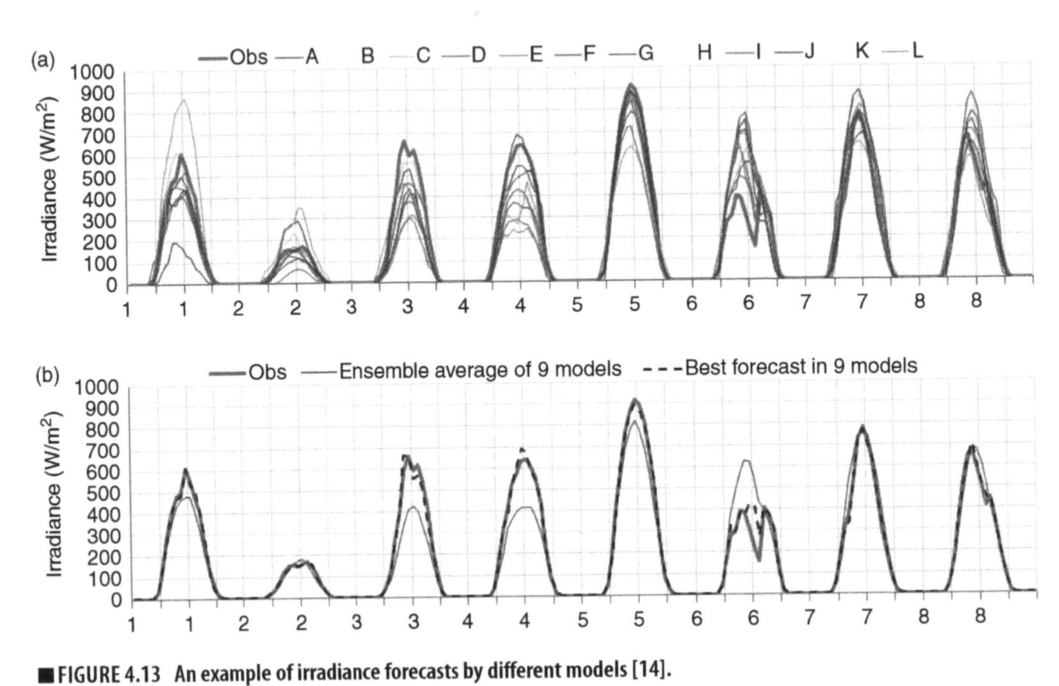

FIGURE 4.13 An example of irradiance forecasts by different models [14].

models such as GFS and global spectral model (GSM)/JMA, the combination of different postprocessing approaches, etc.

Fig. 4.13a shows the spatial average irradiance forecasts in the Chubu region in Japan by nine different day-ahead forecast models [14]. The forecasts by three models, that is, models B, H, and K are not included in Fig. 4.13a. Some models calculate the irradiance forecasts by using WRF with postprocessing, in which the boundary conditions are given by MSM/JMA. Some models are statistical approach using the forecasts of various weather elements by MSM/JMA. As shown in Fig. 4.13a, the irradiance forecasts are different among models. The RMSE for a month (May 2012) is ranging from 100–133 W/m^2. In Fig. 4.13b, the thin line shows the simple average of irradiance forecasts of nine models, and the dashed line shows the best forecasts among nine models in each hour. The RMSE of simple average reduces to 90 W/m^2. If the best forecast can be chosen as an ideal situation, RMSE is only 27 W/m^2. The result suggests that further significant improvement of forecast accuracy is expected.

4.3.4 Satellite cloud motion vector approach

Irradiance forecasting based on cloud motion vectors from satellite images shows good performance for the time-horizon ranging from 30 min up to several hours. Earlier contributions have shown that satellite-derived cloud motion tends to outperform NWP models for forecast horizons up to 4–5 h ahead depending on location [7]. High quality satellite images are available from various weather satellites such as Multifunctional Transport Satellite (MTSAT) in Japan, the Geostationary Operational Environmental Satellites (GOES) in the United States, etc., and have been used extensively in solar resource mapping. Because the irradiance at the earth surface highly depends on the cloud optical depth, the clearness index, which is defined as the ratio of the horizontal global irradiance to the corresponding irradiance available out of the atmosphere, can be calculated accurately based on the reflectance measured for each pixel in the visible satellite images. Fig. 4.14 shows an example of estimated irradiance calculated based on a radiative transfer model [15].

The spatial resolution of geostationary satellite images is 1 km for example in MTSAT-2. Therefore, satellite images are useful to detect large and thick clouds such as stratocumulus and cumulonimbus. On the other hand, detection of cirrus or cirrostratus at the high altitude may be difficult. The sampling interval of full disk images is 30 min in MTSAT-2 and GOES-14. The sampling interval of latest MTSAT in Japan, which was launched in Sep 2014, is improved to 10 min for full disk images and 2.5 min for Japanese local images.

Based on these features, one of the advantages of satellite image-based forecasting compared with NWP models is higher accuracy in a several hours-ahead forecasting. In many methods, the irradiance is forecasted by

■ **FIGURE 4.14 An example of irradiance estimation based on satellite image.**

estimating cloud motion vectors as follows. First, the same feature points are detected in successive images in the last few hours. Then, based on the spatial and temporal difference of feature points in successive images, cloud motion vectors can be estimated. Finally, by assuming that cloud feature points and their motion vectors do not change for next few hours, irradiance is forecast based on the motion vectors of the clouds getting closer to the target location.

4.3.5 All Sky images

Ground-based all-sky images have much higher spatial and temporal resolution compared with the satellite images, though the field-of-view is much smaller than that of satellite images. Therefore, irradiance forecasting based on all sky image analysis is suitable measures for a single point irradiance forecasting of shorter time-horizon shorter than several 10 min [16]. All sky image based forecasting is also useful for the irradiance distribution forecasting and nowcasting within a supply territory of power distribution network. All sky-image-based forecasting is performed by estimating cloud motion vectors by using successive images. With the significant improvement of charge coupled device (CCD) devices, all sky images are available using low cost fish-eye camera [17]. Due to the limited field-of-view of sky imager, the forecast horizon would be shorter than about 15 min. Fig. 4.15 shows an example of forecast output by Sky Imager developed at the University of California San Diego San Diego (UCSD) [18]. Top left is raw HDR image, cropped to remove static objects near horizon. Top center shows the red–blue ratio image. Top right is the cloud decision image (blue: clear sky, light gray: thin cloud, dark gray: thick cloud). Bottom left shows a shadow map over the UCSD domain, showing predicted cloud shadows from images taken 10 min ago. Ground stations are marked by solid black squares. The cloud field mean velocity vector is indicated by the solid black arrow extending from the center, with magnitude indicating predicted distance traveled in 30 s. Bottom right shows the universal sky image (USI) GHI forecast issued at current time for a 15 min horizon (dashed red), USI GHI forecast time series for constant 10 min forecast horizon (solid black), and corresponding measured GHI (solid green). In the bottom right graph, the first vertical dashed line indicates forecast issue time, while the second vertical dashed line shows the 10 min forecast horizon (solid black line must equal red dashed line at that point).

4.3.6 Statistical models

For an irradiance forecast of very short time horizon (up to a few hours), statistical models based on regression analysis using online irradiance measurement data are accurate and practical approach. The simplest model utilizes the irradiance measurements data only. If relevant exogenous data

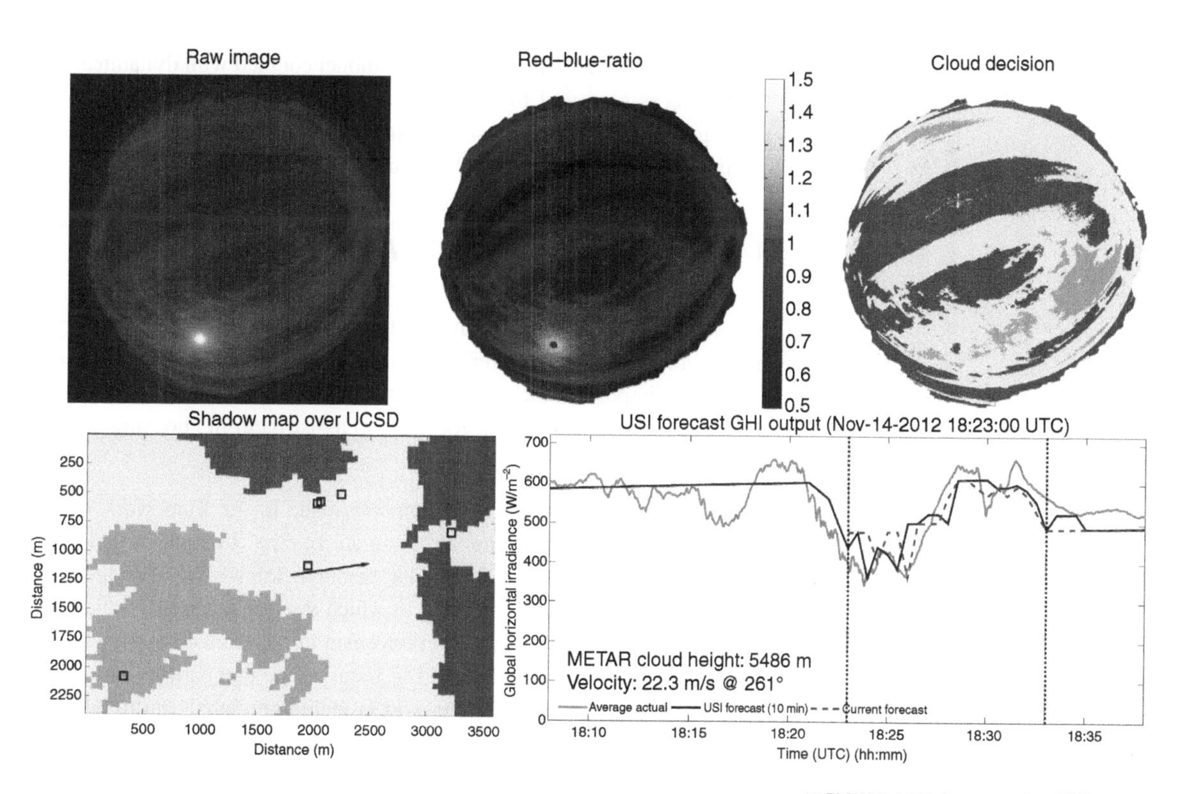

FIGURE 4.15 An example of USI forecast output using all-sky image.

such as all sky images, satellite images, various weather forecasts of NWP models, and other meteorological observations are available, the accuracy and reliability can be significantly improved. Because the irradiance strongly depends on the solar zenith angle, it may be favorable to treat the nondeterministic atmospheric extinction by excluding the influences of the deterministic solar geometry when using statistical models. For this purpose, clearness index, which is the ratio of measured irradiance at ground level to extraterrestrial irradiance is utilized. The clear sky index, which is the ratio of measured irradiance at ground level to estimated irradiance in clear sky conditions at ground level, is also utilized if a clear sky model and information on atmospheric input parameters are available.

Autoregressive moving average (ARMA) model is one of the most popular statistical tools for time series data analysis and is very useful to predict the future value of a specified time series. ARMA consists of two parts, that is, lagged past values (autoregression) and error terms (moving average). One major requirement for ARMA model is that the time series must be stationary. The autoregressive integrated moving average (ARIMA) model can be applied when the time series are nonstationary but their differences are stationary. In addition to the conventional regression models such as

ARMA and ARIMA, an autoregressive model coupled with dynamical system model is proposed for one hour ahead forecasting of irradiance. The combination of dynamical system model contributes to improve the forecast accuracy especially in mostly cloudy days [19].

4.4 EXAMPLES OF FORECASTED RESULTS

As described above, various research organizations have developed different methods to forecast irradiance or PV system power output. The performance comparison of different models in a standardized evaluation methodology is important for researchers to further improve their models. The comparison is also important for users to assist them in choosing suitable models for their purpose.

One of the performance comparison was made in the framework of The Solar Heating and Cooling Programme of the International Energy Agency (SHC IEA) Task 36 "Solar resource knowledge management" (http://archive.iea-shc.org/task36/), in which three independent validations of GHI multiday forecast models performed in the United States, Canada, and Europe were compared [20]. The focus of the comparison was on the end-use accuracy of the different models including global, multiscale, and mesoscale NWP models as a basis. In addition, different postprocessing techniques to derive site-specific every hour forecasts such as very simple interpolation and advanced statistical method were compared. The models considered for this evaluation are listed below:

1. Satellite image as a reference.
2. Persistence model.
3. The Global Environmental Multiscale (GEM) model from Environment Canada in its regional deterministic configuration.
4. An application of the ECMWF model.
5. A model based on cloud cover predictions from the US NDFD proposed.
6. Several versions of WRF model initialized with GFS forecasts from the US NOAA/NCEP (NCEP).
7. The advanced multiscale regional prediction system model.
8. The mesoscale atmospheric simulation system model.
9. The regional weather forecasting system Skiron operated and combined with statistical post-processing based on learning machines at Spain's National Renewable Energy Center (CENER).
10. The high resolution limited area model operational model from the Spanish weather service (AEMet) combined with a statistical post-processing at Centro de Investigaciones Energéticas, Medioambientales y Tecnológicas (CIEMAT).

11. BLUE FORECAST: statistical forecast tool of Bluesky based on the GFS predictions from NCEP.
12. Forecasts based on meteorologists' cloud cover forecasts by Bluesky.

GEM and ECMWF are directly based on global NWP systems. Some models are tested in two operational modes: with and without MOS postprocessing. The MOS process consists of integrating ongoing local irradiance measurements, when available, to correct localized errors from the numerical weather prediction process. The last four models are not used in the US case.

The main results are as follows. RMSE composite values of seven sites in the US show a considerable spread for the different models ranging between 32 and 47% for day 1 forecasts as shown in Fig. 4.16. Lowest MAE and RMSE values (or highest accuracy) are found for the global model ECMWF and GEM irradiance forecasts. All considered mesoscale model forecasts as well as the NDFD based forecasts show larger forecast errors. This indicates some drawbacks of irradiance and/or cloud schemes in the selected mesoscale models. On the other hand, the GFS model irradiance forecasts were found to have a similar performance to those of the ECMWF model when applying a simple postprocessing. There is a considerable variation of accuracy in terms of RMSE and MAE for the different sites and climates in the United States. For an arid climate (Desert Rock, United States) with many sunny days, relative RMSEs in the range of 20–25% for day 1 forecasts are considerably smaller than for the other sites for all investigated models, where the RSME values exceed 30%. Largest day 1 RMSE values between 38 and 48% are found for Penn state with the lowest mean irradiance.

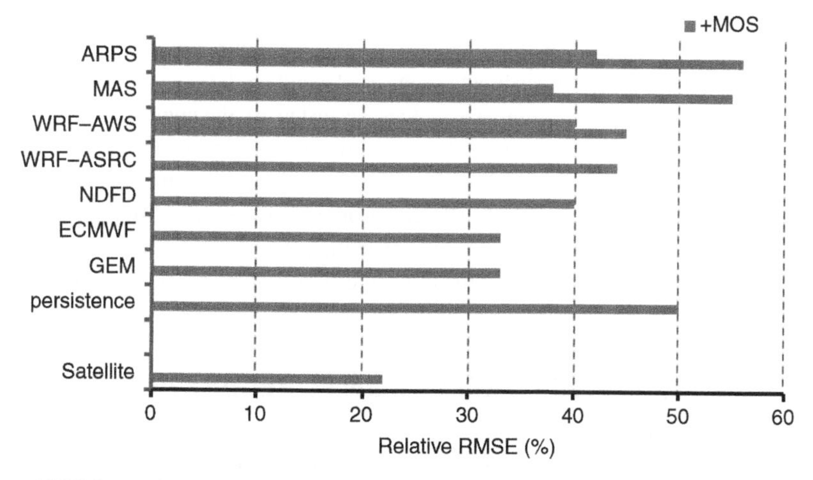

■ **FIGURE 4.16 Composite RMSE of different forecast models (US case).**

Extending the model comparison from the United States to Canada and Europe, the finding that ECMWF based irradiance forecasts show a higher accuracy than irradiance forecasts with WRF and the other investigated mesoscale models is confirmed. For Canada, like for the United States, the performance of the Canadian GEM model is similar to the performance of the ECMWF model.

Focusing on the difference in machine learning techniques, the irradiance forecasting methods were compared [21]. Popular nonlinear techniques such as neural networks and some rather new methods such as Gaussian Processes and support vector machines were evaluated against simple methods like the autoregressive linear model and reference models like scaled persistence. The performances of the following models were compared in terms of several hours-ahead forecasting of historical GHI data measured on three French islands.

1. Climatological mean (mean historical value of clearness index, which is independent of the forecast horizon).
2. Clear sky index persistence model (SC-pers).
3. Autoregressive process (AR) model.
4. Neural network (NN) model.
5. Gaussian process (GP) model.
6. Support vector machine (SVM) model.

The first two models are called the native model. AR model is a linear model. The last three models are nonlinear models. Fig. 4.17 shows relative RMSE of the different methods for each forecasting time horizon for the case of Reunion Island. The nonlinear methods such as NN, GP, and SVM perform better than the scaled-persistence and the linear model. The advantage increases with the forecasting horizon. For hours ahead forecasting, the picture is less clear and seems to depend on the sky conditions. For stable clear sky conditions (clear skies for instance), the nonlinear methods slightly improve the scaled-persistence. For unstable sky conditions, the discrepancy between the machine learning methods and the simple models is more pronounced with a 2% rRMSE difference in average. Similar results are obtained for the other sites. Because the performance of the three nonlinear methods is practically the same, the choice of the method will depend on the skill and experience of the modeler.

4.5 SMOOTHING EFFECT ON FORECAST ACCURACY

Because of a so-called smoothing effect of irradiance forecast errors at different locations, the forecast accuracy would be higher for the spatial average irradiance than the single point irradiance. Higher impact of smoothing

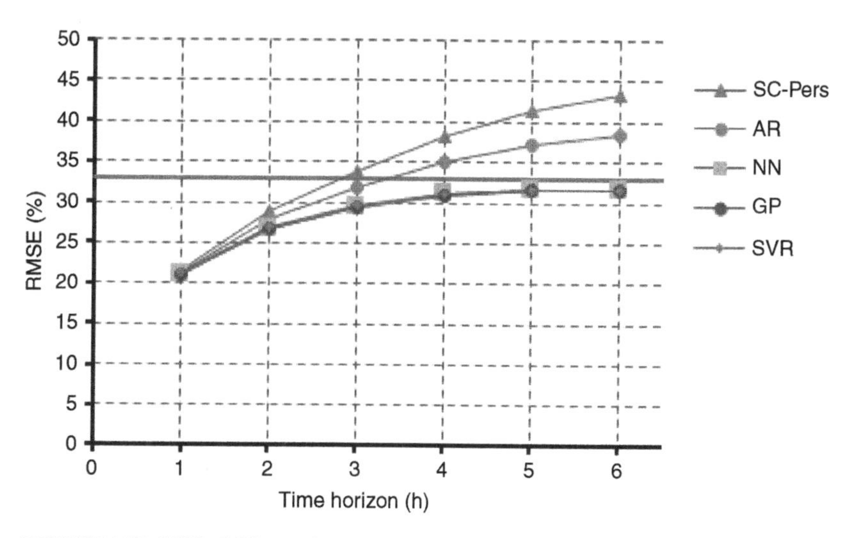

■ **FIGURE 4.17 RMSE of different forecast models as a function of prediction time horizon.**

effect is expected for mesoscale model with hourly values and a finer grid resolution. Quantifying the accuracy improvement by smoothing effect seems difficult because it depends on various factors such as climate diversity within the region, PV system distribution, capacities, etc. Nevertheless, some case studies regarding the forecast accuracy improvement by smoothing effect are available in the technical literature.

For example, in the case of ECMWF forecasts, RMSE of a day-ahead forecasting is 13% for the ensemble average irradiance of more than 200 stations in the complete German region with a size of 9 × 10°, while overall RMSE is 37% for single sites [22]. The best results are achieved for average values of 4 × 4 grid points corresponding to a region of 100 × 100 km. An analysis of the GFS model and NAM model showed that 100 × 100 km as a suitable irradiance forecasts [23]. Similarly, in the case of forecast of the average irradiance of 10 ground stations across Canada and the United States, RMSE is about 67% lower than individual RMSE of the irradiance at individual ground stations [24].

4.6 POWER SYSTEM OPERATION CONSIDERING PV POWER OUTPUT FLUCTUATIONS

In order to maintain the electric power system reliability, electric power utility or Independent System Operator (ISO) must continuously control the electricity supply to meet the demand on a second-to-second basis.

Historically, the ISO has controlled conventional thermal power plant units. With the growing penetration of renewable energy resources, there are higher levels of noncontrollable, variable generation resources in a power system. In some countries, renewable power generations increasingly satisfy the electricity demand in certain times of the year. As a result, the requirements to manage a power system are changing due to the high penetration of intermittent and unstable renewable energy resources.

The time series of residual electricity load, which is the difference between the actual electricity demand and the electricity supply from renewable energy sources, is changing to quite different form from the current load curve. In certain times of the year, the residual load curves produce a "bally" appearance in the mid-afternoon that quickly ramps up to produce an "arch" similar to the neck of a duck as shown in Fig. 4.18 [25]. Such a residual load curve is called as "duck curve." The first ramp in the downward direction occurs in the morning starting around 7:00 am as the power output of PV systems increases steeply. As a result, online conventional generation is replaced by supply from PV power output, producing

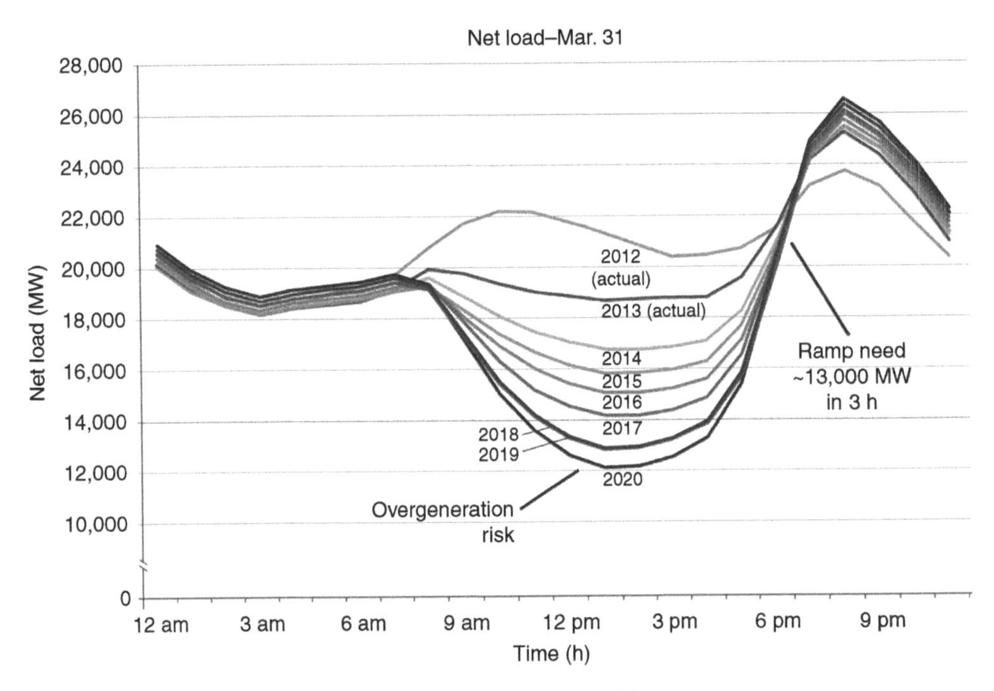

■ **FIGURE 4.18 An example of change in residual electricity load called Duck curve.**

the belly of the duck. As the PV power output starts to decrease around 4:00 pm, the ISO must dispatch resources that can meet the second significant upward ramp (the arch of the duck's neck). Immediately following this steep ramp up, as demand on the system decreases into the evening hours, the ISO must reduce or shut down that generation to meet the final downward ramp. Moreover, as shown in Fig. 4.18, surplus electricity supply or overgeneration happens in the daytime when more electricity is supplied by PV systems than is needed. Similarly, overgeneration may happen during night when the electricity supply from wind power generations is large but the demand is small.

The residual load curves represent the variable portion that the ISO must meet in real time. Important parts of power system operations affected by duck curve include regulation reserves requirements and subhourly dispatch. How the system operations can be changed to more economically integrate large amounts of PV power is an open question currently being considered by many utilities and ISO. To understand changing grid conditions, the CAISO, the independent system operator in California performed detailed analysis for every day of the year from 2012–2020 as follows [25]. The analysis shows that the ISO requires a resource mix that can react quickly to adjust electricity production to overcome several emerging conditions due to the high penetration of renewable resources, that is, the ramp up/down of residual electricity load, the surplus power supply exceeding the demand, insufficient capacity to maintain the system frequency. To ensure reliability under changing grid conditions, the ISO needs resources with ramping flexibility and the ability to start and stop multiple times per day and the flexibility to change output levels as dictated by real-time grid conditions. At the same time, the ISO needs to increase interconnection capabilities to neighboring system, and require the curtailment of renewable generation.

For the best use of these flexibilities, an accurate and reliable forecasting of renewable generation is essential. Because forecasting of PV system power output is a cutting edge research area, it has only recently been introduced into the electricity system operation in some countries such as the United States, Germany, and Spain. Probabilistic or ensemble forecasting methods are used for probabilistic unit commitment of electric power system with high penetration PV system.

For example, the CAISO uses a day-ahead forecast and several hours ahead forecast as follows [26]. A day-ahead forecast is submitted at 5:30 prior to the operating day, which begins at midnight on the day of submission and covers (on an hourly basis) each of the 24 h of that operating day. Therefore, a day-ahead forecast is provided 18.5–42.5 h prior to the forecasted operating day. An hour-ahead forecast is submitted 105 min prior to each operating hour. It also provides an advisory forecast for the 7 h after the operating hour. CAISO is considering the implementation of intra-hour forecasts at 5 min intervals.

4.7 ENERGY MANAGEMENT EXAMPLES OF SMART HOUSE WITH PV

The concept of the "Smart house" is an intelligent house that incorporates advanced automation systems to provide the inhabitants with sophisticated monitoring and controlling of building's functions such as lighting, room temperature, security, and etc. Thanks to the cost reduction of PV system and storage battery, a recent smart house often has a home energy management system (HEMS). In the last decade, a number of demonstration projects of smart house have been conducted throughout the world.

4.7.1 United States/Japan demonstration smart grid project in Los Alamos [27]

The New Energy and Industrial Technology Development Organization, Japan, undertakes the United States/Japan demonstration smart grid project in Los Alamos. A smart house equipped with 2 MW PV system, 8.3 MWh storage batteries, and a micro energy management system was demonstrated together with regional microgrid. The HEMS applied in this project controls them that allow for electric demand in the house to be responsive to smart grid signals, minimizing electricity costs and preserving the comfort of residential usage patterns. Based on time of use (ToU) electricity price, PV power output forecast, load power performance, and hot water usage results, the applied optimization method runs a linear programming to minimize the electricity cost of next three days by optimizing the battery charge/discharge and hot water storage of every 15 min.

Fig. 4.19 shows a configuration of PV power output forecasting system. The forecasting method is a combination NWP model, PV power output estimation model, and statistical model. The time-horizons are both short, up to 24 h, and long, up to 1 week. At first, NWP model forecast irradiance, wind speed, and temperature. The grid point data of mesoscale calculated by the JMA is used as the initial conditions of NWP model. Then, PV power output model translates the forecasted weather elements to the PV power output based on the PV system parameters such as installation conditions, specifications of PV module and power conditioner, etc. Finally, the statistical

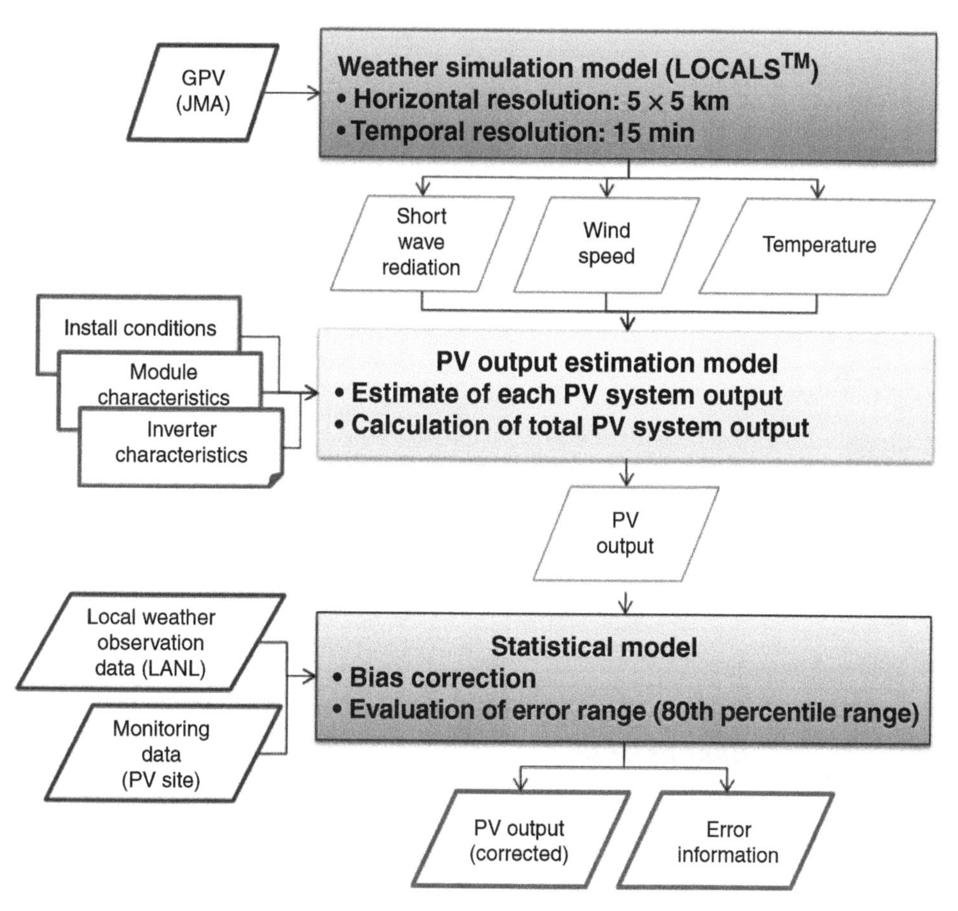

■ **FIGURE 4.19 A configuration of PV power output forecasting system.**

model, which was developed in advance based on the observed values of PV power output and local weather observation data, corrects the statistical bias error of calculated values of PV power output. In addition to the bias correction, the statistical model calculates the forecasting error range (confidence interval).

Fig. 4.20 shows an example of HEMS operation in a day. HEMS controls the storage battery so as to charge with grid power when the TOU price is low during mid night. As a result, the SOC reaches 90% before 5:00 am. Because the daytime electricity demand is small, the excess PV power output is reversed to the grid. Then, after the noon when the ToU price is the highest, HEMS controls the storage battery so as to discharge the electricity and sell it to the grid.

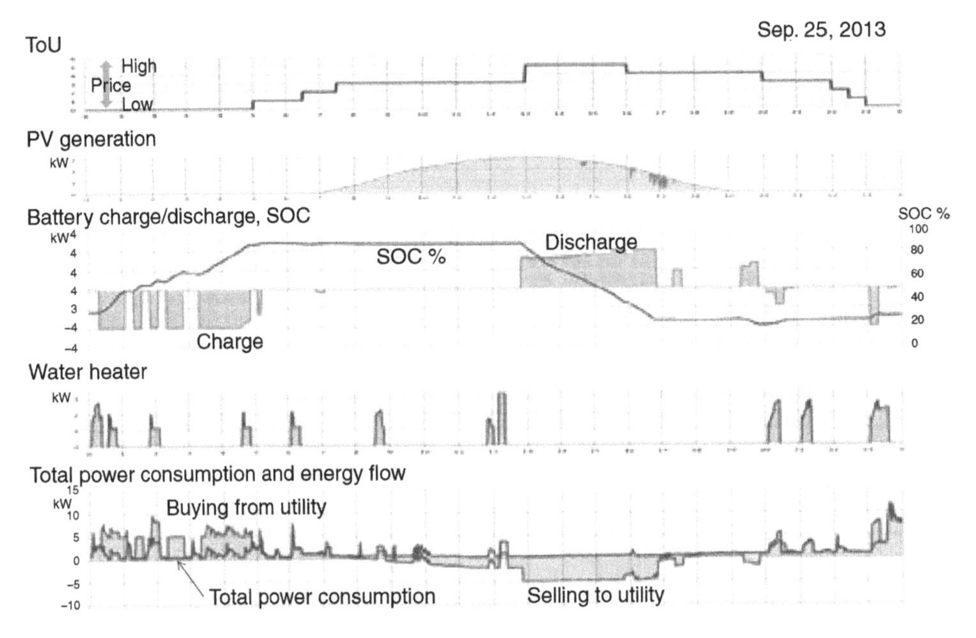

■ **FIGURE 4.20 An example of HEMS operation in a day.**

REFERENCES

[1] IEA/PVPS. TRENDS 2014 in photovoltaic applications, Survey Report of Selected IEA Countries between 1992 and 2013; 2014.

[2] Dierauf T, Growitz A, Kurtz S, Cruz JLB, Riley E, Hansen C. Weather-corrected performance ratio, NREL/TP-5200-57991; 2013.

[3] Kato T, Kumazawa S, Suzuoki Y, Honda N, Koaizawa M, Nishino S. Evaluation of long-cycle fluctuation of spatial average insolation in electric utility service area, Proceedings of The 2012 IEEE PES General Meeting, PESGM2012-001461; 2012, CD-ROM.

[4] Nagoya H, Komami S, Ogimoto K. A method for presuming total output fluctuation of highly penetrated photovoltaic generation considering mutula smoothing effect. IEEJ Trans. EIS 2011;131-C (10), 1688–1696 (in Japanese).

[5] Kato T, Kumazawa S, Honda N, Koaizawa M, Nishino S, Suzuoki Y. Spatial average irradiance fluctuation characteristics in Chubu Region considering smoothing effect around observation points. IEEJ Trans. PE 2013;133-B(4), 373–382 (in Japanese).

[6] Kato T, Manabe Y, Funabashi T, Matsumoto K, Kurimoto M, Suzuoki Y. A study on pattern classification of diurnal variation of spatial average irradiance. Proceedings of the twenty-sixth annual conference of power & energy society. IEE of Japan; 2015, 170 (in Japanese).

[7] Perez R, Kivalov S, Schlemmer J, Hemker K Jr, Renne D, Hoff TE. Validation of short and medium term operational solar radiation forecasts in the US. Sol Energy 2010;84:2161–72.

[8] JMA web site. http://www.jma.go.jp/jma/en/Activities/nwp.html.
[9] WRF web site. http://www.wrf-model.org/index.php.
[10] Shimada S, Liu YY, Yoshino J, Kobayashi T, Wazawa Y. Solar irradiance forecasting using a mesoscale meteorological model, Part II: Increasing the accuracy using the Kalman filter. J Jap Sol Energ Soc 2013;39(3):61–7. (in Japanese).
[11] Marquez R, Coimbra CFM. Forecasting of global and direct solar irradiance using stochastic learning methods, ground experiments and the NWS database. Sol Energy 2011;85:746–56.
[12] Kataoka Y, Kato T, Suzuoki: Y. A study on spatial average insolation forecast in electric utility service area using mesoscale model grid point value. IEEJ Trans PE 2013;133-B(6):548–54. (in Japanese).
[13] Fonseca JGS Jr, Oozeki T, Takashima T, Koshimizu G, Uchida Y, Ogimoto K. Use of support vector regression and numerically predicted cloudiness to forecast power output of a photovoltaic power plant in Kitakyushu, Japan. Prog Photovolt Res Appl 2012;20(7):874–82.
[14] Kato T. Report on competition of day-ahead forecasting of irradiance and wind power output. The 2015 Annual Meeting Record I.E.E. Japan, No. 4-S13–17; 2015.
[15] Takenaka H, Nakajima TY, Higurashi A, Higuchi A, Takamura T, Pinker RT, Nakajima T. Estimation of solar radiation using a neural network based on radiative transfer. J Geophys Res 2011;116:D08215.
[16] Chow CW, Urquhart B, Kleissl J, Lave M, Dominguez A, Shields J, Washom B. Intra-hour forecasting with a total sky imager at the UC San Diego solar energy testbed. Sol Energy 2011;85:2881–93.
[17] Gauchet C, Blanc P, Espinar B, Charbonnier B, Demengel D. Surface solar irradiance estimation with low-cost fish-eye camera. ES1002 Workshop; 2012.
[18] Yang H, Kurtz B, Nguyen D, Urquhart B, Chow CW, Ghonima M, Kleissl J. Solar irradiance forecasting using a ground-based sky imager developed at UC San Diego. Sol Energy 2014;103:502–24.
[19] Huang J, Korolkiewicz M, Agrawal M, Bolandl J. Forecasting solar radiation on an hourly time scale using a coupled autoregressive and dynamical system (cards) model. Sol Energy 2013;87:136–49.
[20] Perez R, Lorenz E, Pelland S, Beauharnois M, Knowe GV, Hemker K Jr, Heinemann D, Remund J, Müller SC, Traunmüller W, Steinmauer G, Pozo D, Ruiz-Arias JA, Lara-Fanego V, Ramirez-Santigosa L, Gaston-Romero M, Pomares LM. Comparison of numerical weather prediction solar irradiance forecasts in the US, Canada and Europe. Sol Energy 2013;94:305–26.
[21] Lauret P, Voyant C, Soubdhan T, David M, Poggi P. A benchmarking of machine learning techniques for solar radiation forecasting in an insular context. Sol Energy 2015;112:446–57.
[22] Lorenz E, Hurka J, Heinemann D, Beyer HG. Irradiance forecasting for the power prediction of grid-connected photovoltaic systems. IEEE J Sel Topics Appl Earth Observ Remote Sens 2009;2(1):2–10.
[23] Mathiesen P, Kleissl J. Evaluation of numerical weather prediction for intra-day solar forecasting in the continental united states. Sol Energy 2011;85:967–77.
[24] Pelland S, Galanis G, Kallos G. Solar and photovoltaic forecasting through post-processing of the global environmental multiscale numerical weather prediction model. Prog Photovolt Res Appl 2013;21:284–96.

[25] California I.S.O. What the duck curve tells us about managing a green grid. Fast Facts; 2013.
[26] Kleissl J. Solar energy forecasting and resource assessment. Oxford: Elsevier; 2013.
[27] NEDO. Japan–US New Mexico Smart Grid Collaborative Demonstration Project; 2015.

Chapter 5

Prediction of wind power generation output and network operation

Ryoichi Hara
Graduate School of Information Science and Technology, Hokkaido University, Hokkaido, Japan

CHAPTER OUTLINE

5.1 Need for forecasting wind power output in electric power systems 110
5.2 Power output fluctuation characteristics 112
5.2.1 Fundamentals 112
5.2.2 Maximum variation 115
5.2.3 Umbrella curve 115
5.2.4 Standard deviation 115
5.2.5 Power spectral density 118
5.3 Power output smoothing control 120
5.3.1 Application of energy-storage system 120
5.3.2 Kinetic energy of wind turbines 121
5.3.3 Pitch angle control 121
5.4 Forecasting methods 122
5.4.1 Difficulties 122
5.4.2 Physical approach 123
5.4.3 Statistic approach 123
5.4.4 Regional forecasting 124
5.4.5 Probabilistic forecast 125
5.5 Examples of forecasted results 125
5.6 Forecasting applications 128
5.6.1 Scheduled generation of wind farms and solar power plants with energy-storage systems 128
5.6.2 Suppression of ramp variation of wind output 129
References 130

Integration of Distributed Energy Resources in Power Systems

5.1 NEED FOR FORECASTING WIND POWER OUTPUT IN ELECTRIC POWER SYSTEMS

Wind power generation, which can convert the kinetic energy of wind into electric energy without serious environmental damages, is regarded as one of the most promising distributed energy resources in the world. Relatively cheap installation cost accelerates the installation of wind power generation in the world. The Global Wind Energy Council reported that the annual global installed capacity in 2014 exceeded 50 GW and the global cumulative wind power generation capacity has grown exponentially as shown in Fig. 5.1 [1].

Wind power generation appeals because of its merits in cost, ecological compatibility, sustainability, enormousness, and ubiquity natures; however, disadvantages such as intermittency, variability, and uncertainty still remain technological issues. As introduced in Section 5.2.1, power generated from wind power varies depending on the wind speed variation.

In a typical electric power systems, the total generation output of conventional generators such as thermal, hydro, and/or nuclear power plants must meets the total demand (electricity consumption) at every moment to maintain the system frequency. The system operator achieves adequate frequency regulation by the weekly/daily generation scheduling including unit commitments, the online economic load dispatch, and the frequency control based on the speed–droop characteristics, which cover different time domains. The purpose of generation scheduling is to find the most economical generation schedule that can meet the forecasted demand in the coming week/day and satisfy the static and dynamic operational constraints such as capacity limits, voltage limits, procurement of regulation margin, and other criteria on stability, reliability, and security. Here, the demand forecast considered in the scheduling process should be the forecasts for demand that is to be supplied by the conventional generators, or the net demand (the actual demand minus the total output of renewable energy generation systems). That is, the wind power output during the targeted period must be forecast with the same or higher temporal resolution (typically 30 min in daily scheduling process) in the power system with mass penetration of wind power generations. Since the forecast data used in the generation scheduling contains errors, the system frequency deviates even though the generators are running as scheduled. The online economic load dispatch revises the generation schedule based on the short-term (from several minutes to several hours ahead, depending on the country and region) demand and wind power forecasts. Fluctuation of net demand during the scheduling interval also affects the system frequency. As wind power generation grows, the short-term net demand variation is also widened. This short-term

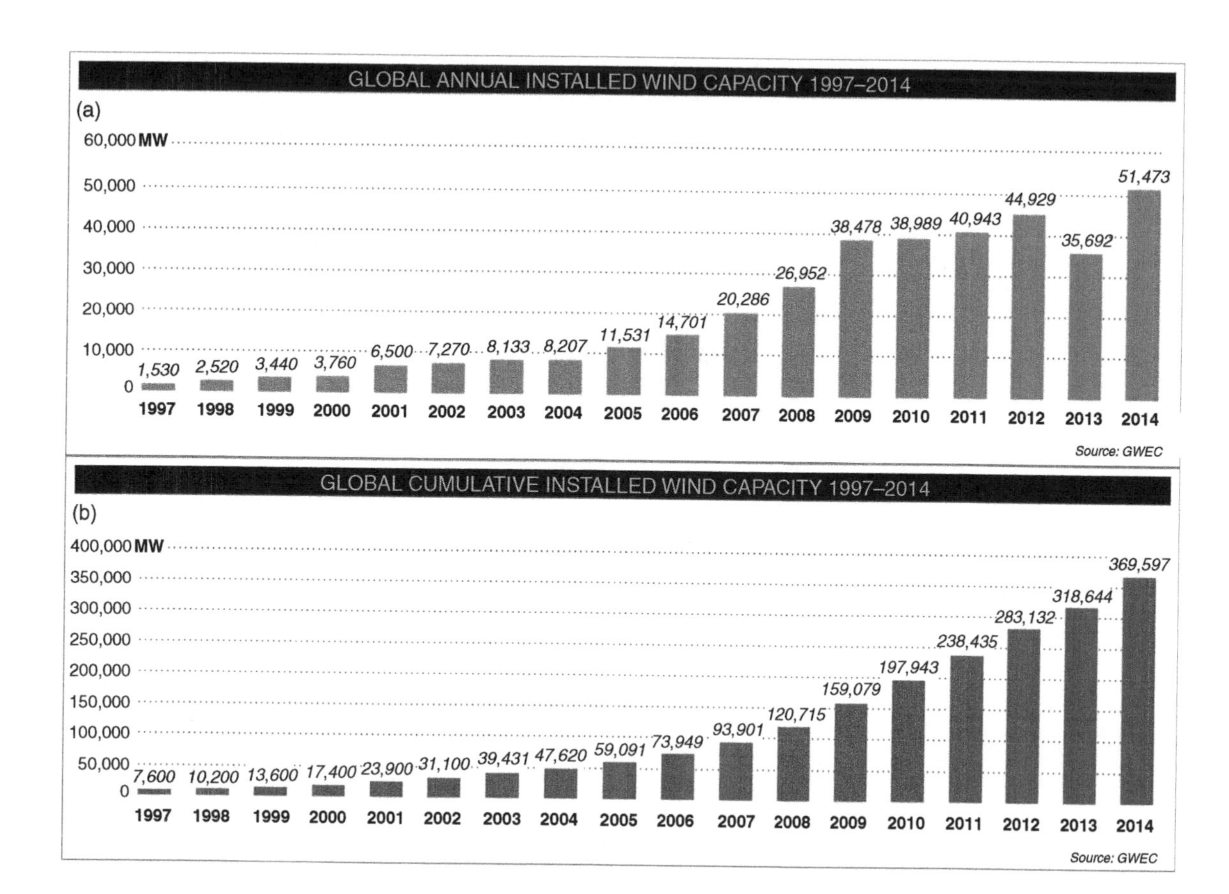

FIGURE 5.1 Global growth of wind power generation. (a) Annual installed capacity; (b) cumulative installed capacity.

frequency deviation is covered by means of the primary and secondary load frequency controls, in which the generator output is automatically regulated within the primary and secondary reserve margin procured in the generation scheduling process. That is, understanding the characteristics of wind power output fluctuation is also an important factor to estimate the adequate level of reserve margins.

As described above, the wind power output forecast is becoming an important technology for the stable operation of power system with mass wind power generations. Actually, some regional system operator (RSOs)/ independent system operator (ISOs) and transmission system operator (TSOs) such as Bonneville Power Administration, Electric Reliability Council of Texas, and New York ISO in the United States, Alberta Electric System Operator in Canada, 50Hertz in Germany, EirGrid in Ireland, Energinet in Denmark, Tohoku-EPCO in Japan, and so on, have integrated the wind power output forecasting function into their generation scheduling and/or online economic load dispatch operations [2].

Another need for wind power output forecast is emerging from the wind farm owner/investor side. In some countries and regions, the wind farm is allowed to sell their power at the day-ahead electricity market, to where the market participants have to show their generation schedule in the next day. For wind farms, a generation schedule should be planned based on the day-ahead wind power forecast. If the intraday market becomes more popular, needs for short-term forecast would also emerge.

5.2 POWER OUTPUT FLUCTUATION CHARACTERISTICS

5.2.1 Fundamentals

Theoretically, the available power, P_a (W), in the wind can be expressed as follows:

$$P_a = \frac{1}{2} A \rho V^3 \tag{5.1}$$

where, A is the area wind passing through (which is considered perpendicular to the direction of the wind) (m²), ρ is the density of air (kg/m³), and V is the wind speed (m/s). The Eq. (5.2) represents that P is equal to the kinetic energy possessed by the volume of air passing through A per every second. The actual power generation by single turbine generator is, needless to say, less than the above available power. Reasons for this reduction is not only energy conversion losses that consists of mechanical loss associated with bearing and gearbox and cupper and iron losses of generator, but also the electromechanical characteristics of wind power generator; cut-in and cut-out wind speeds and rated power. The cut-in wind speed is the minimum wind speed at which the turbine blades overcome friction and begin to rotate. The cut-out speed is the wind speed at which the pitch angle of turbine blades are regulated to flat to avoid damage from high pressure of wind and the generation is stopped. The actual wind power generation, P (kW), in the steady state is then expressed as follows:

$$P = \begin{cases} 0 & (V < V_{\text{cut-in}}) \\ \eta P_a & (V_{\text{cut-in}} \le V < V_{\text{rated}}) \\ P_{\text{rated}} & (V_{\text{rated}} \le V < V_{\text{cut-out}}) \\ 0 & (V \le V_{\text{cut-out}}) \end{cases} \tag{5.2}$$

where, $V_{\text{cut-in}}$ and $V_{\text{cut-out}}$ are the cut-in and cut-out wind speeds (m/s), V_{rated} is the wind speed (m/s) at which the wind power output reaches to the rated power P_{rated} (W), and η is the total conversion efficiency of wind power generator and its associated mechanical transmissions (note that

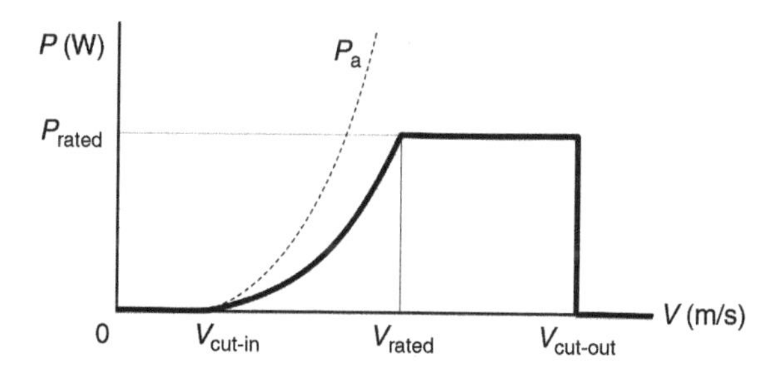

FIGURE 5.2 Example of power curve of single wind power generator.

conversion efficiency is also a function of wind speed indeed). For the wind speed $V_{rated} < V < V_{cut\text{-}out}$, the pitch angle control maintains the wind power output at P_{rated}. For example, the relationship between P and V, called the power curve, is illustrated in Fig. 5.2. IEC 61400-12 standardizes the wind power output measurement in detail [3]. As observed in Fig. 5.2, the wind power output strongly depends on the wind speed, and therefore, it would vary moment to moment. Fig. 5.3 shows the actual power output of single 1500 kW wind power generator for 10 days [4].

For a wind farm, where multiple wind power generators are aggregated together and interconnected to the main grid through the common connection point, the fluctuation of total generation output would be smoothed as shown in Fig. 5.4, which shows the total output of six wind power generators

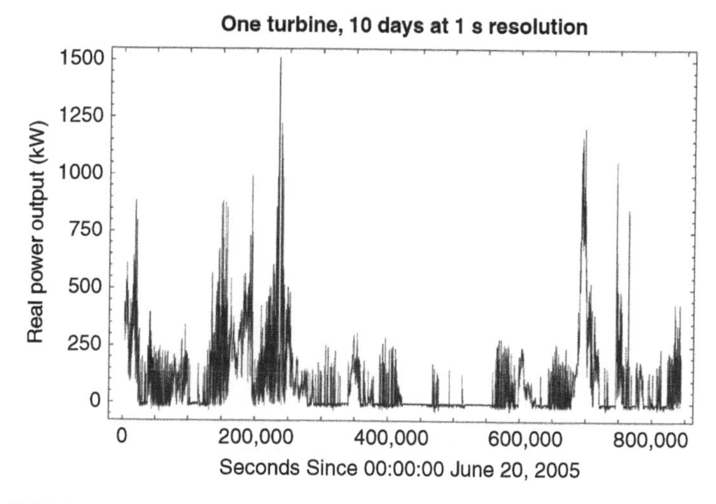

FIGURE 5.3 Real power output (kW) of 1500 kW wind power generator for 10 days [4].

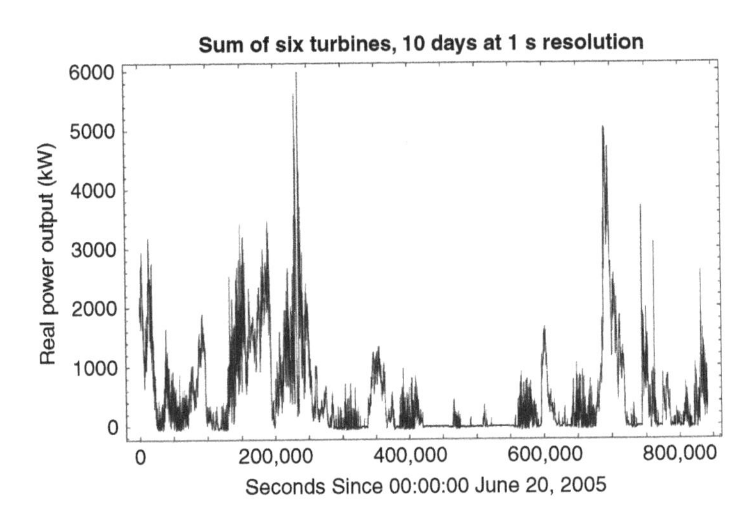

FIGURE 5.4 Real power output (kW) of six wind power generators for 10 days [4].

including the generator shown in Fig. 5.3. In detail, the outputs in both figures are similar in trend, but the jaggy (short-term) fluctuation observed in the single generator output disappears in the aggregated output. This is because the short-term fluctuations among generators are uncorrelated due to the spatial dispersion while the long-term trends of output are correlated and similar. Note that the absolute width of fluctuation of synthetic output is statistically wider than those of individual outputs, but the ratio of the fluctuation width to the total capacity (or the mean output), called the relative fluctuation width, becomes narrower than those of individuals. Due to this smoothing effect brought by the geographical dispersion, the shape of the power curve of a wind farm becomes relatively gentle as shown in Fig. 5.5. It can be generally said that a greater number of wind power generators and

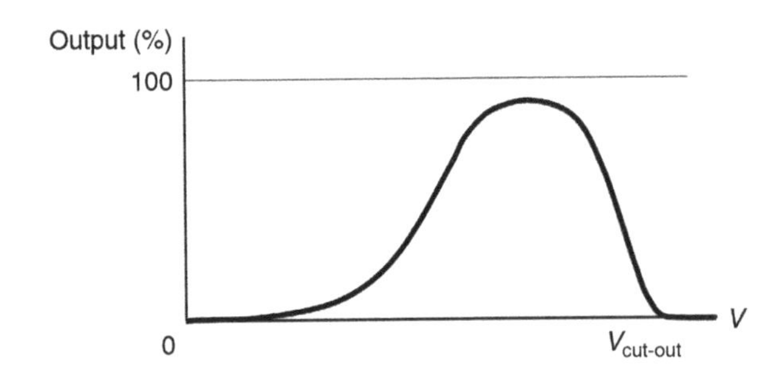

FIGURE 5.5 Example of power curve of a wind farm.

farther placement would reduce the short-term fluctuation width measured in ratio to the capacity.

5.2.2 Maximum variation

One of the indices used in fluctuation measurement is the maximum variation, which is defined as a maximum stepwise change of generation output within the specified duration. The maximum variation, which is measured in kW, can represent the impact to the power system directly and intuitively. For stable and safe operation of the power system, the system operator must procure the adequate capacity of reserve, which can cover the maximum variation (more precisely, larger than the maximum variation of net demand) even though the event is an extreme case. The literature [5] summarizes the maximum variations in some countries as shown in Table 5.1. Even in power systems with a large number of sites, the maximum variation of total wind power output is small (about 10%) in shorter timescale (in the order of minutes). In a longer timescale perspective, however, it would reach to 70% in the worst case within 4 h.

5.2.3 Umbrella curve

While the maximum variation is used to measure the possible fluctuation deterministically, the distribution of fluctuation magnitude with its occurrence probability is also often used to understand the characteristics of output fluctuation. The umbrella curve, presenting the occurrence frequency of changes in average wind power output during a certain period, is widely used. An example of an umbrella curve is shown in Fig. 5.6, which is reported in Ref. [6]. As illustrated in Fig. 5.6, the distribution of output fluctuation is generally different from the normal distribution and spreads wider.

5.2.4 Standard deviation

One of the important statistical indices for output fluctuation is the standard deviation (SD), defined as follows:

$$\sigma = \sqrt{\frac{1}{M}\sum_{m=0}^{M-1}\left(P_m - P_{\mathrm{ave}}\right)^2} \tag{5.2.3}$$

where, P_{m} is the wind power output (W) measured at mth sampling, P_{ave} is the average output (W), and M is the number of samples. For SD of total output from N generators (σ_{total}), the following equation can be derived from Eq. (5.2.3):

$$\sigma_{\mathrm{total}}^2 = \sum_{k=1}^{N}\sigma_k^2 + \sum_{k=1}^{N-1}\sum_{l=k+1}^{N} 2r_{kl}\sigma_k\sigma_l \tag{5.2.4}$$

Table 5.1 Extreme Variations of Large-Scale Regional Wind Power as a Percent of Installed Capacity

			10–15 min		1 h		4 h		12 h	
Region	**Region size (km^2)**	**Number of sites**	**Max decrease (%)**	**Max increase (%)**	**Max decrease (%)**	**Max increase (%)**	**Max decrease (%)**	**Max increase (%)**	**Max decrease (%)**	**Max increase (%)**
Denmark	300 × 300	>100			−23	+20	−62	+53	−74	+79
West Denmark	200 × 200	>100			−26	+20	−70	+57	−74	+84
East Denmark	200 × 200	>100			−25	+36	−65	+72	−74	+72
Ireland	280 × 480	11	−12	+12	−30	+30	−50	+50	−70	+70
Portugal	300 × 800	29	−12	+12	−16	+13	−34	+23	−52	+43
Germany	400 × 400	>100	−6	+6	−17	+12	−40	+27		
Finland	400 × 900	30			−16	+16	−41	+40	−66	+59
Sweden	400 × 900	56			−17	+19	−40	+40		
US Midwest	200 × 200	3	−34	+30	−39	+35	−58	+60	−78	+81
US Texas	490 × 490	3	−39	+39	−38	+36	−59	+55	−74	+76
US Midwest+OK	1200 × 1200	4	−26	+27	−31	+28	−48	+52	−73	+75

Denmark, data 2000–2002 from http://www.energinet.dk; Ireland, Eirgrid data, 2004–2005; Germany, ISET, 2005; Finland, years 2005–2007; Sweden, simulated data for 56 wind sites 1992–2001; United States, NREL years 2003–2005; Portugal, INETI.

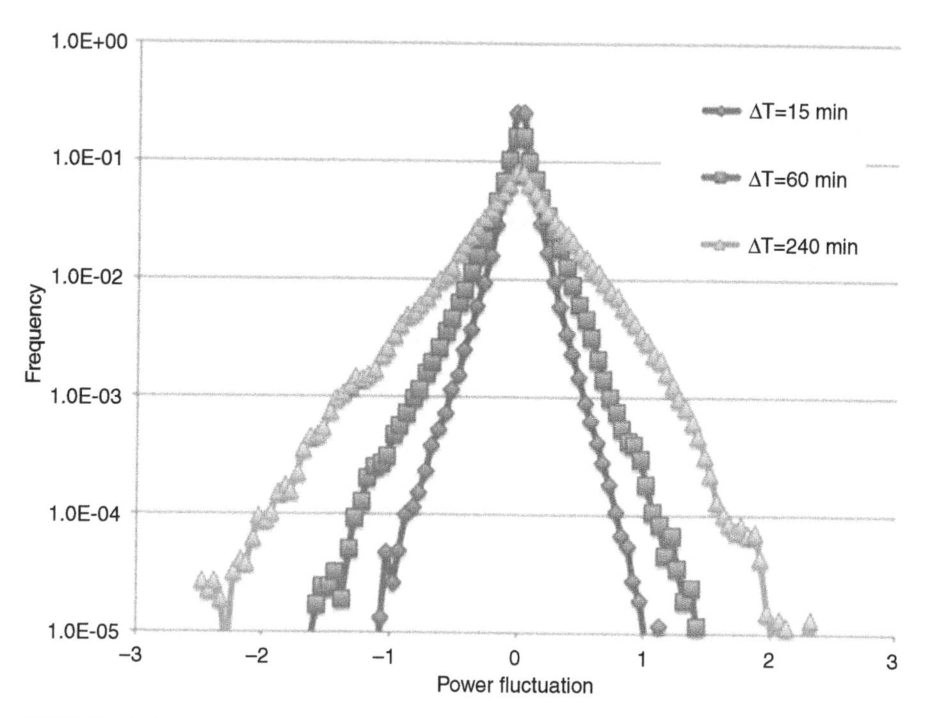

FIGURE 5.6 Frequencies of relative power fluctuation for time period 1/4, 1, and 4 h [6].

where, σ_k is SD of output at kth generator (W), r_{kl} is the correlation coefficient between kth and lth generators. Here, assume that the SD of each generator is identical, that is, $\sigma_k = \sigma_l = \sigma$. When the output from N generators are completely correlated ($r_{kl} = 1$),

$$\sigma_{\text{total}} = N\sigma \tag{5.2.5}$$

On the other hand, in case of total uncorrelated situation ($r_{kl} = 0$), total output can be well smoothed, then

$$\sigma_{\text{total}} = \sqrt{N}\sigma \tag{5.2.6}$$

In most cases, the outputs of multiple generators are partially correlated as discussed in Section 5.2.5; as a result, ratio of total SD to individual SD stays between $\sqrt{N}$ and N. For the aggregation of generators in different size and fluctuation characteristics, the following average correlation coefficient ρ is sometimes used in analysis.

$$\rho = \frac{\sum_{k=1}^{N-1}\sum_{l=k+1}^{N} r_{kl}\sigma_k\sigma_l}{\sum_{k=1}^{N-1}\sum_{l=k+1}^{N}\sigma_k\sigma_l} = \frac{\sigma_{\text{total}}^2 - \sum_{k=1}^{N}\sigma_k^2}{\sum_{k=1}^{N-1}\sum_{l=k+1}^{N} 2\sigma_k\sigma_l} \tag{5.2.7}$$

5.2.5 Power spectral density

In some analysis, the fluctuation of wind power generator/wind farm output is studied as the power spectral density (PSD, or simply called the power spectrum in some literature). The terminology "power spectrum" is not the spectrum of power produced, but the square of spectrum magnitude ($\xi(f)$) obtained by the discrete Fourier transform shown in the following equation:

$$\xi(f) = \sum_{m=0}^{M-1} P_m \exp\left(j\frac{2\pi mf}{Mf_s}\right) \tag{5.2.8}$$

where, f is the frequency (Hz), f_s is the sampling rate (Hz), and j is the imaginary unit ($j^2 = -1$).

Literature [4] revealed that the PSD of wind farm output, shown in Fig. 5.7, shows good agreements with Kolmogorov spectrum that can express the characteristics of turbulence of incompressible fluidities and is proportional to $f^{-5/3}$.

Parseval's theorem assures that the variance of P_i and the sum of PSD are identical; that is, the PSD can be recognized as the decomposition of

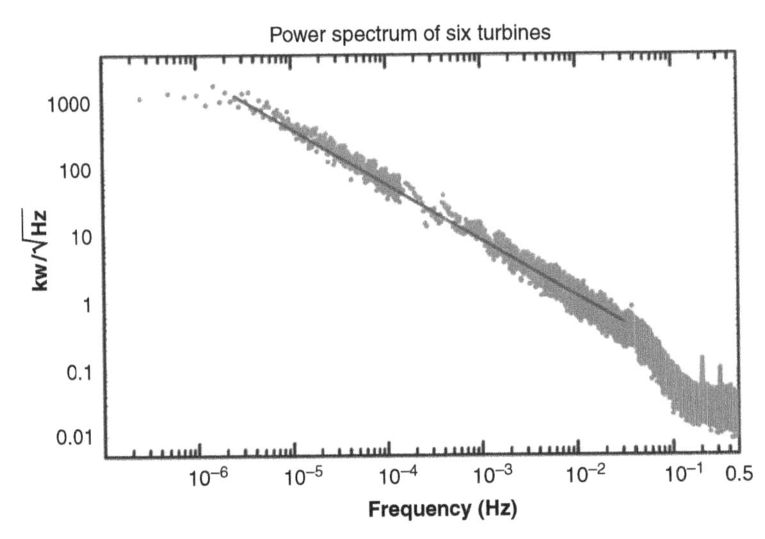

FIGURE 5.7 Power spectral density of wind farm with six turbines (gray points) and f − 5/3 spectrum (solid line) [4].

variance in the frequency domain. This idea naturally evolves Eq. (5.2.4) into the analogical expression for power spectra of N aggregated output and individuals, as follows.

$$S_{\text{total}}(f) = \sum_{k=1}^{N} S_k(f) + \sum_{k=1}^{N-1}\sum_{l=k+1}^{N} 2\gamma_{kl}(f)\cos\phi_{kl}(f)\sqrt{S_k(f)S_l(f)} \qquad (5.2.9)$$

Where, S_{total} and S_k are the power spectra of total and individual outputs, and γ_{kl} and ϕ_{kl} are the coherent and phase angle between two individual outputs. The literature [7] develops a similar extension to Eq. (5.2.7) and reaches to the idea of average coherent (coh_{av}), defined as

$$\text{coh}_{\text{av}}(f) = \frac{\sum_{k=1}^{N-1}\sum_{l=k+1}^{N} \gamma_{kl}(f)\cos\phi_{kl}(f)\sqrt{S_k(f)S_l(f)}}{\sum_{k=1}^{N-1}\sum_{l=k+1}^{N} \sqrt{S_k(f)S_l(f)}} = \frac{S_{\text{total}}(f) - \sum_{k=1}^{N} S_k(f)}{\sum_{k=1}^{N-1}\sum_{l=k+1}^{N} 2\sqrt{S_k(f)S_l(f)}} \qquad (5.2.10)$$

Due to the above definition, the average coherence represents the degree of correlation at the specified frequency and ranges from 0 (uncorrelated) to 1 (correlated). The literature [7] also analyzes the average coherent for the 16 wind farms in Hokkaido, Japan, placed over some hundreds of kilometers apart. The major conclusion is that the average coherence for periods longer than 100 min is different day by day, but is relatively high compared with shorter periods. Especially the average coherent for periods less than 10 min is quite small; that is, the smoothing effect by geographical dispersion can only be expected for periods less than 10 min. Distributions of average coherent in different three days reported in Ref. [7] are shown in Fig. 5.8.

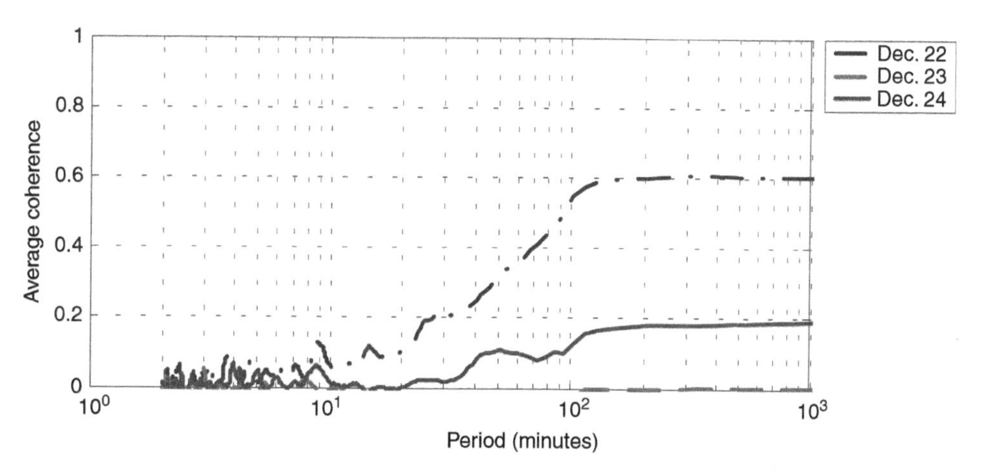

FIGURE 5.8 Example of average coherent; evaluated for 16 wind farms in Hokkaido, Japan [7].

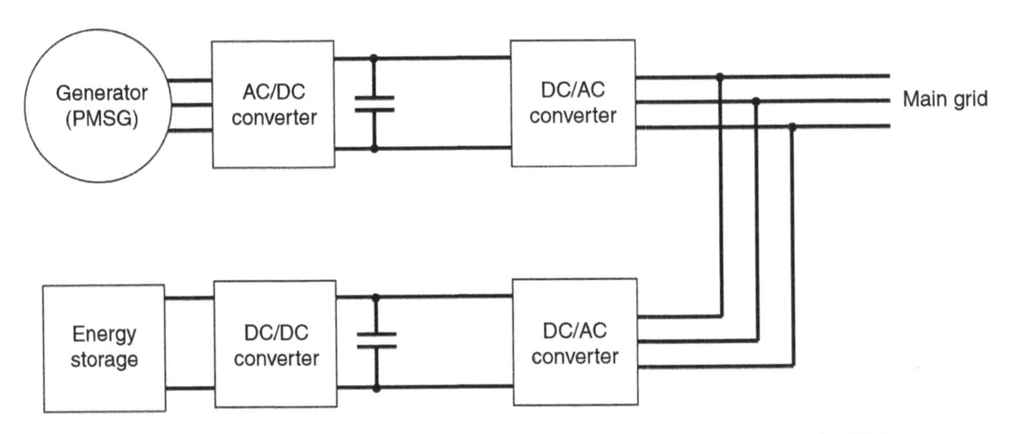

■ **FIGURE 5.9 Circuit topology of wind power generator with energy-storage system connected to AC line.**

5.3 POWER OUTPUT SMOOTHING CONTROL

5.3.1 Application of energy-storage system

One of the major fluctuation smoothing approaches is to use the energy-storage system, such as lead–acid, nickel cadmium, lithium ion, sodium sulfur batteries, and ultra capacitor [8–14]. Most energy-storage systems work with DC, therefore, the bidirectional DC/AC converter is needed in principle. Fig. 5.9 shows a typical circuit topology of wind power generator with energy-storage system. Another choice of circuit topology available for a double-fed induction generator or a permanent magnet synchronous generator, which has a DC-link inside, is to embed the energy-storage system into the DC-link as shown in Fig. 5.10. The advantage of the latter topology is to save the system cost and volume, by sharing the DC/AC converter with the

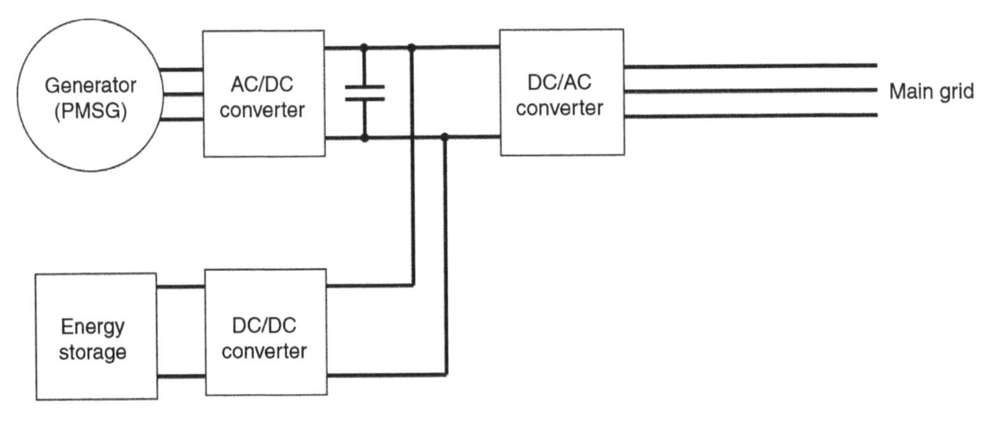

■ **FIGURE 5.10 Circuit topology of wind power generator with energy-storage system embedded at DC-link.**

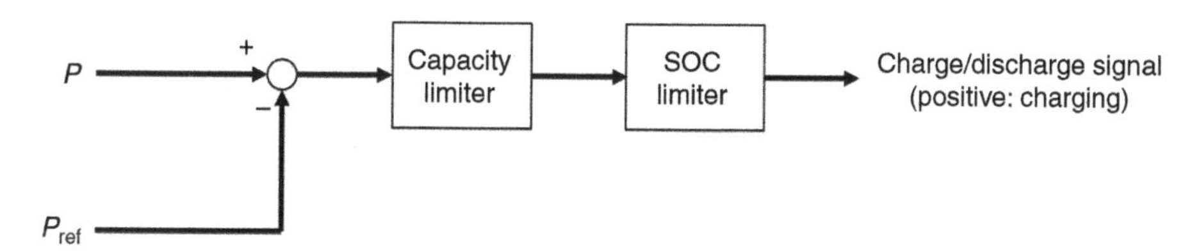

■ **FIGURE 5.11 Control of energy-storage system for fluctuation smoothing.**

generator. For a wind farm consisting of multiple wind power generators, it is beneficial to install single energy system to the farm with the circuit topology as in Fig. 5.9.

A control strategy for an energy-storage system is to charge or discharge the difference between the actual output (P) and the reference output (P_{ref}), as illustrated in Fig. 5.11. The capacity limiter maintains the charge/discharge power with in the inverter capacity. The SOC (state of charge) limiter is also used to avoid overcharging and overdischarging. Several approaches have been developed for deciding reference output; the most major and popular approach is to employ the moving average of actual output. The advantage of moving the average method is the simplicity of the SOC management since the average charged and discharged energy can be balanced automatically in principle. Strictly speaking, the SOC is drifting to empty due to charging/discharging efficiency even if the moving average is applied. For this reason, it is beneficial to implement the SOC feedback control [11] in parallel with Fig. 5.11.

Application of flywheel technology, which stores the energy as a kinetic energy of rotating disk, is also discussed in Refs. [15–17].

5.3.2 Kinetic energy of wind turbines

Needless to say, the wind generator is driven by the wind turbine, whose moment of inertia is relatively large. Thus, the inertia of the wind turbine itself can store energy in the form of kinetic energy. The idea is to store excess wind power by increasing the rotation speed of turbine, and compensate the deficient wind power by decreasing the rotation speed [18,19]. The rotation speed of turbines can be controlled by regulating the generator output while the turbine is receiving the optimal mechanical torque from the wind.

5.3.3 Pitch angle control

Pitch angle control is usually used to maintain the output at rated power for the wind speed higher than the rated speed (V_{rated}), as shown in

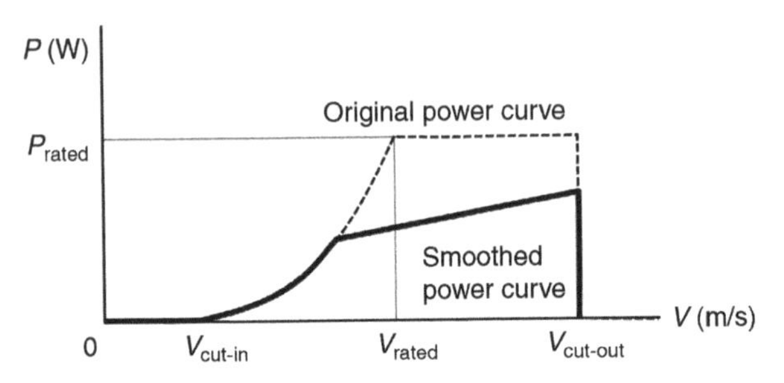

FIGURE 5.12 Image of power curve with pitch angle control for output smoothing.

Fig. 5.2. Implementation of pitch angle control to the wind speed below the rated speed can realize the constant (or smoothed) output control for wider wind speed range, as shown in Fig. 5.12. The merit of the pitch angle is simplicity of implementation since it does not require any additional devices such as energy-storage system. However, the pitch angle control for smoothing is associated with abandoning the available energy of wind; as a result, this method sacrifices the energy production. In order to avoid unnecessary energy loss, sophisticated pitch angle control schemes have been developed [20–22].

5.4 FORECASTING METHODS

5.4.1 Difficulties

Wind power output forecasting is fundamentally the forecasting of wind speed and direction at the location of wind power generator or wind farm during the short time intervals.

For short-term forecasting, which is employed in the online economic load dispatch, atmospheric behavior in a relatively small area must be forecast. Such small area forecasting requires precise and high geographical resolution meteorological measurement data, such as input variables, which are difficult to obtain.

For the day-ahead forecasting, which is used in the generation scheduling, atmospheric behavior in regional size, from mesoscale to synoptic scale, must be forecast. In most region/countries, these are measured, forecast, and provided by a governmental agency; however, their geographical resolution is sometimes coarse to use for the wind generator/farm output forecasting. Furthermore, the length of the temporal horizon (look-ahead period) itself makes forecasting more difficult.

Against the above difficulties, there has been much research and many developments. Of these efforts, there have been two major approaches: physical and statistical. Most wind power forecasting methods fall into one of them, or a combination of both.

Another side to difficulties arises when the system operator tries to forecast the total wind power output in its control area. A straightforward approach is to obtain and sum the output forecasts of all wind farms. In this approach, however, the system operator would handle the mess of the data and require long computation cost. Furthermore, the system operator could not acquire the specifications or online status of some wind farms.

5.4.2 Physical approach

In the physical approach, atmospheric behavior is firstly forecast by numerically solving a set of differential equations representing the state and movement of atmosphere in a the region under consideration, which is modeled by the three-dimensional grid cells with a finite space resolution. In detail, the considered differential equations consist of equations of atmosphere motion (for three dimension), conservation of mass, state equation and thermodynamics of atmosphere, and conservation of water vapor (seven equations) with wind speed (in three dimensions), temperature, pressure, density, and water vapor (seven state variables) for each grid cell. This prediction is called the Numerical Weather Prediction (NWP) and is generally provided by a governmental organization/agency service.

Then, the obtained wind speed at the nearest grid point is scaled considering the hub's height and the property of terrain, such as the roughness, orography, and obstacles. The refined wind speed is then applied to the power curve of the target wind generator to obtain the wind power output forecast. In order to eliminate the effect of systematic error associated with the NWP modeling and improve the forecast accuracy, a postprocessing named Model Output Statistics (MOS) would be applied. For MOS, one of the statistic approaches described in the next section is applied. The whole organization of the physical approach is illustrated in Fig. 5.13 [23].

5.4.3 Statistic approach

Applying fine temporal and special mesh resolutions in the NWP may improve the forecast accuracy; however, the computation burden would become heavier. In order to shorten the computation time, some statistical approaches have been developed. A statistical approach predicts the wind speed or wind power output simply based on the input data. Statistical approaches are used (1) as a MOS in the physical approach, or (2) as a

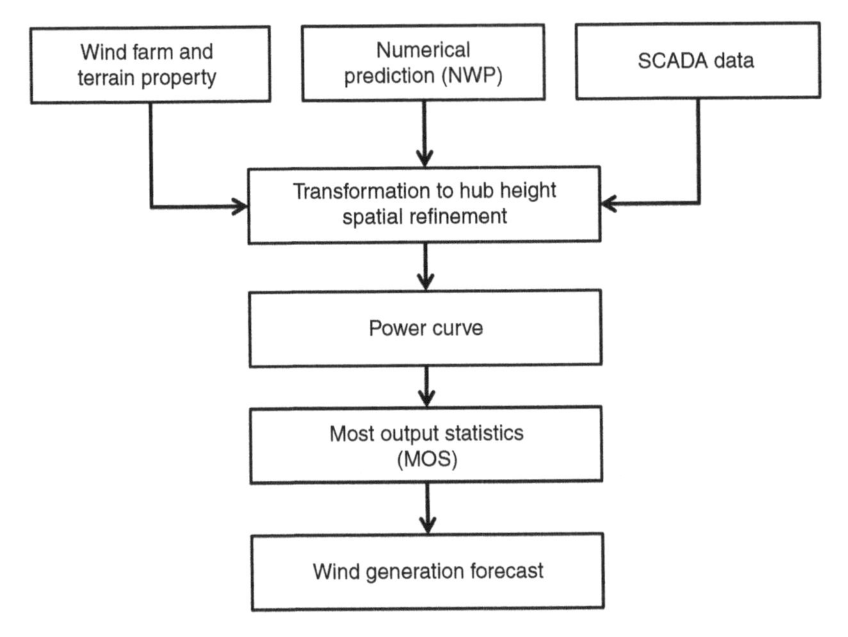

■ **FIGURE 5.13 Structure of physical approach.**

stand-alone predictor that uses the recent wind condition data (speed and direction). For a MOS application, artificial neural networks (ANNs), multivariable regression models, support vector machines, and fuzzy reasoning techniques are major technologies. For a stand-alone predictor, the autoregressive model or ANN is often employed.

Another form of statistical approach is the forecast ensemble method, in which multiple forecast wind speeds or outputs are composed to obtain the final forecast, as shown in Fig. 5.14. The idea of the ensemble method assumes that the error included in individual forecast is unbiased and uncorrelated, and therefore, composite forecast includes fewer errors. Two strategies can be considered to obtain the multiple forecasts: one is to employ the different forecast methods and another is to vary the input data within their range of uncertainty [24].

5.4.4 Regional forecasting

For the system operator, the target of forecasting would be the total wind power outputs in the control area. However, it is not an easy task for the system operator to sum the output forecasts of all wind farms for several reasons (see Section 5.4.1). Thus, for regional forecasting, the output of

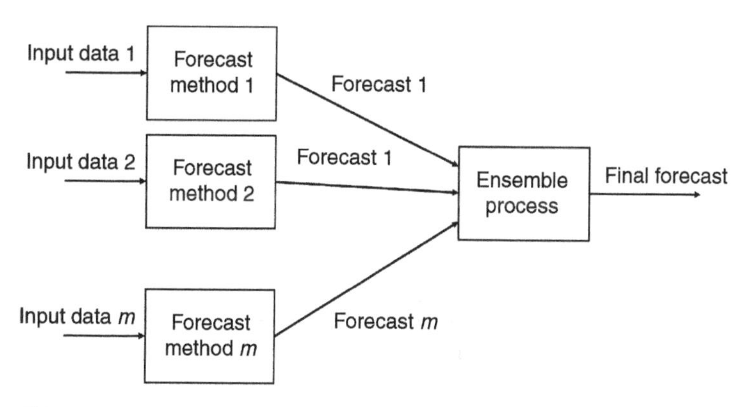

■ **FIGURE 5.14 Structure of ensemble forecast.**

reference wind farms is forecast and then the upscaling technique is applied. The literature [23] summarizes the upscaling frameworks in detail.

5.4.5 Probabilistic forecast

In the preceding examples of forecasting, only a single point value is provided for each time interval, which can be regarded as the mean of the conditional distribution of the wind power. However, recent research has focused on forecasting the probabilistic distribution of wind power during the targeted time interval. This type of forecasting is called probabilistic forecast. Major variations of probabilistic forecast provide one or some of the following properties as a forecast: (1) mean, variance, skewness, and or kurtosis, (2) value at risk (quantiles), (3) confident interval, and (4) probability density function or cumulative distribution function. More detailed explanations are found in [23].

5.5 EXAMPLES OF FORECASTED RESULTS

The Ministry of the Environment of Japan promoted a demonstration project named "wide-area operation systems for multiple renewable energy power plants" in financial year 2012–2014. The primary purpose of this project is the development of a new control system for smoothing the total generation output of multiple wind and solar power plants with the help of weather forecasts and energy storage [25]. The targeted three wind farms are (1) 1980 kW, (2) 2200 kW, and (3) 1200 kW in size and are located in the Hokkaido area of Japan. Two 1000 kW solar power plants are also considered. All five renewable energy power plants are sited at different locations in Hokkaido, and the longest distance is hundreds of kilometers. In this

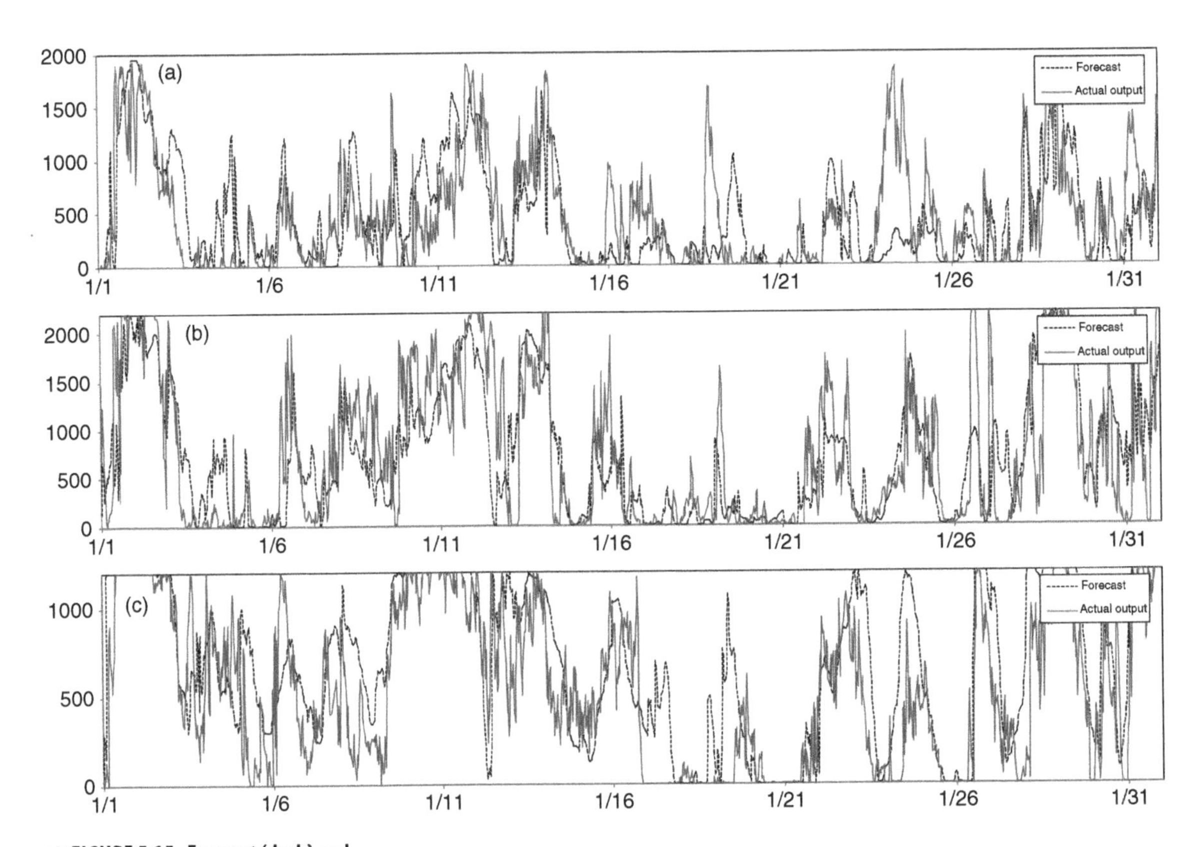

■ FIGURE 5.15 Forecast (dash) and actual (solid) wind farm outputs (kW in vertical) in January. (a) Site A; (b) site B; (c) site C.

project, day-ahead wind output forecasting system has been developed and run for 1 year to evaluate the forecast accuracy. The developed day-ahead forecasting system can provide average wind power outputs during 30 min intervals in the next day (48 intervals) at 4:00 pm. That is, look-up period ranges from 8–32 h. The developed forecasting system has adopted the physical approach that employs SYNFOS-3D (developed by Japan Weather Association, the spatial resolution is 5 km and the temporal resolution is 1 h) with correction based on the observation data [26].

Fig. 5.15 shows the obtained forecasts and the actual outputs at each wind farm for 1 month. Trends of forecast and actual output are similar, but some large error can be still observed (Jan 17th at site C, for example). A histogram of the normalized forecast errors at each site, shown in Fig. 5.16, presents the quite large (greater than 90%) errors that could be included in the forecast. When all three wind farms results are combined together, as shown in Fig. 5.17, the agreements between the forecast and actual output are improved. This is because the forecast errors at individual sites are not correlated with each other. In statistics, the normalized root mean square

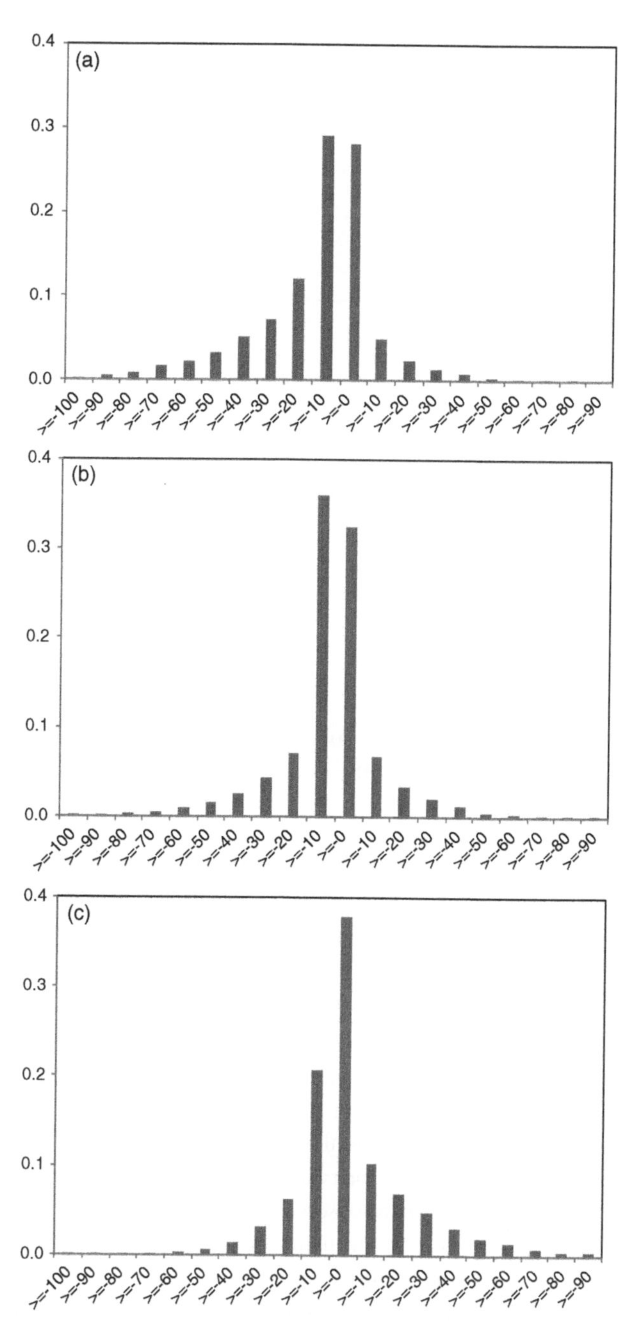

■ FIGURE 5.16 Normalized frequency (vertical) of the normalized forecast errors (percentage in horizontal) counted through 1 year. (a) Site A; (b) site B; (c) site C.

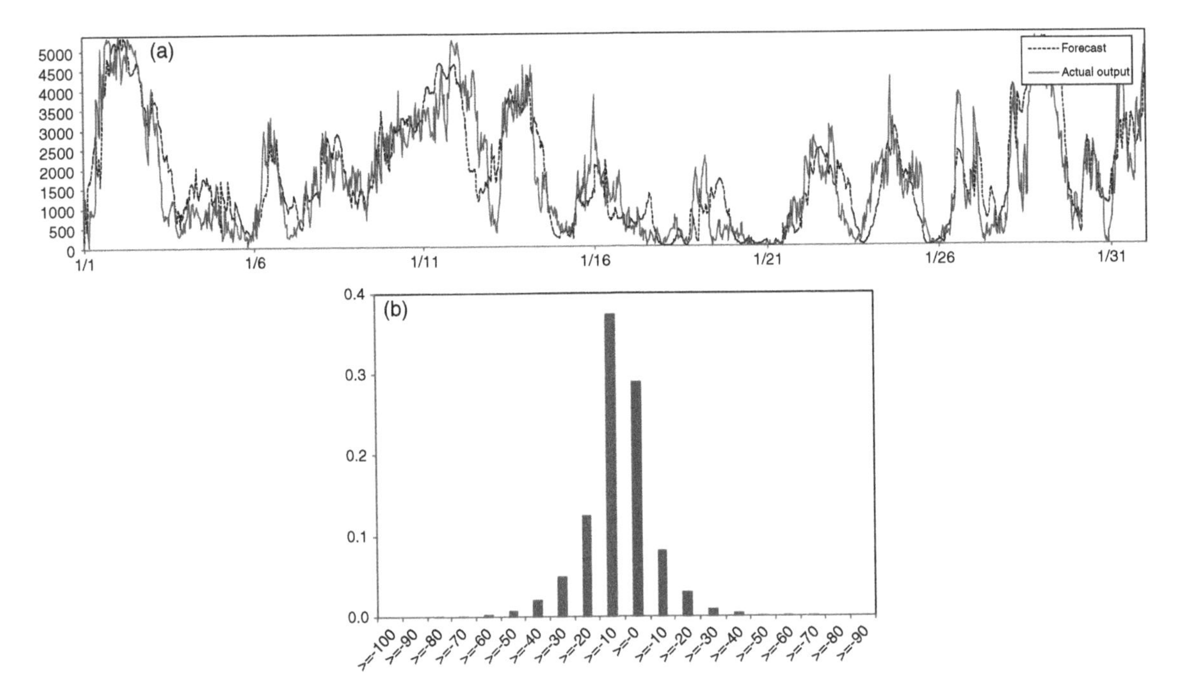

■ **FIGURE 5.17** (a) Forecast and actual total output in January and (b) the normalized frequency of forecast errors for total output (counted through 1 year).

errors (RMSE) of each site are 19.5, 19.1, and 18.3%, but in the sum of three sites, the RMSE is reduced to 13.1%.

5.6 FORECASTING APPLICATIONS

5.6.1 Scheduled generation of wind farms and solar power plants with energy-storage systems

The literature [25] investigates the scheduled generation operation, in which the total generation output is compensated by the energy-storage systems is regulated to meet the generation schedule, specified 1 day ahead. In this scheduled generation, the generation schedule is made with a 30-min interval based on the day-ahead output forecasts. Both the 30-min energy imbalance and the standard deviation of instantaneous total output during the 30-min interval are used as performance indices. Computer simulations based on the observed field data and output forecasts reveal that, it is not the short-term fluctuation but the error in forecast that has a significant impact on the performance of scheduled generation. As described in Section 5.5, aggregation of wind farms has the potential to mitigate the relative forecast error; therefore, the wide-area forecast and operation are possible solutions in the future. Another option to reduce the effect of forecast error is to introduce the probabilistic forecast. Confident interval or quantile information

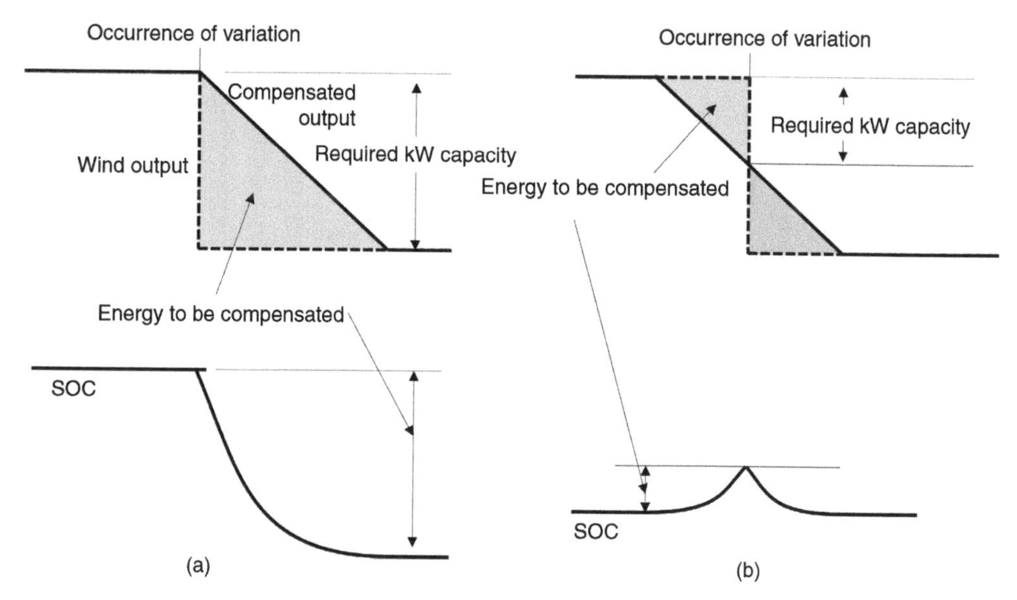

■ FIGURE 5.18 Image of compensation of step-wise variation of wind power output by energy-storage system. (a) After the variation detection and (b) before the variation with the support of output forecast.

may provide useful information for more safety side (pessimistic) generation scheduling.

5.6.2 Suppression of ramp variation of wind output

The New Energy and Industrial Technology Development Organization of Japan has launched the demonstration project named "R&D Project on Grid Integration of Variable Renewable Energy: Mitigation Technologies on Output Fluctuations of Renewable Energy Generations in Power Grid" since financial year 2014 (5-year project). One of the primary purposes of this project is to mitigate the wind power output fluctuations, especially the ramp variation (relatively slow output variation, from tens of minutes to several hours) with the help of energy-storage systems. For this kind of application, it is expected that the wind power forecast could contribute considerably. When the ramp variation cannot be forecast, the energy-storage system must begin its compensation behavior right after the detection of ramp variation, as shown in Fig. 5.18a. In this case, the energy-storage system must store or release the energy corresponding to the hatch area. If the occurrence of ramp variation can be anticipated in advance and the compensation by energy-storage system can be initiated earlier, as shown in

Fig. 5.18b, the necessary compensation power and energy could be reduced significantly (in Fig. 5.18b, only 50% of the kilowatt capacity and 25% of the energy capacity are required compared with Fig. 5.18a).

REFERENCES

[1] Global Wind Energy Council. http://www.gwec.net.

[2] Lawrence E.J. Strategies and decision support systems for integrating variable energy resources in control centers for reliable grid operations – global best practices, examples of excellence and lessons learned. Executive Summary of DE-EE0001375, U.S. Government; 2009.

[3] IEC61400-12-1: 2005(E) Wind turbines Part 12–1, Power performance measurements of electricity producing wind turbines; 2005.

[4] Apt J. The spectrum of power from wind turbines. J Power Sources 2007;169: 369–74.

[5] Holttinen H, Meibom P, Orths A, van Hulle F, Lange B, O'Malley M, Pierik J, Ummels B, Tande JO, Estanqueiro A, Matos M, Gomez E, Söder L, Strbac G, Shakoor A, João RJ, Smith C, Milligan M, Ela E. Design and operation of power systems with large amounts of wind power. Final Report IEA Wind Task, vol. 25; 2006–2008.

[6] The Institut für Solare Energieversorgungstechnik e.V. (ISET). Windenergy Report Gearmany 2008; 2008.

[7] Nanahara T, Asari M, Maejima T, Sato T, Yamaguchi K. Japan Weather Association, Sapporo, Japan, Masaaki Shibata: Smoothing Effects of Distributed Wind Turbines. Part 2. Coherence Among Power Output of Distant Wind Turbines. Wind Energy 2004;7:75–85.

[8] Abedini A, Nasiri A. Applications of super capacitors for PMSG wind turbine power smoothing, 34th annual conference of IEEE IECON; 2008, pp.3347–3351.

[9] Abbey Chad, Joos Géza. Supercapacitor energy storage for wind energy applications. IEEE Trans Ind Appl 2007;43(3):769–76.

[10] Muyeen SM, Shishido S, Hasan AM, Takahashi R, Murata T, Tamura J. Application of energy capacitor system to wind power generation. Wind Energy 2008;11: 335–50.

[11] Yoshimoto K, Narahara T, Koshimizu G. A control method of charging level for battery energy storage system for smoothing output fluctuation of wind power generation. IEEJ Trans Power Energ 2009;129(5):605–13. (in Japanese).

[12] Khalid M, Savkin AV. A model predictive control approach to the problem of wind power smoothing with controlled battery storage. Renew Energ 2010;35:1520–6.

[13] Teleke S, Baran ME, Bhattacharya S, Huang AQ. Optimal control of battery energy storage for wind farm dispatching. IEEE Trans Energy Convers 2010;25(3): 787–94.

[14] Li Q, Choi SS, Yuan Y, Yao DL. On the determination of battery energy storage capacity and short-term power dispatch of a wind farm. IEEE Trans Sustain Energy 2011;2(2):148–58.

[15] Cardenas R, Peña R, Asher G, Clare J. Power smoothing in wind generation systems using a sensorless vector controlled induction machine driving a flywheel. IEEE Trans Energy Convers 2004;19(1):206–16.

[16] Jerbi L, Krichen L, Ouali A. A fuzzy logic supervisor for active and reactive power control of a variable speed wind energy conversion system associated to a flywheel storage system. Electr Pow Syst Res 2009;79:919–25.

[17] Ghedamsi K, Aouzellag D, Berkouk EM. Control of wind generator associated to a flywheel energy storage system. Renew Energ 2008;33:2145–56.

[18] Sato D, Saitoh S H. Smoothing control of wind farm output by using kinetic energy of variable speed wind power generators. IEEJ Trans Power Energ 2009;129(5):580–90. (in Japanese).

[19] Howlader AM, Urasaki N, Senjyu T, Yona A, Uehara A, Saber AY. Output power smoothing of wind turbine generator system for the 2 MW permanent magnet synchronous generators. IEEE international conference on electrical machine and systems; 2010, pp. 452–457.

[20] Sakamoto R, Senjyu T, Kinjo T, Urasaki N, Funabashi T, Fujita H, Sekine H. Output power leveling of wind turbine generator for all operating regions by pitch angle control. IEEE Trans Energy Convers 2006;21:467–75.

[21] Chowdhury MA, Hosseinzadeh N, Shen W. Fuzzy logic systems for pitch angle controller for smoothing wind power fluctuations during below rated wind incidents. IEEE PowerTech 2011;2011:1–7.

[22] Chowdhury MA, Hosseinzadeh N, Shen WX. Smoothing wind power fluctuations by fuzzy logic pitch angle controller. Renew Energ 2012;38:224–33.

[23] Monteiro C, Bessa R, Miranda V, Botterud A, Wang J, Conzelmann G. Wind power forecasting: state-of-the-art 2009. Argonne National Laboratory, ANL/DIS-10-1; 2009.

[24] Zack J. Overview of wind energy generation forecasting. TrueWind Solutions LLC and AWS Scientific, Inc.; 2003.

[25] Manabe Y, Hara R, Kita H, Takitani K, Tanabe T, Ishikawa S, Oomura T. Cooperative control of energy storage systems and biogas generator for multiple renewable energy power plants. Proceedings of the 18th power systems computation conference, no. 288; 2014, pp.1–7.

[26] Website of Japan Weather Association. http://www.jwa.or.jp/english/technology.html.

Chapter 6

Energy management systems for DERs

Atsushi Yona

Department of Electrical and Electronics Engineering, Faculty of Engineering, University of the Ryukyus, Okinawa, Japan

CHAPTER OUTLINE

6.1 Basic concepts of home energy management systems 132
6.2 Control strategies for energy storage systems 137
6.3 Control strategies for EVs as storage 145
6.4 Use of smart meter data 146
References 155

When connecting power plants of distributed energy source to existing power systems, we should consider the energy management. This chapter describes application methods about energy management concentrating on home energy management systems (HEMS), energy storage systems, electric vehicles (EVs), and smart meter data. This chapter starts by explaining the basic concepts of HEMS. In Section 6.2, control strategies for energy storage systems are described. Section 6.3 introduces application example of EVs. Finally, Section 6.4 summarizes the use of smart meter data.

6.1 BASIC CONCEPTS OF HOME ENERGY MANAGEMENT SYSTEMS

Due to a fluctuating power from distributed energy sources such as renewable energy sources and loads, supply/demand balancing of power system becomes unstable. Kenichi Tanaka et al. [1] proposed a methodology for optimal operation of a smart grid to minimize the interconnection point power flow fluctuation. The system consists of basic concepts of HEMS including photovoltaic (PV) generator, heat pump (HP), battery, solar collector (SC), and load. For further research and dissections, Refs. [2–9] review work for basic concepts of HEMS. Marc Beaudin et al. [2] have written a review of modeling and complexity of HEMS. A. Tascikaraoglu et al. [3] introduced

Integration of Distributed Energy Resources in Power Systems

the concept for a demand side management strategy of smart home system in Turkey. Yumiko Iwafunea et al. [4] discussed, in considerable detail, cooperative home energy management using batteries for a PV system. Rim Missaoui et al. [5] explained in detail the performance analysis of a global model-based anticipative building energy management. Hanife Apaydın Özkan [6] provided a real-time solution to reduce the electricity cost and to avoid the high peak demand problem simultaneously for a smart home. Phani Chavali et al. [7] presented distributed energy scheduling algorithm as a demand response for the smart grid, and Mohammad Chehreghani Bozchalui et al. [8] proposed hierarchical control approach for energy management of greenhouse. The proposed model incorporates weather forecasts, electricity price information, and the end-user preferences to optimally operate existing control systems. Amjad Anvari-Moghaddam et al. [9] applied a multiobjective mixed integer nonlinear programming model to optimal energy in a smart home considering a balance between energy saving and a comfortable lifestyle.

This section summarizes Ref. [1] as an example of HEMS. In the Ref. [1], the authors present an optimal operation method of the direct current (DC) smart house group with the controllable loads in the residential houses as a smart grid. The DC smart house consists of a SC, a PV generator, a HP, and a battery. The HP and the battery are used as controllable loads in this research. The proposed method has been developed in order to achieve the interconnection point power flow within the acceptable range and the reduction of max–min interconnection point power flow error as low as possible to smooth the supply power from distribution system. Power consumption of the controllable load is determined to optimize the max–min interconnection point power flow error based on the information collected from power system through communication system. By applying the proposed method, we can reduce the interconnection point power flow fluctuation, and it is possible to reduce electricity cost due to the reduction of the contract fee for the electricity power company. Also, by using battery as the power storage facility, which can operate rapidly for charge and/or discharge, the rapid output fluctuations of DC load and PV generator are compensated. The smart grid model is shown in Fig. 6.1. The smart grid has six smart houses, and is connected to the power system and control system through a transmission line and communication infrastructures. The control system sends required control signals to the smart house group which response to the system's conditions. Each smart house determines the operation of controllable loads. The interconnection point power flow is the power flow from the power system to the smart grid in Fig. 6.1. The DC smart house model is shown in Fig. 6.2, which consists of a DC load, a PV generator, an SC, an HP, and a battery. The HP and the battery are used as controllable loads.

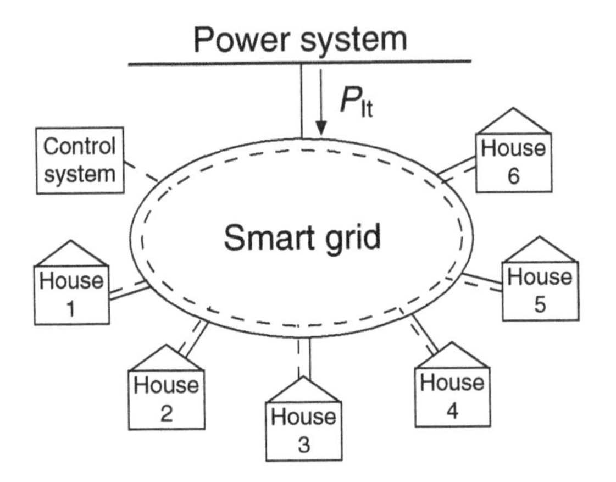

■ **FIGURE 6.1 Smart grid model [1].**

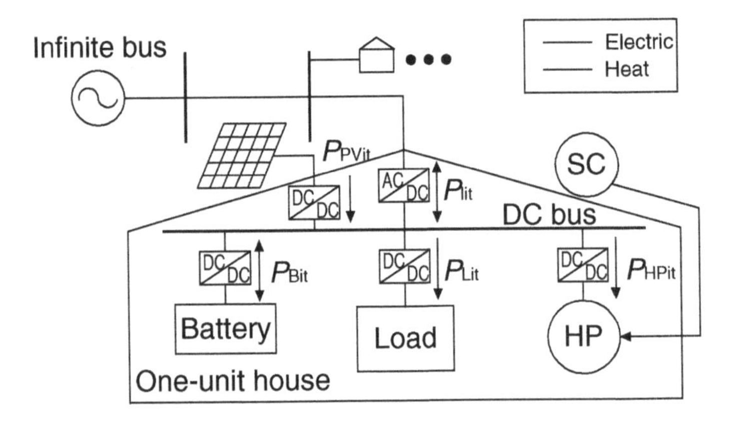

■ **FIGURE 6.2 DC smart house model [1].**

Fig. 6.3 shows the numerical model of SC system. It is assumed that sum of total solar radiation will be falling on the SC and array, and it does not consider the incidence angle of SC array. Optimal operation of smart grid is determined to minimize the interconnection point power flow fluctuations. Due to reduce the interconnection point power flow fluctuation, it is possible to suppress the harmful effects to power system, and it is possible to reduce electricity cost.

Furthermore, the proposed configuration of electricity price as shown in Fig. 6.4 assumes the smart grid system in the future. If the interconnection point power flow within the bandwidth (Region A) has an electricity purchase cost of 10 Yen/kWh, and if the interconnection point power flow

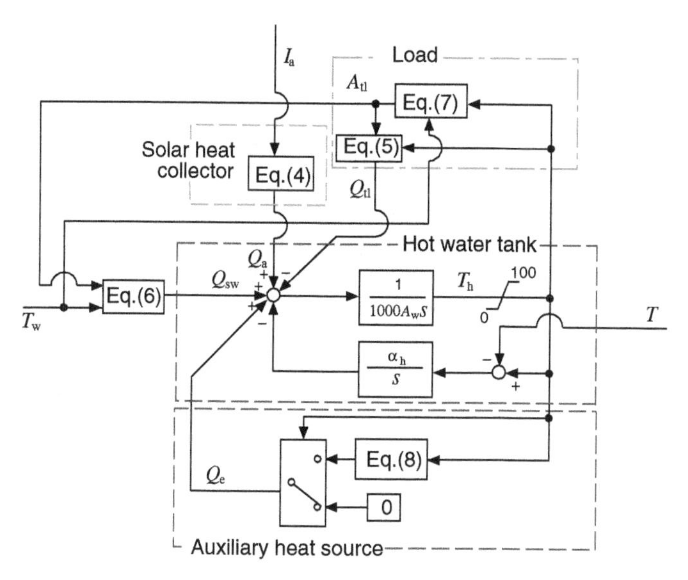

■ FIGURE 6.3 Model of SC system [1].

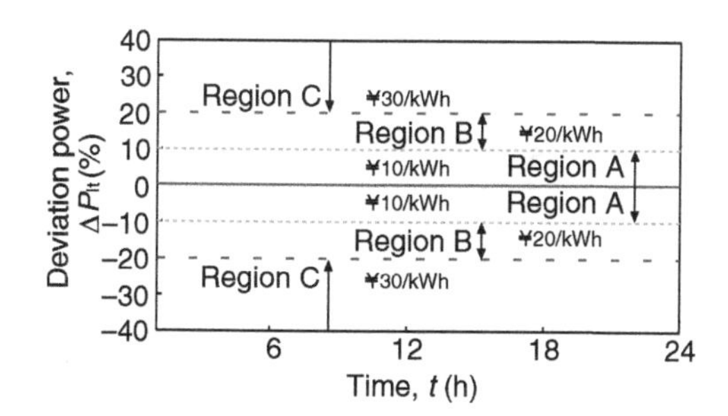

■ FIGURE 6.4 Electric price [1].

departs from the bandwidth (Regions B and C), the electricity purchase costs are 20 and 30 Yen/kWh, respectively. Moreover, the electricity selling cost to the power system is 10 Yen/kWh. The bandwidth is given by power system to smart grid as power reference. Therefore, it is important that the customer follows the power reference and interconnection point power flow within the bandwidth.

Tabu Search (TS) with local search methodology, which always moves neighborhood solution, is used as the optimal technique to address the problems of determining the charge/discharge power of battery and HP operation time for each smart house. It is possible to determine heating time of an HP by assuming power consumption, except controllable loads and heat load which could be forecast. Therefore, the charge/discharge power of the battery and HP operation time for each smart house are calculated by using TS under the objective function and constraints. Simulation results for sunny, cloudy, and rainy weather conditions are shown in Figs. 6.5, 6.6, and 6.7, respectively. Assuming power consumption except controllable loads and PV output power in each weather conditions are shown in Figs. 6.5(a), 6.5(b), 6.6(a), 6.6(b), 6.7(a) and 6.7(b), respectively. These figures show power consumption except controllable loads and PV output power in smart grid and each smart house, respectively.

It is possible to reduce the cost with the proposed control because the interconnection point power flow is controlled within the bandwidth by operating controllable loads. In sunny weather conditions, PV power output and SC heat generation are high due to sufficient solar radiation. Therefore, it is not necessary to heat by HP and the power consumption for each house is not

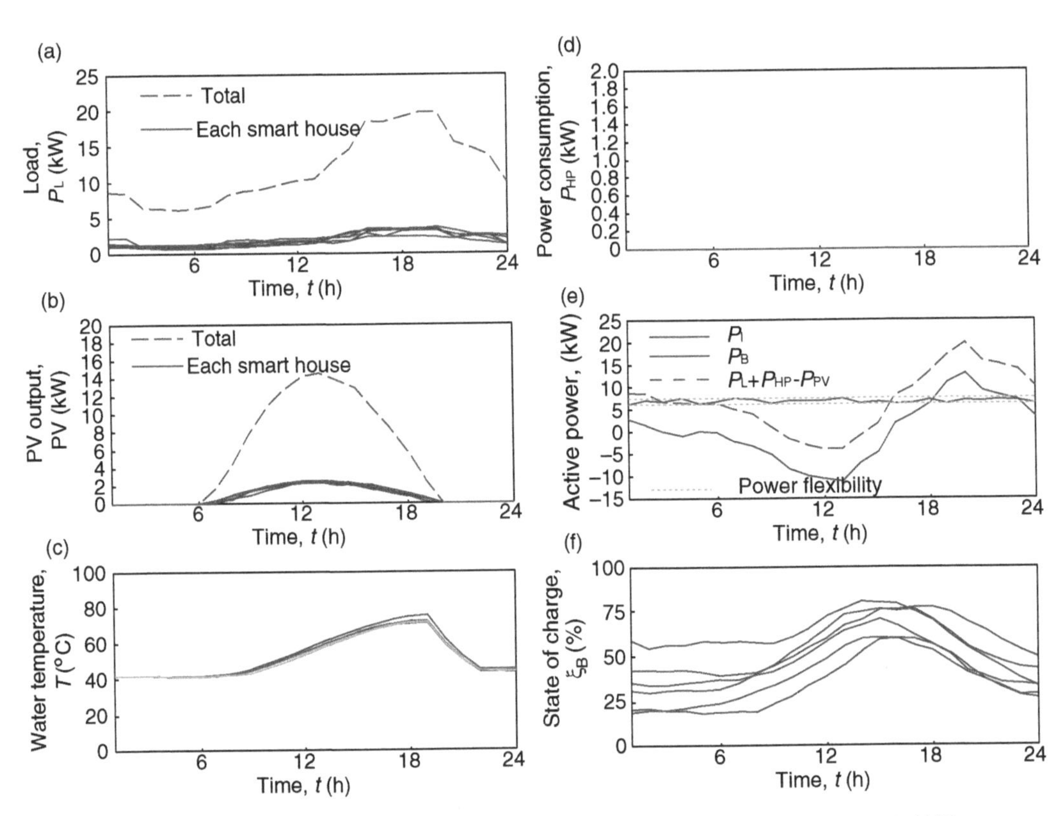

FIGURE 6.5 Simulation results in sunny weather condition [1]. (a) Power consumption except controllable loads; (b) PV output power; (c) water temperature of HP; (d) power consumption of HP; (e) supplying power from infinite bus; (f) remaining energy capacity of battery.

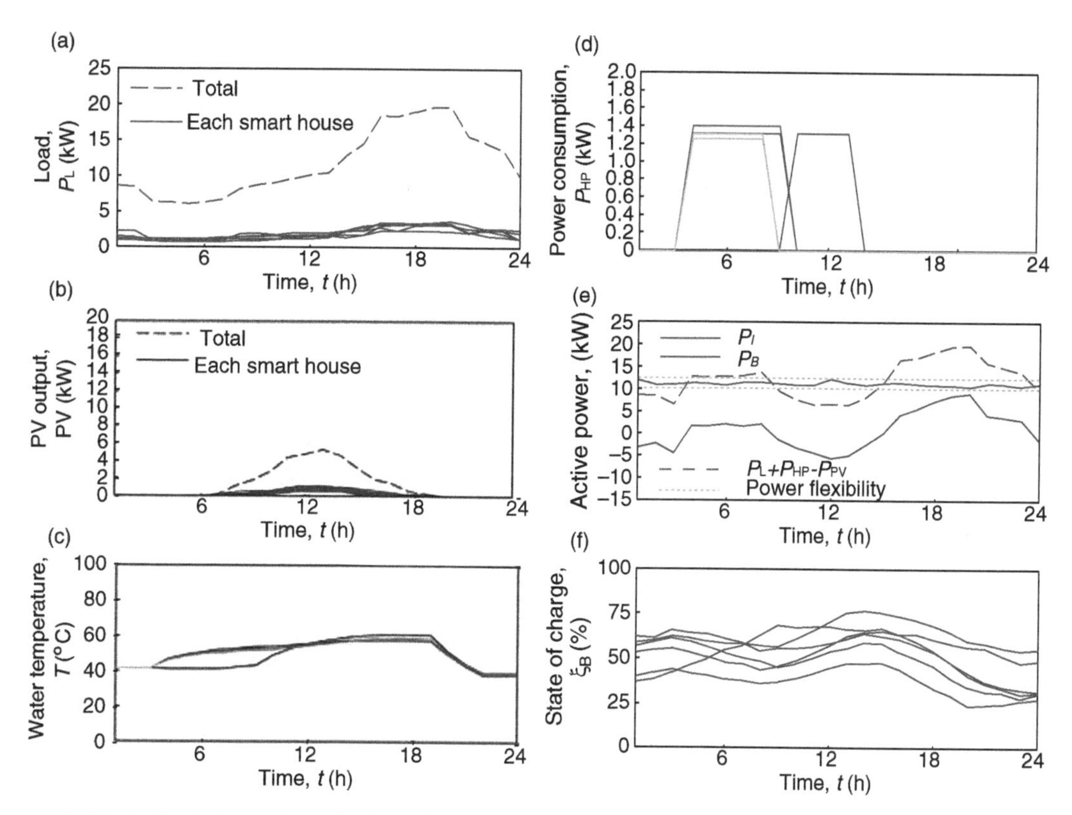

FIGURE 6.6 Simulation results in cloudy weather condition [1]. (a) Power consumption except controllable loads; (b) PV output power; (c) water temperature of HP; (d) power consumption of HP; (e) supplying power from infinite bus; (f) remaining energy capacity of battery.

increased any more. So, the sunny weather condition is the lowest cost in the three cases. Power consumption in smart grid is smoothed by achieving the proposed method, so we can suppress the impact of PV against the power system. Consequently, we can expect high quality power supply and reduce the cost by cooperative control in the smart grid.

6.2 CONTROL STRATEGIES FOR ENERGY STORAGE SYSTEMS

When renewable distributed generators (DGs) are connected to the power system, there is a concern about their harmful effects as most of their power fluctuates with weather conditions. Therefore, many researchers have been working on practical application of microgrids. Kyohei Kurohane et al. [10] proposed a DC microgrid with renewable energy. The proposed method is

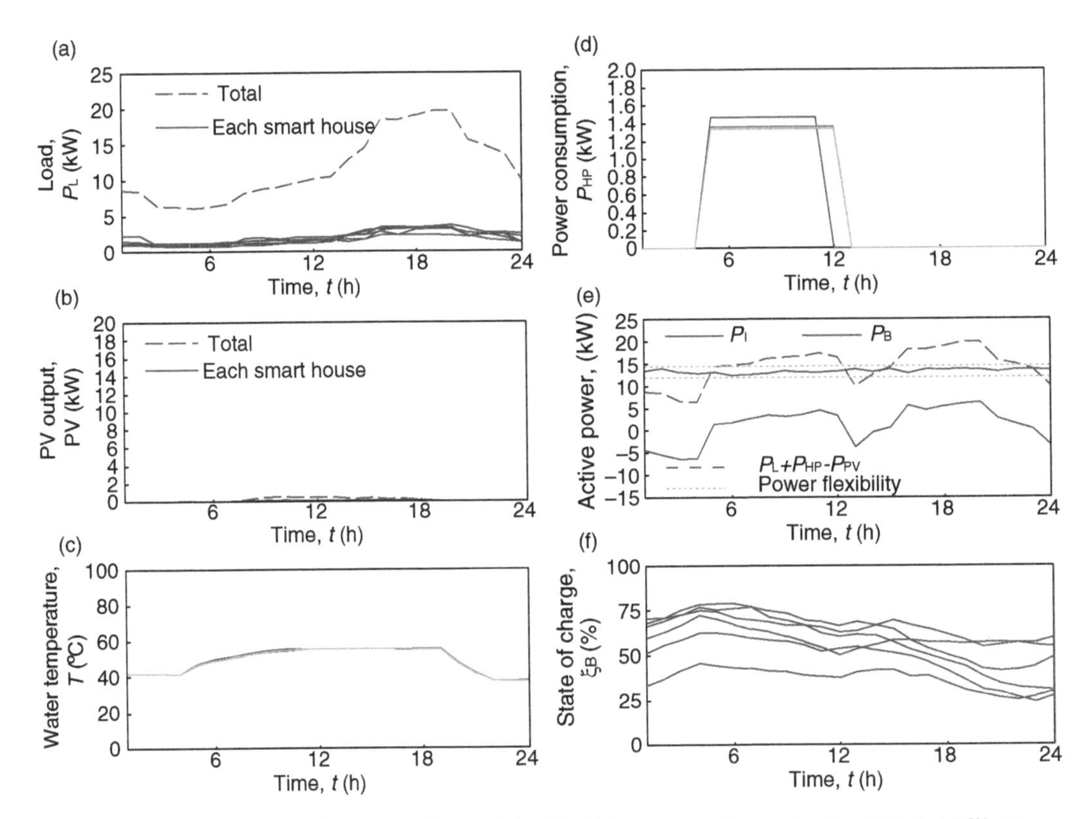

FIGURE 6.7 Simulation results in rainy weather condition [1]. (a) Power consumption except controllable loads; (b) PV output power; (c) water temperature of HP; (d) power consumption of HP; (e) supplying power from infinite bus; (f) remaining energy capacity of battery.

composed of a gearless wind power generation system and a battery in a DC distribution system. The battery helps to avoid DC overvoltages by absorbing the power of the permanent magnet synchronous generator (PMSG) during line fault. The proposed system presents a control strategy based on the maximum power point tracking (MPPT) control to generate the maximum power for the variable wind speed and a pitch angle control to smooth the output fluctuation at low wind speed. The system consists of control strategies for energy storage systems. For further research and dissections, Refs. [11–15] discuss control strategies for energy storage systems. Yu Zheng et al. [11] proposed battery energy storage system (BESS)-based energy acquisition model for the operation of distribution companies in regulating price considering energy provision options, and the results show that the capacity requirement is reduced. Sathish Kumar Kollimalla et al. [12]

introduced the concept of a hybrid energy storage system. In the method, batteries are used to balance the slow changing power surges, whereas supercapacitors are used to balance the fast changing power surges. Ayman B. Eltantawy et al. [13] discussed in method for increasing the capacity of small-scale renewable distributed generation sources that can be installed in distribution systems based on both a technical and economic assessment. Xinda Ke et al. [14] discussed in detail the control algorithms and sizing strategies for using energy storage to manage energy imbalance for variable generation resources. Zhaoyu Wang et al. [15] presented a decentralized power dispatch model for the coordinated operation of multiple microgrids and a distribution system.

This section summarizes Ref. [10] as an example of energy storage systems. The DC distribution system has the following advantages over the alternating current (AC) distribution system. (1) Each power generator connected to the DC distribution system can easily be operated in coordination because it controls only the DC bus voltage. (2) When the AC grid system is faulty, the DC distribution system is disconnected from the AC grid. It is then switched to the stand-alone operation in which the generated power is supplied to the loads connected to the DC distribution system. (3) The system cost and loss can be reduced because only a single AC grid-side inverter unit is needed. With the rapid growth of the wind turbine generator (WTG) systems, it is difficult to stabilize the operation of the power system by disconnecting WTGs when there are line faults. Under the line fault, the dispersed power sources are disconnected from the power system, requiring much more time and energy to compensate for the power supply–demand balance. In addition, with the recovery of the power system, disconnected WTGs need to be restarted. Thus, the frequency of the power system rise as many WTGs return to the system. As a countermeasure, in Europe, large WTGs are required to remain connected to the power system under line fault, and supply power to the power system after fault clearing. This requirement is called fault ride-through. Moreover, under the line fault, the DC bus voltage in the DC distribution system experiences overvoltage. Therefore, unstable power supply from DC distribution system and an overvoltage problem of semiconductor devices of the power converter occur. It is important to solve these problems of the stable operation for DC distribution system. Stable power supply strategies for DC distribution system and stable control strategies for PMSG under the line fault are proposed. The proposed method uses a battery for the DC distribution system. Under the line fault, a chopper circuit is used to avoid DC bus overvoltage by absorbing energy from the PMSG and by supplying to the battery. By means of the proposed method, stable operation of the DC power system under the line fault becomes possible, and a

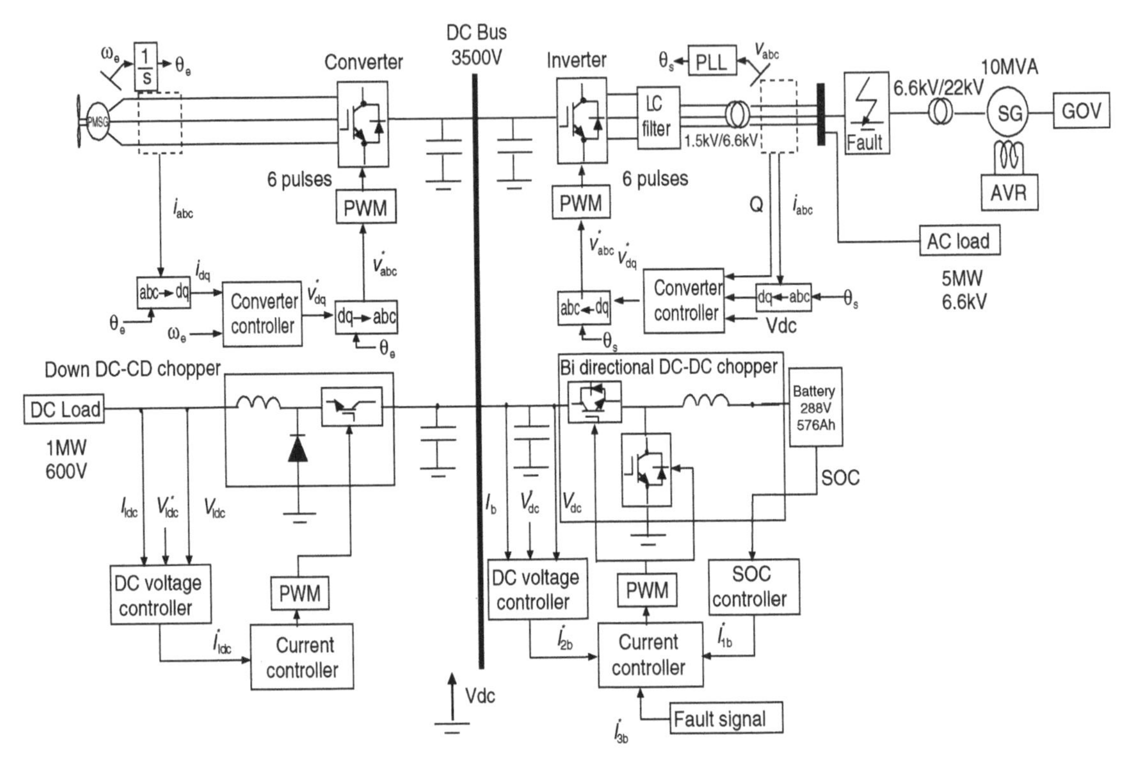

■ **FIGURE 6.8 DC power system [10].** PLL, phase locked loop; GOV, governor.

highly reliable power supply can be achieved from the grid-side inverter to the AC grid after the line fault is cleared.

The DC power system used in this study is shown in Fig. 6.8. The wind power generator is a gearless 2 MW PMSG. The PMSG has a simple structure and high efficiency. In addition, the DC distribution system consists of a gearless 2 MW PMSG, a grid-side inverter, a 576 Ah battery and 100 kW DC loads. The DC system is connected to a 10 MVA diesel generator and 5 MW AC loads through the grid-side inverter. Wind power energy obtained from the windmill is sent to the PMSG. In order to generate maximum power, the rotational speed of the PMSG is controlled by the pulse width modulation (PWM) converter and the generated power is leveled by a pitch angle control. This power is then supplied to the DC load. The rest of the power is supplied to the AC load through the grid-side inverter.

The power converter control systems are shown in Figs. 6.9 and 6.10. Generator-side converter achieves variable speed operation by controlling rotational speed of the PMSG. On the other hand, the grid-side inverter supplies

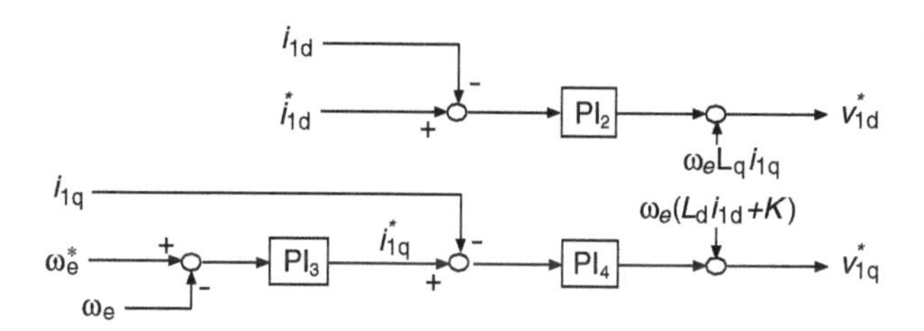

■ **FIGURE 6.9 Wind generator-side converter control system [10].**

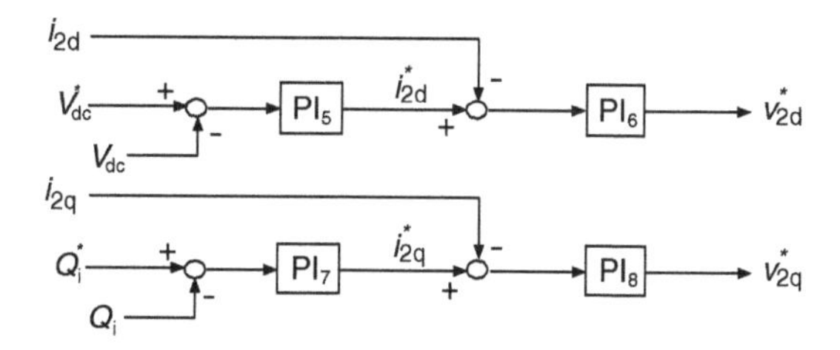

■ **FIGURE 6.10 AC grid-side inverter control system [10].**

electrical power and its frequency is synchronized with the frequency of the power system. Each of the power converters is a standard three-phase two-level unit, is composed of six insulated gate bipolar transistors (IGBTs), and is controlled by the triangle wave PWM law. In addition, the DC distribution system includes a battery, in order to avoid DC bus overvoltages under line fault.

The battery model is considered for battery's discharge and charge characteristics. In this study, the authors consider a lithium ion battery. The state of charge (SOC) is calculated by the integration of the discharge and charge power of the battery. The proposed constant DC bus voltage control is achieved by using a bidirectional DC chopper in connection with a battery. The control systems are shown in Figs. 6.11 and 6.12. The bidirectional DC chopper controls the duty ratio for normal operation (SW_{normal}) or line fault (SW_{fault}). The switching determinations for SW_{normal} and SW_{fault} are performed by considering the AC grid voltage, v_t. When the AC grid voltage is $v_t \geq 0.8$ pu, the bidirectional DC chopper performs the normal operation. If the AC grid voltage is $v_t < 0.8$ pu, the bidirectional DC chopper performs under the fault operation. Under the line fault, this chopper circuit helps to avoid DC bus overvoltage by absorbing energy from PMSG and supplying it to the battery, keeping the DC bus voltage constant. The rapidly rising DC bus voltage is difficult to keep constant using the only proportional–integral (PI) controller of the AC grid side inverter. The control system under the line

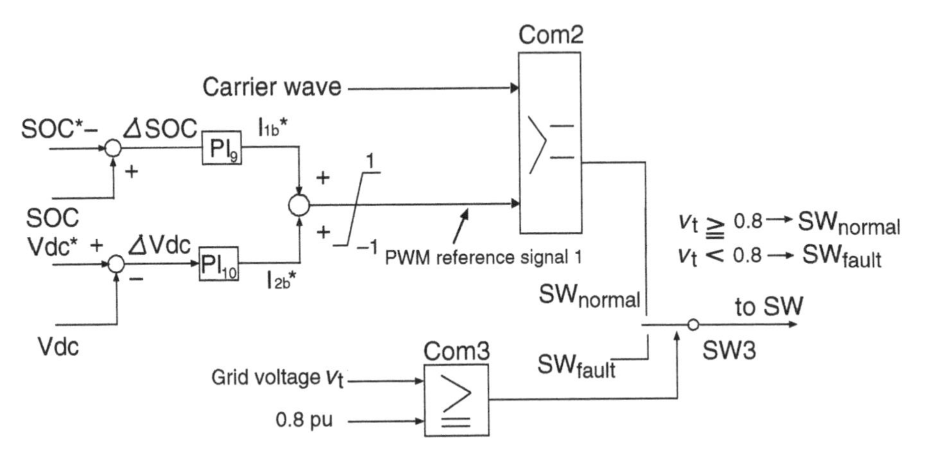

■ FIGURE 6.11 Battery control system [10].

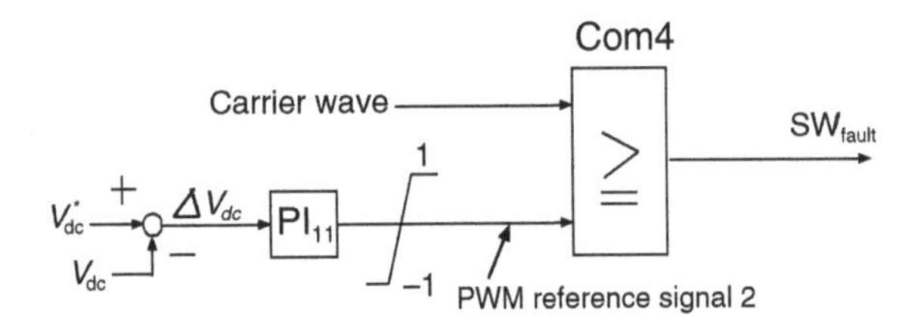

■ FIGURE 6.12 DC bus constant voltage control system for fault [10].

fault is shown in Fig. 6.12. In this system, the PWM reference signal 2 is determined by the output of PI_{11} controller. The output of the comparator 4 depends on the comparison of PWM reference signal 2 and carrier wave signal. The carrier wave signal is 1 when the carrier wave signal is greater than the reference signal, and is 0 when the carrier wave is less than the reference signal. IGBTs are used as switching devices for the DC chopper circuit.

The effectiveness of the proposed method is examined by a switching simulation with the system model shown in Fig. 6.8. This simulation considers that the three-line to ground fault occurs at the middle of transmission line of Fig. 6.8 and electrical power supply to the load is shut down. The sequence of simulation is described further:

1. At $t = 5.0$ s: The three-line to ground fault occurs at the middle of transmission line.
2. When the AC grid voltage is within $v_t < 0.8$ pu, the gate signals for grid-side inverter are stopped.

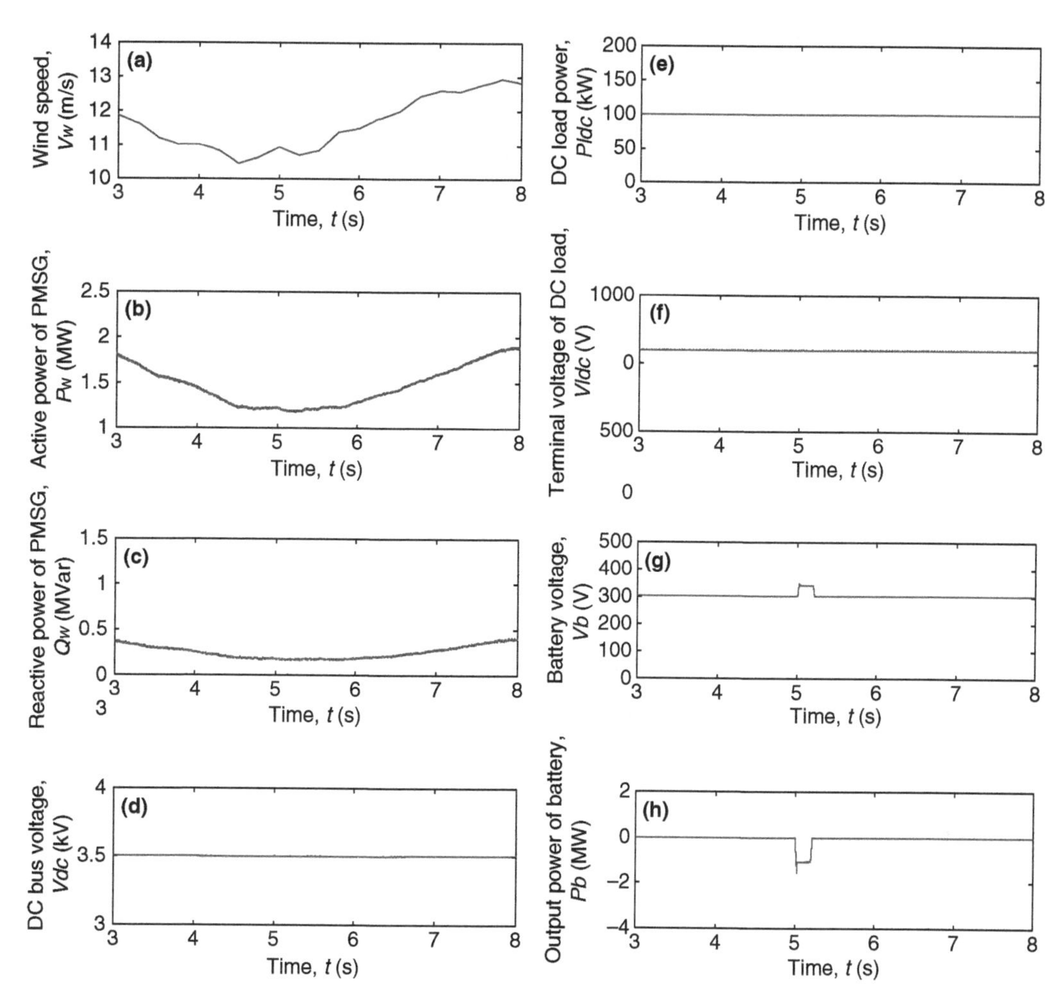

■ FIGURE 6.13 Simulation results during three-line to ground-fault [10]. (a) Wind speed; (b) active power of PMSG; (c) reactive power of PMSG; (d) DC bus voltage; (e) DC load; (f) terminal voltage of DC load; (g) terminal voltage of battery; and (h) output power of battery.

3. At $t = 5.1$ s: The line fault is cleared.
4. When the AC grid voltage is within $v_t \geq 0.8$ pu, the gate signals for grid-side inverter are restarted.

The simulation results are shown in Figs. 6.13 and 6.14.

In normal operation, the proposed system presents a control strategy based on the MPPT control to generate the maximum power for the variable wind speed and a pitch angle control to smooth the output fluctuation at low wind

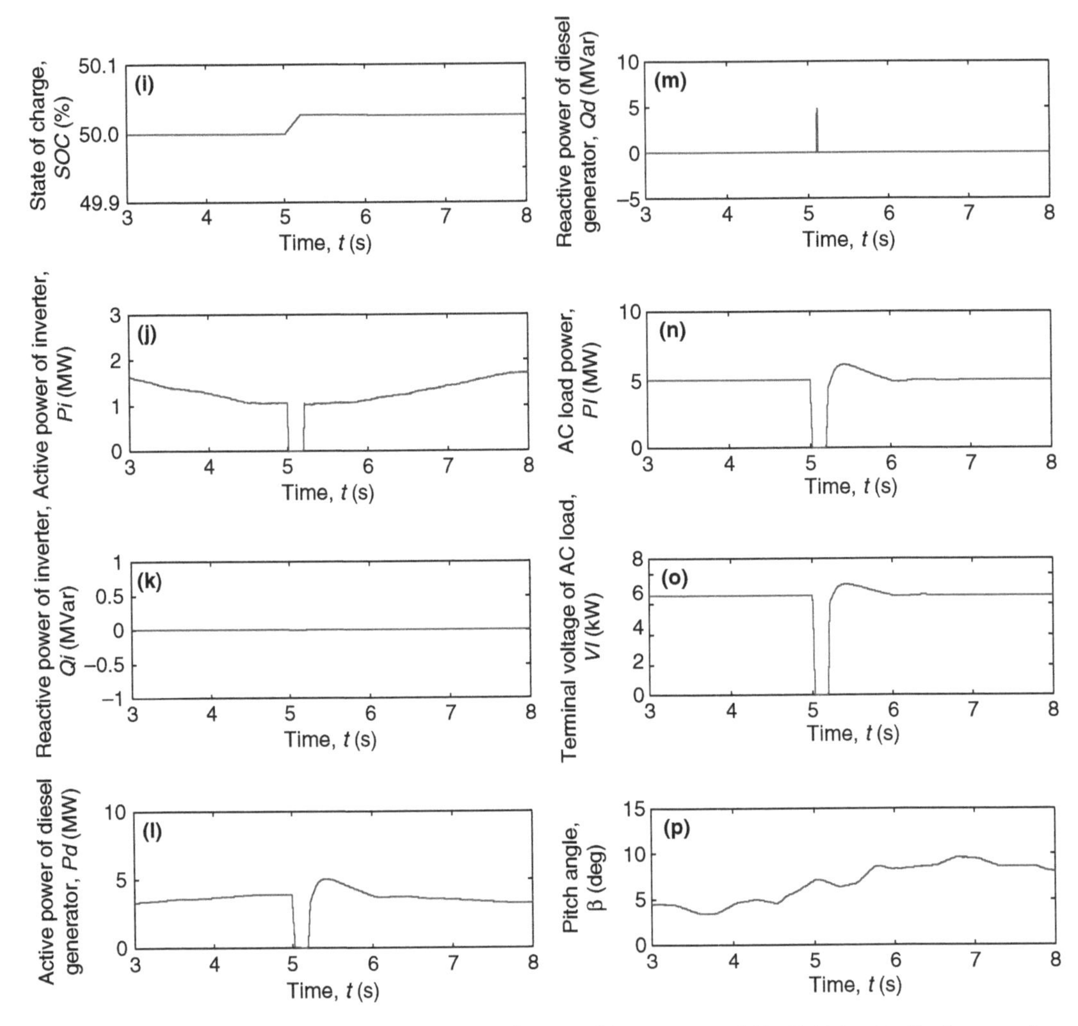

FIGURE 6.14 (i) State of charge; (j) active power of AC grid-side inverter; (k) reactive power of AC grid-side inverter; (l) active power of diesel generator; (m) reactive power of diesel generator; (n) AC load; (o) terminal voltage of AC load; and (p) pitch angle [10].

speed. Besides, at high wind speeds, it is possible to control the output power of the PMSG by the pitch angle control system. In addition to these control strategies, the DC bus voltage is controlled by using a bidirectional chopper and battery under the line fault. From the simulation results, it was confirmed that the DC distribution system with the proposed method can stabilize power system operation under the line-fault and can supply stable power from the grid-side inverter to the power system after the line-fault clearing.

6.3 CONTROL STRATEGIES FOR EVs AS STORAGE

In the installation area of a large renewable energy sources, such as a PV system, appropriate operation is required, and it is better to minimize the size of the battery and its capital cost. Atsushi Yona et al. [16] proposed an optimization approach to determine the operational planning of the power output for a large PV system. The approach includes the method of determination of the charge/discharge amount for the battery of an EV as a demand response. The method aims to obtain a more beneficial deal with the sale of electrical power. The operation consists of control strategies for EVs as storage. For further research and dissections, Refs. [17–21] are related with control strategies for EV as storage. Mingrui Zhang et al. [17] proposed a pricing scheme through a fuzzy control method to facilitate the energy management for EVs and battery swapping station based on islanded operation of the microgrid. Preetham Goli et al. [18] proposed a unique control strategy based on DC link voltage sensing with integration of PV and plug-in EVs for efficient transfer energy. Tan Ma et al. [19] have written in detail the economic analysis of a real-time power flow control algorithm for charging a large-scale plug-in EVs network. Kan Zhou et al. [20] introduced decentralized random access framework with charging plug-in EVs to avoid bus congestion and large voltage drop in the distribution grid. Uwakwe Christian Chukwu et al. [21] performed to investigate the real-time management of power systems with vehicle-to-grid (V2G) facilities.

This section summaries Ref. [16] as an example of control strategies for EVs as storage. The authors described an optimal operational planning strategy applying demand response (DR) for a large PV/battery system. The authors assume that the objective function is to maximize the obtained benefit of electrical power selling from the combined power output from the PV system developer and demand from EVs users. At present, the contribution of EVs as a DR is quantified. The required electric power for the initial SOC of the battery is described by simulation, with the result that the purchased power at nighttime is assumed to be available. The forecast power output errors of the PV system for the case of over/under fitting are assumed. The PV power output by MPPT is restrained, considering the variation of combined power output, inverter capacity, and its conversion efficiency, and then the combined power output is smoothed according to the command for charge/discharge of the BESS. A genetic algorithm is applied as the optimization technique to determine the combined power output and initial SOC of BESS. In order to decrease the battery capacity, the proposed technique should be implemented by the aggressive introduction of EVs. The simulation results show the potential of DR as an option to compensate for the PV power output forecast errors for optimal operational planning.

6.4 USE OF SMART METER DATA

Due to global warming and the depletion of fossil fuels, we are required to reduce CO_2 emissions and energy consumption. To solve these problems, the installation of PV and SC systems in residential houses has been proposed for the demand side. Moreover, the introduction of the HP is also expected as an efficient water-heating appliance. The general electric utility company sets a time-of-use electricity rate system to absorb peak power demand and to operate the thermal power generator at the highest efficiency. Much research has been carried out regarding this matter, pertaining to decreasing peak power demand and to the control of house loads. Consequently, houses with an HP have benefited from the electricity pricing system in residential areas. In addition, if a fixed battery is introduced to houses under this electricity pricing system, the electricity cost for the houses would further decrease because it would be possible to recharge the battery using cheap electricity in the middle of the night. Another recent trend is the increase of occupants who have EVs and who are able to implement the vehicle to home system, which can compensate the power consumption of the houses with the EV battery by attaching it to the grid (V2G). Thus, it is expected that users will be likely to install EVs and fixed batteries with effective utilization strategies in the coming decades. Unfortunately, the investment cost for these appliances is quite high. Thus the installation cost to the consumer is likely to increase. Akihiro Yoza et al. [22] proposed an expansion-planning model of PV and battery systems for the smart house. The expansion-planning period is 20 years and ranges from 2015 to 2035 in Japan. The operation consists of using smart meter data. The proposed method clarifies the optimal installation year, capacity, and appliances during the 20-year period considering variable characteristics such as investment cost, selling price, and purchasing price which change year by year. For further research and dissections, Refs. [23–26] discuss smart meters. Waleed Aslam et al. [23] discussed in detail the smart meter technology and applications across residential, commercial and industrial sectors. Sabine Erlinghagen et al. [24] discussed in considerable detail the identifying wired and wireless communication standards for smart metering in Europe, and described a comprehensive set of criteria for standard selection compared with the existing standards. Myriam Neaimeh et al. [25] have written a probabilistic method to combine two datasets of real EV charging profiles and residential smart meter load demand. Franklin L. Quilumba et al. [26] presented the efforts involved in utilizing the advanced metering infrastructure data to improve the load forecasting accuracy at the system level.

This section shows summary of the Ref. [22] as an example of use of smart meters data. The authors proposed a model of optimal expansion planning of PV and battery systems in a smart house for a 20-year period. The optimization method employs the TS algorithm which is known to be metaheuristic and a model to reduce calculation time is devised. It is assumed that the expansion-planning period ranges from 2015 to 2035. In advance, the smart house contains HP and SC systems. It is assumed that the occupant in the smart houses has a plan to install PV and battery systems. By solving this optimization problem, the optimal installed year, and the most beneficial capacity for the appliances are revealed for 20-year period, and the minimal total payment cost during the expansion period for the consumer is calculated. The yearly varying electricity purchasing price for the consumer and the selling price for the general utility company are considered. In order to verify the effectiveness of the proposed system, MATLAB® is used for simulations.

The smart house system model is shown in Fig. 6.15. The smart house model, which consists of SC, HP, PV, and battery, is used for the expansion problem of the PV and battery on the residential side. The authors assumed the occupant has a plan to introduce PV and a fixed battery system in the house within 20 years, where the HP and the battery are used as controllable loads. From Fig. 6.15, P_{It}, P_{Lt}, P_{Bt}, P_{PVt}, and P_{HPt} represent power flow from the power supply to the smart house, power consumption other than controllable loads, charge/discharge power of the battery, PV output power, and power consumption of the HP respectively.

The HP is used for a hot water supply and the standard is a 370 L type with 1.0/4.0 kW heater with a coefficient of performance of 4.0. Moreover, the examined capacity of the PV varies from 0 to 10 kW, and the examined

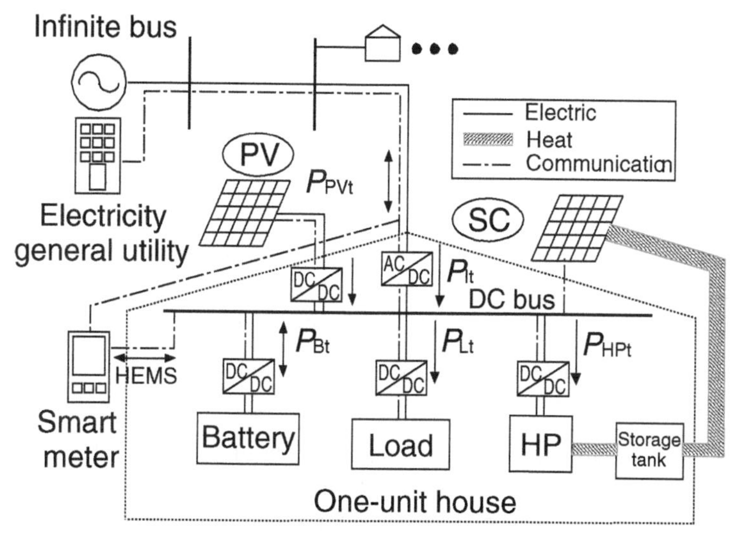

■ **FIGURE 6.15 Smart house model [22].**

inverter capacity and battery capacity vary from 0 to 3 kW and 0–10 kWh, respectively. It is assumed that the simulation is carried out in Okinawa in Japan where the temperature is comparatively warm and the solar intensity is strong. In addition the season is summer-like in October. Weather data of the Japan Meteorological Agency is used.

The objective function and the constraints for optimization of expansion planning of the appliances are described as follows.

The total cost during an expansion period of 20 years for the occupants is defined as summation of investment and operational cost in the smart house. It is assumed that the expansion-planning period ranges from 2015 to 2035. The optimization problem is to solve for when, what capacity, and which appliances (PV and/or battery) are installed during the two-decade period. Thus, the objective function is set to minimize the total cost and described as follows.

Objective function:

$$\mathrm{Min}\,\mathrm{TC} = C_{\mathrm{INVEST}} + C_{\mathrm{OPE}} \sum_{t_e=1}^{T} \left\{ \begin{array}{l} \left[E_{\mathrm{PV}}(t_e)C_{\mathrm{PV}}(t_e) + E_{\mathrm{BA}}(t_e)C_{\mathrm{BA}}(t_e)\right] + C_{\mathrm{PAY}}(t_e) \\ -\left[E_{\mathrm{PV}}(t_e)C_{\mathrm{PS}}(t_e) + E_{\mathrm{BA}}(t_e)C_{\mathrm{BS}}(t_e)\right] \end{array} \right\}$$

where TC: total cost during 20 years; C_{INVEST}: investment cost for the appliances; C_{OPE}: operational cost for the consumer; T: expansion planning period ($T = 20$, inaugural year is 2015); t_e: time index (by year); $E_{\mathrm{PV}}(t_e)$: binary variable for PV to be installed in year t_e; $E_{\mathrm{BA}}(t_e)$: binary variable for battery to be installed in year t_e; $C_{\mathrm{PV}}(t_e)$: investment cost for PV in year t_e; $C_{\mathrm{BA}}(t_e)$: investment cost for battery in year t_e; $C_{\mathrm{PAY}}(t_e)$: electricity payment for the consumer in year t_e; $C_{\mathrm{PS}}(t_e)$: benefit of sold by PV output power in year t_e; and $C_{\mathrm{BS}}(t_e)$: benefit of sold by battery discharge power in year t_e.

The first and second terms mean the investment cost, which is existent according to binary variable representing whether or not the PV or battery system has been installed in year t_e. The third term means the electricity cost which the occupant pays the utility company in year t_e. The fourth and fifth term mean the benefit obtained by sold power generated by the PV and battery in year t_e according to the binary variable, respectively.

The objective function of the constraints for optimal scheduling is depicted as follows.

For the objective function of the optimization of expansion planning, if the installation year is determined, the investment cost can be calculated as C_{INVEST} for the appliances in year t_e for a short while. However, the operational cost, C_{OPE}, which the occupant of the smart house pays during a 20-year period must be simulated at hourly time, t. Thus, the following objective function means minimization of operational cost during a 20-year period in the simulation at hourly time, t.

Objective function:

$$\text{Min}\, C_{\text{OPE}} = \sum_{t_e \in T} \sum_{t \in T_Y} \sum_{t \in T_M} \sum_{t=1}^{24} \left(U_{\text{P}(t,T_M,T_Y,T)} \cdot C_{\text{P}(t,T_M,T_Y,T)} - U_{\text{S}(t,T_M,T_Y,T)} \cdot C_{\text{S}(t,T_M,T_Y,T)} \right)$$

where T: expansion planning period (T = 20, inaugural year is 2015); t_e: time index (by year); T_Y: represents the four seasons in year; T_M: days in 1 month including fair, cloudy and rainy day; t : time index (by hour); $C_P(t,T_M,T_Y,T)$: cost of purchased power in hour t, month T_M, season T_Y, year T; $C_S(t,T_M,T_Y,T)$: benefit of sold power in hour t, month T_M, season T_Y, year T ; $U_P(t,T_M,T_Y,T)$: unit price for purchased power in hour t, month T_M, season T_Y, year T ; and U_S (t,T_M,T_Y,T): unit price for sold power in hour t, month T_M, season T_Y, year T.

The TS is a metaheuristic global optimization method discovered by Glover, which has been effectively used for a combinational optimization such a scheduling problem. TS can find the optimal solution by carrying out the iteration step until a criterion is achieved. However, this simple iteration may cause the cycling in which the search moves within the same loop among local solutions. In order to prevent this cycling from occurring, a memory system called the tabu list, which records the latest moves, is applied in the searching procedure. The utilization of the tabu list can reduce the probability of going into a local iteration since the highest evaluated solution is selected by referring to the tabu list for the next solution x_{nex}. The following implementation parameters are used for the TS. The global iteration max is 4000 and the length of the tabu list is 500. The optimization problem is encoded into the program based on the algorithm and simulated on MATLAB.

The simulation time becomes very long because an optimal hourly scheduling must be solved for many optimization variables and the operational cost of the smart house during 20 years is calculated with the optimization procedure in hourly steps for a long-term simulation of 20 years. A solution methodology for shortening the simulation time is cubic spline interpolation. The function which calculates the operational cost of smart house for 1 year is used in order to shorten the simulation time. The procedure to derive the function is described in this subsection. First, an optimization for hourly scheduling of controllable loads in the house is solved for many variables with a specific small interval. The variables are the capacity of the PV, battery, electricity purchasing price, and selling price from consumer to utility. The minimum operational costs of the house are obtained from the optimal scheduling of appliances for a year. However, an operational cost which has not been investigated cannot be obtained in combinations of small intervals for these variables. Thus, operational costs for the variables which have not been investigated are estimated with cubic spline interpolation, which is a useful technique to estimate the unknown data from the obtained data. The cubic spline interpolation uses piecewise polynomials such that three order polynomials are each divided into small intervals, continuing for a finite interval.

The optimization flow chart is depicted in Fig. 6.16. At first, the minimal operational cost in various cases is revealed in determining the optimal scheduling of the appliances. After that, the function is then derived as piecewise continuous

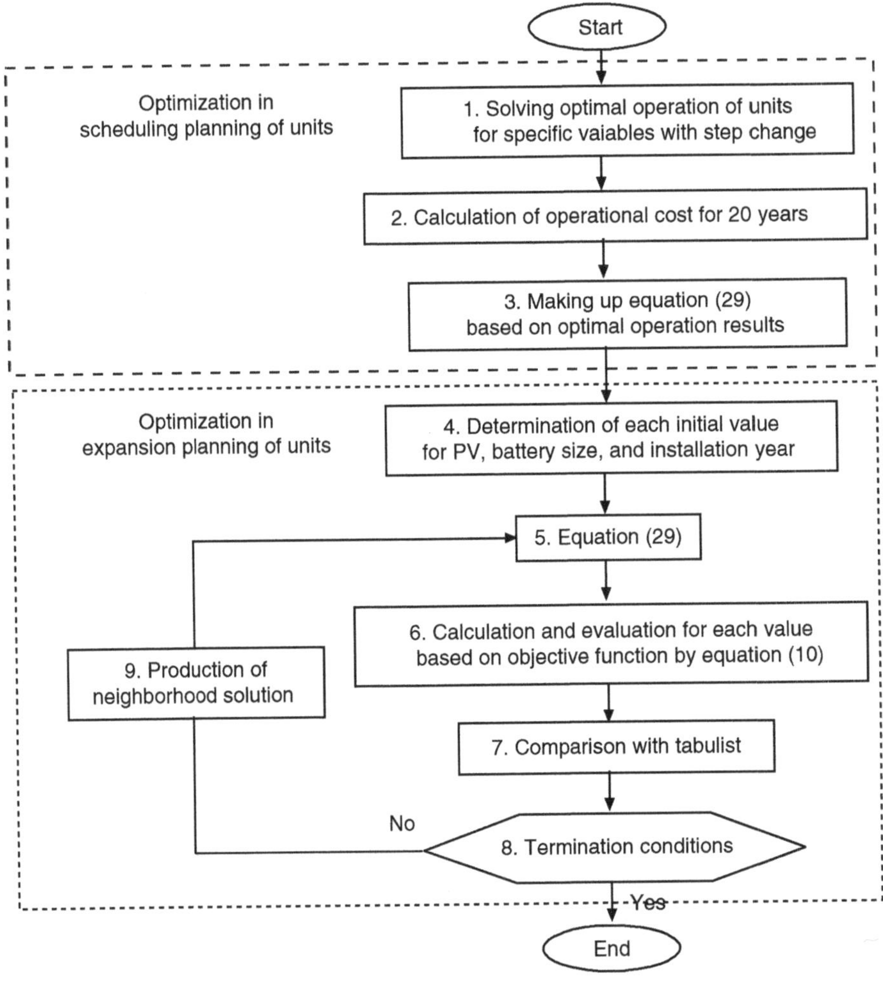

FIGURE 6.16 Flow chart for optimal scheduling and expansion planning [22]. Equation (29) is operational cost (C_{OPE}) for the consumer; equation (10) is objective function (TC) for total cost during 20 years.

polynomial by cubic spline interpolation. Next, the solution of the optimization problem of expansion planning for the PV and battery is determined by TS.

Four simulation cases are carried out for the optimization of expansion planning for PV and battery systems. Case 1 has neither PV nor battery installed during the planning period (20 years). Case 2 is the scenario in which the occupant has a plan to install only the PV system during the planning period. Case 3 is the scenario in which the occupant has a plan to install only a battery during the planning period. Case 4 includes both PV and battery installation during the planning period. The detailed simulation conditions are described as follows for the optimization problem of expansion planning.

- The selling unit price by feed in tariff (FIT) is kept the same from the year installed until after 10 years, that is, when the contract is finished. After that, the selling unit price of the year is applied and is guaranteed for 10 years.
- Once the appliances are installed in the smart house, they are kept as set. Furthermore, the same appliances are not reintroduced (such as two or three times) during the expansion-planning period.
- It is assumed that three people use 30 L and 150 L of hot water for showers from 7:00 to 8:00 and from 19:00 to 22:00, respectively, in a smart house. The HP is operated in such a manner that the temperature does not fall below 50°C in the storage tank.

Fig. 6.17 shows that the operational cost of the smart house for each year, which is calculated by solving for optimal scheduling of the appliances at time index hour, t, in advance before the optimization for expansion planning is investigated. It is observed that the operational cost has a trend of increasing from 2015 to 2034 since the purchased electricity unit price becomes expensive. Moreover, if the battery is installed, the operational cost rises since in this case, the sold electricity unit selling price is lower than the unit price of PV due to price system of double generation which means the selling price for a PV only owner is higher than that for the PV and battery owner or battery only owner. This price rate is actually applied in Japan, even if one battery is set in the smart house. Thus, in the case which includes a battery with a capacity of 0 kWh and PV capacity of 10 kW, the operational cost is very beneficial.

Figs. 6.18–6.21 depict the optimal installation year, capacity, investment cost, operational cost during 20 years, and variation during a period of 20 years, respectively. In Fig. 6.20 and Fig. 6.21, a positive cost means that the occupant pays the cost, and a negative cost means that the occupant obtains a benefit. In Case 1, although the investment cost is 0 Yen (see Fig. 6.19), the operational cost becomes expensive during 20 years, and in

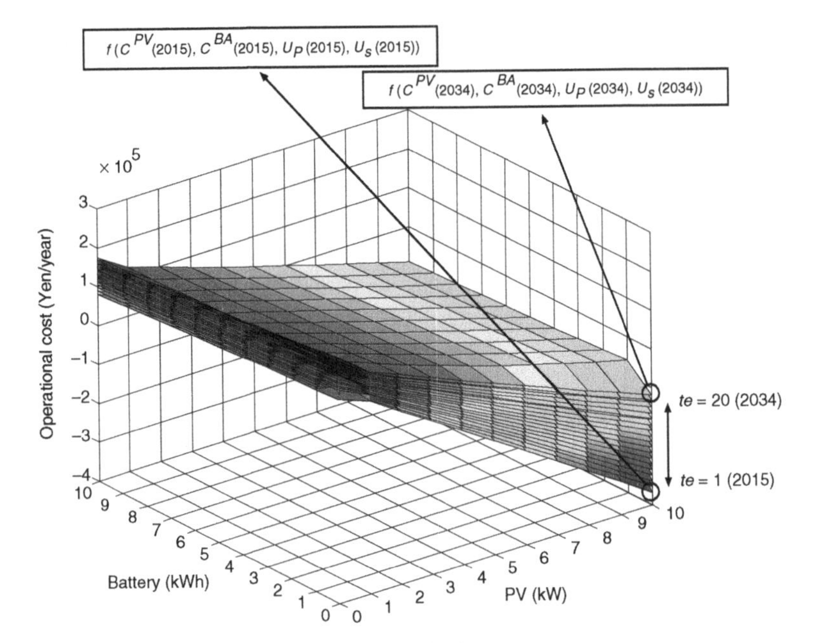

FIGURE 6.17 Operational cost for each year [22].

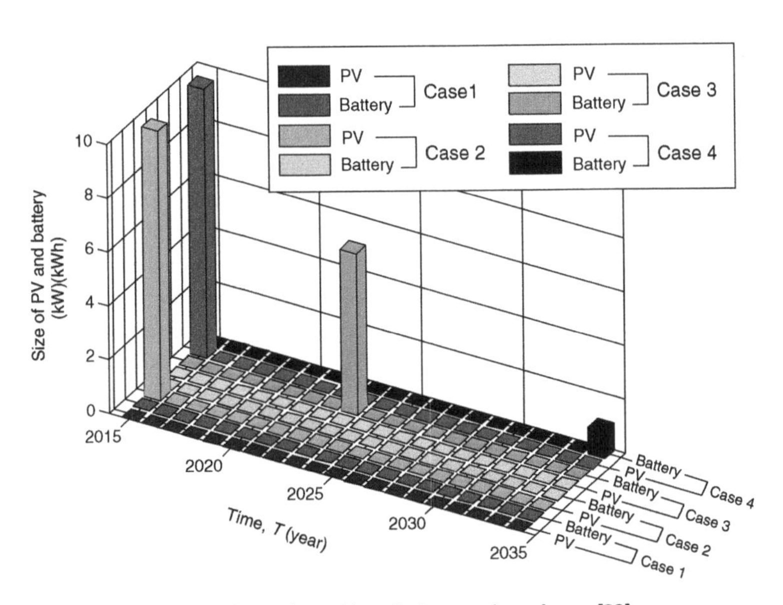

FIGURE 6.18 Optimal capacity and installation year in each case [22].

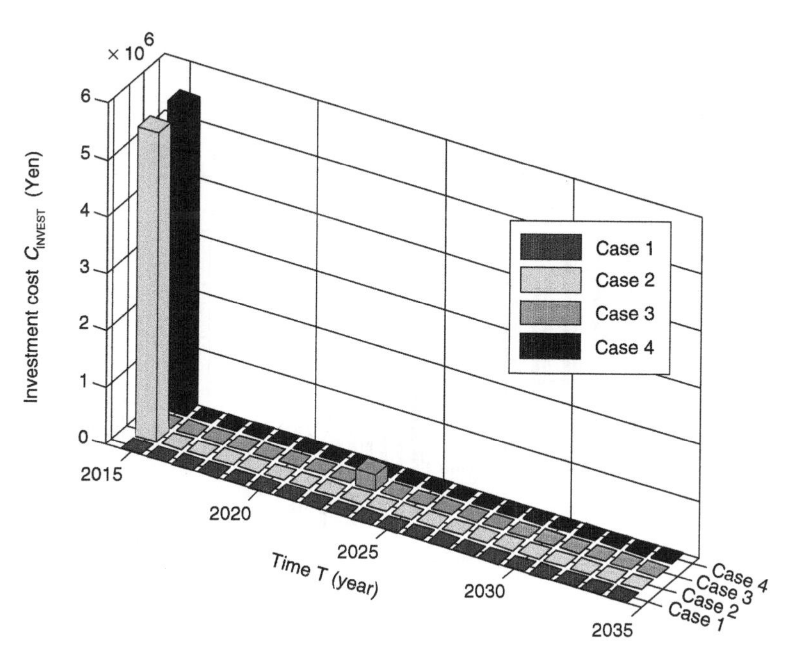

■ **FIGURE 6.19 Investment cost during 20 years in each case [22].**

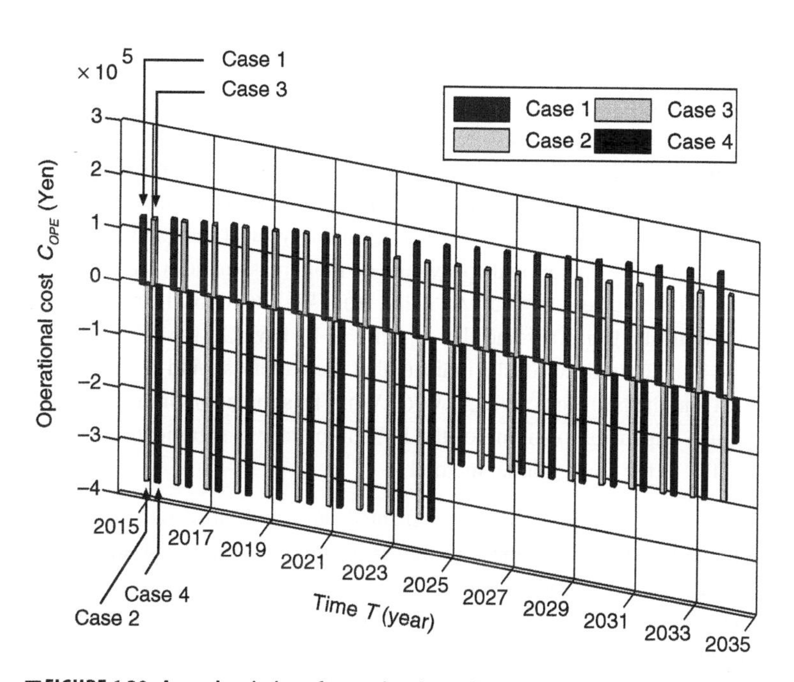

■ **FIGURE 6.20 Annual variation of operational cost during 20 years in each case [22].**

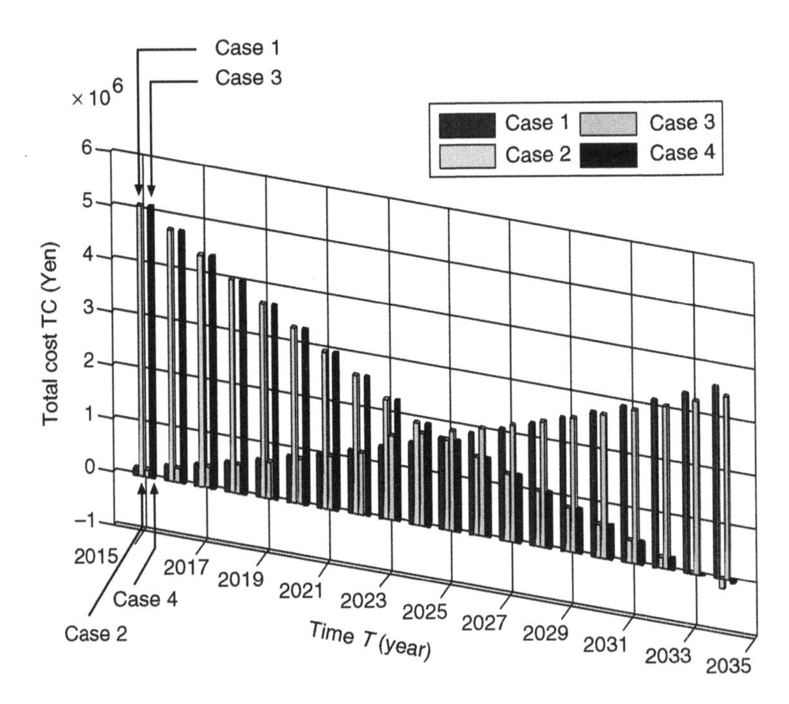

■ **FIGURE 6.21 Annual variation of total cost during 20 years in each case [22].**

Fig. 6.20 the total cost is more expensive than any other case (see Fig. 6.21). It can be seen that the case in which the total cost is the least expensive among Cases 2–4 is Case 2 (see Fig. 6.21). The reason for this is that only the PV is installed and the electricity unit selling price is high. In addition, if the PV is installed in 2015, which is beginning year of the expansion planning period, the operational cost would be more reasonable since the selling price decreases year by year. Actually, from Figs. 6.18–6.21, in Case 2 the PV is installed in 2015 and the capacity is 10 kW. On the other hand, in Case 3 the optimized results indicate that the battery should be installed in 2024 and the best capacity is 3 kW/6 kWh. The simulation time to solve the optimal expansion planning of the PV and battery is significantly reduced by 50% to the conventional method.

The optimization problem was separated into two parts, and the optimal scheduling of the appliances is solved in time indices 1 h in advance. In expansion planning optimization, the cost function is derived by scheduling optimization by cubic spline interpolation, and TS is employed as an optimization tool. This research clarified the optimal installation year, capacity, and type of appliances which make a better choice, and solved for the total cost during a period of 20 years in each case for the consumer side.

REFERENCES

[1] Tanaka K, Yoza A, Ogimia K, Yona A, Senjyu T, Funabashi T, Kim C-H. Optimal operation of DC smart house system by controllable loads based on smart grid topology. Renew Energ March 2012;39(1):132–9.

[2] Beaudin M, Zareipour H. Home energy management systems: a review of modelling and complexity. Renew Sust Energ Rev 2015;45:318–35.

[3] Tascikaraoglu A, Boynuegri AR, Uzunoglu M. A demand side management strategy based on forecasting of residential renewable sources: a smart home system in Turkey. Energ Buildings 2014;80:309–20.

[4] Iwafunea Y, Ikegami T, da Silva Fonseca JG Jr, Oozeki T, Ogimoto K. Cooperative home energy management using batteries for a photovoltaic system considering the diversity of households. Energ Convers Manage 2015;96:322–9.

[5] Missaoui Rim, Joumaa Hussein, Ploix Stephane, Bacha Seddik. Managing energy smart homes according to energy prices: analysis of a building energy management system. Energ Buildings 2014;71:155–67.

[6] Apaydın Özkan H. A new real time home power management system. Energ Buildings 2015;97:56–64.

[7] Chavali P, Yang P, Nehorai A. A distributed algorithm of appliance scheduling for home energy management system. IEEE Trans Smart Grid 2014;5(1):282–90.

[8] Bozchalui MC, Cañizares CA, Bhattacharya K. Optimal energy management of greenhouses in smart grids. IEEE Trans Smart Grid 2014;6(2):827–35.

[9] Anvari-Moghaddam A, Monsef H, Rahimi-Kian A. Optimal smart home energy management considering energy saving and a comfortable lifestyle. IEEE Trans Smart Grid 2015;6(1):324–32.

[10] Kurohane K, Uehara A, Senjyu T, Yona A, Urasaki N, Funabashi T, Kim C-H. Control strategy for a distributed DC power system with renewable energy. Renew Energy 2011;36(1):42–9.

[11] Zheng Y, Yang Dong Z, Ji Luo F, Meng K, Qiu Jing, Po Wong K. Optimal allocation of energy storage system for risk mitigation of discos with high renewable penetrations. IEEE Trans Power Syst 2014;29(1):212–20.

[12] Kumar Kollimalla S, Kumar Mishra M Sr, Narasamma NL. Design and analysis of novel control strategy for battery and supercapacitor storage system. IEEE Trans Sustain Energy 2014;5(4):1137–44.

[13] Eltantawy AB, Salama MMA. Management scheme for increasing the connectivity of small-scale renewable DG. IEEE Trans Sustain Energy 2014;5(4):1108–15.

[14] Ke X, Lu N, Jin C. Control and size energy storage systems for managing energy imbalance of variable generation resources. IEEE Trans Sustain Energy 2015;6(1): 70–8.

[15] Wang Z, Chen B, Wang J, Begovic MM, Chen C. Coordinated energy management of networked microgrids in distribution systems. IEEE Trans Smart Grid 2015;6(1):45–53.

[16] Yona A, Senjyu T, Funabashi T, Mandal P, Kim C-H. Optimisation strategy for an operational planning of a large photovoltaic system with enhanced electrical vehicles. Int J Sustain Energ 2015;34(1):10–22.

[17] Zhang M, Chen J. The energy management and optimized operation of electric vehicles based on microgrid. IEEE Trans Power Deliver 2014;29(3):1427–35.

[18] Goli P, Shireen W. PV integrated smart charging of PHEVs based on DC Link voltage sensing. IEEE Trans Smart Grid 2014;5(3):1421–8.

[19] Ma T, Mohammed OA. Economic analysis of real-time large-scale PEVs network power flow control algorithm with the consideration of V2G services. IEEE Trans Ind Appl 2014;50(6):4272–80.

[20] Zhou Kan, Cai Lin. Randomized PHEV charging under distribution grid constraints. IEEE Trans Smart Grid 2014;5(2):879–87.

[21] Chukwu UC, Mahajan SM. Real-time management of power systems with V2G facility for smart-grid applications. IEEE Trans Sustain Energy 2014;5(2):558–66.

[22] Yoza A, Yona A, Senjyua T, Funabashi T. Optimal capacity and expansion planning methodology of PV and battery in smart house. Renew Energ 2014;69:25–33.

[23] Aslam W, Soban M, Akhtar F, Zaffar NA. Smart meters for industrial energy conservation and efficiency optimization in Pakistan: Scope, technology and applications. Renew Sust Energ Rev 2015;44:933–43.

[24] Erlinghagen S, Lichtensteiger B, Markard J. Smart meter communication standards in Europe – a comparison. Renew Sust Energ Rev 2015;43:1249–62.

[25] Neaimeh M, Wardle R, Jenkins AM, Yi J, Hill G, Lyons PF, Hübner Y, Blythe PT, Taylor PC. A probabilistic approach to combining smart meter and electric vehicle charging data to investigate distribution network impacts. Applied Energy, 2015, doi:10.1016/j.apenergy.2015.01.144.

[26] Quilumba FL, Lee W-J, Huang H, Wang DY, Szabados RL. Using smart meter data to improve the accuracy of intraday load forecasting considering customer behavior similarities. IEEE Trans Smart Grid 2015;6(2):911–8.

Chapter 7

Protection of DERs

Raza Haider*, Chul-Hwan Kim**

**Department of Electrical Engineering, Balochistan University of Engineering and Technology, Khuzdar, Pakistan;*
***School of Electronics and Electrical Engineering, Sungkyunkwan University, Republic of Korea*

CHAPTER OUTLINE

7.1 Introduction 157
7.2 Protection in distribution system 160
7.2.1 General protection 160
7.2.2 IEEE standards for protection 163
7.2.3 What to do under fault conditions 164
7.2.4 Fault currents change 165
7.2.5 Smart protection 166
7.3 Power system disturbances 168
7.3.1 Power quality issues 168
7.3.2 Faults 170
7.3.3 Consequences of electric faults 172
7.3.4 Application based three-phase fault analysis 173
7.4 Impact of DER on protection system 173
7.4.1 Protection failure 174
7.4.2 Hosting capacity 175
7.4.3 Loss of coordination 176
7.4.4 Protection issues of DER 178
7.5 Protection schemes for distribution systems with DER 180
7.5.1 Islanded operation 181
7.5.2 The protection equipment for DER networks 182
7.5.3 Recent technological trends in DER protection 188
7.6 Conclusions 190
References 191

7.1 INTRODUCTION

Electric power systems are at their most popular since their invention and are now the most common constraint. They play a key role in technological reforms, infrastructure development, and economic growth. The well-linked

Integration of Distributed Energy Resources in Power Systems

system of electrical power distribution is the conventional one, that is, generation in bulk quantity, transmission at high voltage levels, and distribution via grid stations with the facility of end-user's level voltage.

Experts worldwide are concerned over the grab of energy resources on the atmosphere and their regular supply in the coming future. In the past 10 years, a large fluctuation in oil prices has been recorded, and the scarcity of petroleum products prompted the energy experts to think of alternative and sustainable energy resources for a clean and green environment. Distributed energy resources (DERs) were promoted as an idea for instant electricity generation and connectivity to the main electric grid. A rapid growth of DERs has occurred as it provides clean power generation with less environmental impact, reliability assurance, and insertion of renewable energy sources for modern power scenario requirements [1].

The complexity of electrical power systems is mainly based on the smart and case sensitive protection schemes. A well-designed protection scheme is the backbone of any electrical system. DER is a recent idea that was developed for decentralizing the power system. The Electrical Power System industry comprises generation, transmission, and distribution, working closely with centralization. DER has brought new distribution technologies that not only give the opportunity to utilize the indigenous renewable energy resources at door steps but also make generation level small and less complex. It is a reliable system of distribution that changed the conventional model of power generation as promising and smart integrated system. There are numerous issues regarding fuel prices, genuine or artificial shortages, global warming, and increasing electricity need, which are big challenges for traditional power generation technologies to cope. DER with microgrids enable technologies to get more attention for their efficiency, durability, and improved reliability. There are some issues relating to DER protection which are being addressed and certain models have been developed to minimize and control power system disturbances.

The main purpose of DER is to utilize the energy resources that can serve at a cost of no transmission line, as the generation is close to the load center. The electrical power transmission system, with variable losses and high cost, has made DER a more reliable and promising option that shares load jamming on conventional utility and assists localized power generation. The DER system has the delicate intensity to control and can easily be handled by the distribution utilities.

The sustainable approach behind DER is to integrate it with the ongoing power system utilities by applying novel protective measures. The conventional distribution system is designed as a submissive network in which huge alternative energy penetration may cause bidirectional power flow, and

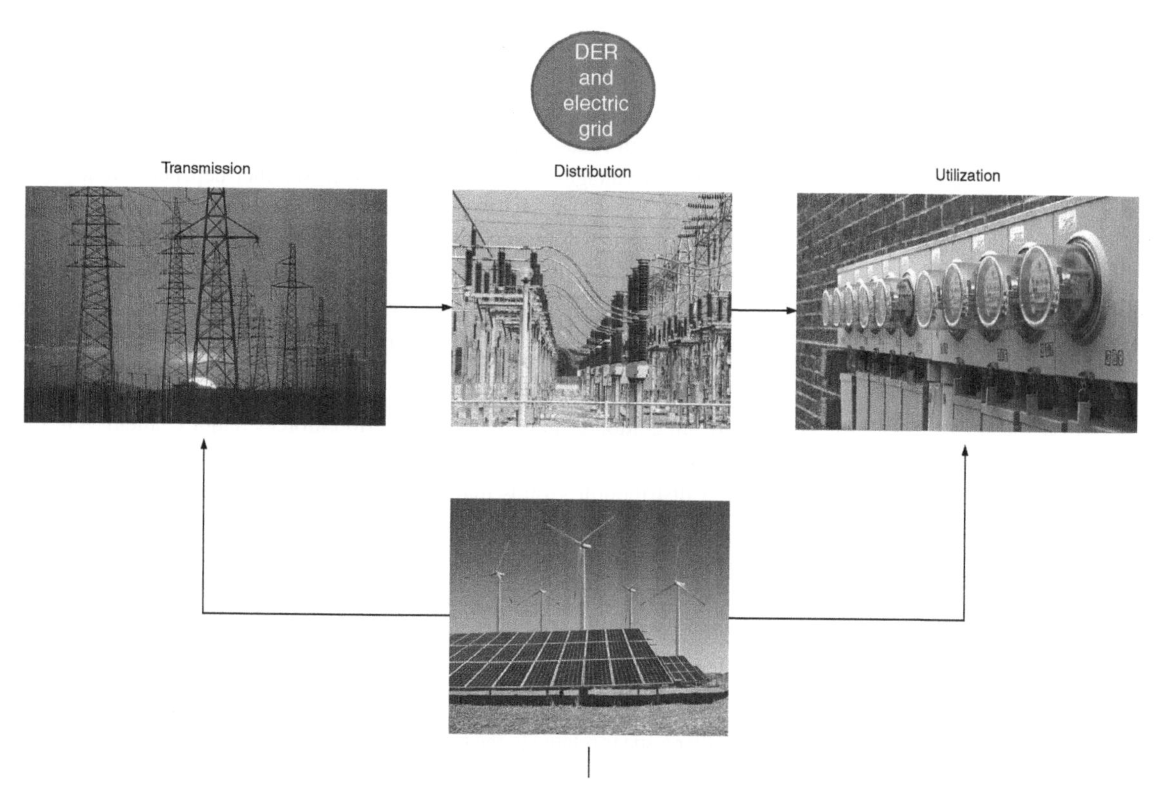

■ **FIGURE 7.1 Distribution energy resource layout.**

as a result there are variations observed in voltage, current, or frequency. In addition, power system oscillations due to sudden fault currents could affect the protective apparatus. This chapter mainly focuses the different protection issues in DER, needs of protection, and reliable protection schemes.

The Institute of Electrical and Electronics Engineers (IEEE) P1547-2003 is a benchmark model for interconnecting DERs with conservative electrical power systems and it also provides guidelines for general interconnection of electric utility. This involves the response to abnormal conditions when functioning, power quality issues, and safety condition together with operation in utility grid connected and islanded mode [2]. Fig. 7.1 shows DER placement in an electrical power system network. This includes multiple power generation sources in a mode of bidirectional or islanded operation, and power flow coordination in a single network.

DERs became mainstream after advanced renewable energy technologies had been explored and implemented. Renewable energy equipment brings a helping hand for grid-integrated electric supply and benefits the suppliers, operators, as well the customers. Keeping environmental concerns in view, the conventional power generation technologies that use to produce more carbon is reduced. At present, researchers are working to suppress voltage fluctuations in renewable energy harnessing technologies because the frequency and voltage always fluctuate due to weather-dependent renewable energy resources. The energy storage devices with state-of-the-art technologies are gaining more importance to cope the fluctuation problems. These devices help for smooth voltage distribution and reduce the technical hurdles in distribution system [3,4]. The mode of transmission of energy through the main electric grid has now being changed by DERs and facilitating consumers to have flexible energy utilization.

DER integration to the main electric grid enforces major challenges in the protection systems. The conventional system of distribution is planned for radial distribution groups that contain feeders at one end with high fault currents. The fault currents always flow downstream in a radial system of distribution, from the utilization service towards the faulty point. Because of the single source of power generation, high fault currents are generated by the utilization service, triggering protective apparatus next to the feeder corridor. Therefore, despite the several advantages, there are many technological challenges behind control and protection mechanism for DERs [5].

Another issue is related to the number of units installed with DER, which connect the main electric grid in order to share supply line and the accessibility of adequate level of short-circuit current in the islanded operating mode. This level of DER may significantly fall down after a detachment from the main electrical grid while keeping express track. Actually, the working conditions of DERs continually vary because of the alternating energy resources, that is, wind, tidal, or solar, and intermittent load distinction. Such conditions may lead to overcurrent protection failure and will not guarantee any discriminating action for all possible faults. Thus it is necessary to check the selected setting for overcurrent protective relays that take into account the grid connectivity, transformed position, and the type and amount of power generation. The overcurrent relays (OCRs) and circuit breaker coordination, and sequential operating conditions at fast tracks are the challenges to be addressed.

7.2 PROTECTION IN DISTRIBUTION SYSTEM

7.2.1 General protection

The prime objective of the power system protection is to ensure safe and reliable electrical power supply to consumers. The system is based on heavy duty and expensive equipment with high electrical power concentration, thus it should be of major concern.

In every system of electrical power protection, control of all electrical quantities is compulsory. Normally the values of voltage, current, and frequency should be under control and limiting these electrical quantities ensures the power system protection reliability.

The distributed energy protection system depends upon the mode of distribution, that is, radial, ring main, or network. The most commonly used system of distribution is the radial one that depends on the supply at one end through a main source. This system is normally very simple to operate and is generally employed with circuit breakers, fuses, and relays. But with the distributed generation injected to the main electricity system, it requires more attention, and can weaken the protection coordination. Therefore, a model and coordination has to be developed that can protect the overall system in a sound manner and allow fault diagnosis to be easily achieved.

Faults on overhead electrical power transmission or distribution systems are characterized as symmetrical and unsymmetrical. In an overhead system of transmission line the faults between any two phases of a three-phase system are rare, but bad weather conditions affect the line and give rise to faults. The most common faults occurring are single phase to ground, which are due to the insulator flashover and breakdown. In a distribution system, the overhead lines are subjected to different kinds of faults because the distance is closer to the utility end. Hence the overall protection scheme and the equipment operation should be linked with grid-integrated DERs. Transients and harmonics are rising as big problems in DERs that are already present in normal distribution systems. The harmonics are the distortion in voltage or current that occurs at a numeral multiple of the fundamental frequency, and the transients are the sudden changes occurred at distribution system depending upon the nature of load connected. One of the drawbacks of utilizing an alternating current (AC) source is that it produces transients and harmonics, because most of the loads in daily life are inductive in nature. There is much research going on to address this challenge and make transfer of AC supply constant and smooth.

Nowadays, electrical energy system engineering is going through a major transformation due to many alternative and renewable energy supplies and better quality of power distribution by the utility. This is going to be supportive in order to ignore disorder of manufacturing or other practices caused by voltage drop, swell, or distort conditions when a short circuit fault takes place in the distribution system [6].

The protection system of distribution network in general is used for power flows from the grid supply point to the downstream low voltage system. It is generally based on OCRs with preferred settings to pledge discrimination between upstream and downstream relays. The protection coordination should be arranged so that a fault on a downstream part must be cleared by the relaying system at the source end of the main feeder. At this moment, the operation of any other relaying system on an upstream part would result in a blackout and the system would be seriously affected [7].

There are several studies into DER protection, which are based on different protective algorithms. In one of the schemes, protection is provided by dividing the whole distribution network in different operating zones, which are known to be an islanding operation. A multilayer perceptron network is utilized for determining the faults by installing a computer-based relaying system. It is the most common neural network model because it gives the desired output mapping sets of all input data. This method is best suited to isolating system fault within a defined range and protecting rest part of the system [8].

In the integrated system of distribution, there are many cases that deal with different characteristics of overcurrent protection. This includes definite or inverse time and inverse definite minimum time characteristics. There are drawbacks of each overcurrent relaying protection, such as the time lag of the unknown nature of short circuits and coordination with the additional load. The protection scheme should be designed in manner that the fault clearing time is less and without causing disturbance for the end users [9]. When the DER is connected to the distribution substation, it makes the overall system complex because the new system will be having its own dynamics. Therefore, in order to synchronize DER with the ongoing distribution system, the corrective protection measures have to be taken. Furthermore, each protective device must be capable to differentiate this characteristic for all disturbing or faulty positions, the nature of faults, different load levels, and possible supportive supply design.

As shown in Fig. 7.2, a DER has to be connected to the ongoing power system bus. According to the fundamental principles of power system operation, the DER should have common electrical parameters for electric grid interconnection. The most important conditions that have to be fulfilled are:

1. The frequency of the DER and the existing system should be same, and
2. The voltages at the terminals of DER and the existing system should be same.

The DER interconnection is complex in the present scenario of main electric grid; it creates technical challenges in feeder design, where one or more generators connect to presented feeder. The areas of common concern are voltage regulation, relay desensitization, islanding, voltage flickers, transformer substitution, and resonant conditions.

The energy generating source, having some governor control mechanism, controls the real power supplied by the source to the electric grid whereas

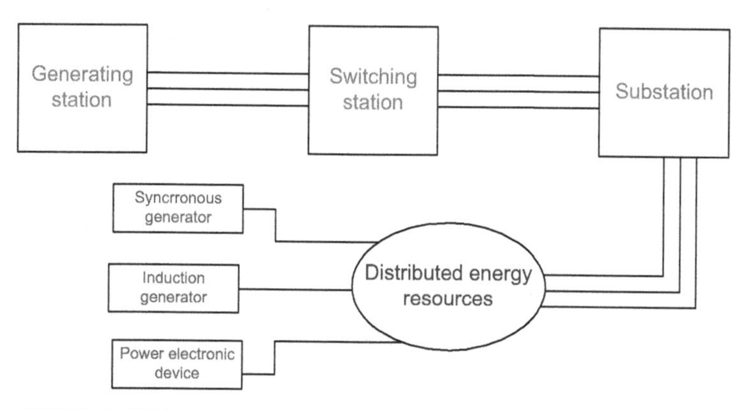

■ FIGURE 7.2 DER interconnection.

the field circuitries decide the amount of reactive power drawn from it. If the DER is connected to the existing system of power generation and running under lower frequency, this will make the DER consume electrical power instead of supplying it. Therefore, some control mechanism is important for the joint operation and control of DER and main electric grid. This can be achieved by a governor controller, which is a control mechanism with different set points to adjust the system frequency.

The prime objective of the utility companies is to supply electricity to consumers with protective standards and keeping the economical means in view. The protection of DER is achieved by protective relaying with the adaptive and new available schemes.

7.2.2 IEEE standards for protection

The recommended practice for protection of domestic, commercial, and industrial power systems given by the IEEE provides information regarding short circuit current, standard power system voltages, sensing device characteristics, and various interruptions in power systems. The general principles [10] for the relevant protection of DER specified in the IEEE standards are mentioned further.

7.2.2.1 *Voltage requirements*

The DER that has to be connected to the main electric grid will not be responsible for voltage fluctuation and may not cause variations in the area electric power system. Any mishandling could cause the severe unbalancing and affect the distribution grid auxiliary equipment operation.

7.2.2.2 *The grounding*

The grounding scheme should be designed in a manner that the connection to the main electric grid may not cause any over voltage that go beyond the rating of the equipment connected to the area of electrical power system. And the overall coordination of the ground-fault protection on the area may not be affected.

7.2.2.3 *Synchronization*

DER has brought new issues on the stability and the transient function of the main electric grid. The penetration of such resources into the grid may cause synchronization problems because the sudden detachment or reconnection gives rise to transients or harmonics. The voltage fluctuation at the point of common coupling around ±5% of the existing voltage level at the main electric grid is allowed for DER integration.

7.2.2.4 *Energization area*

The DERs, when integrated to the main electric grid, do not energize the area power system. The standard is about the DER units that stop to energize the area electrical power system for faults on the circuit to which they are connected. If a fault is detected in a circuit and has been de-energized by its protection system in order to clear the fault, there should be no fixed source on the circuit to energize it.

7.2.3 What to do under fault conditions

DER contains different generation units and capability of each unit to detect fault depends upon the technology utilized for generation and the nature of fault. The synchronous generators operating as a unit with DER produce fault currents during absolute periods of time because of their capability curve characteristics. The initial fault occurring can reach more than six times the generator full-load current and can decompose over a number of seconds under generator full-load current as the generator field disintegrates. Under such conditions, the voltage on the generator considerably comes down and helps to detect and analyze the fault. Different fault analysis techniques are employed based on the fault current and voltage sag on the synchronous generator. The commonly used techniques are OCRs with voltage control and voltage restraint with unlike inverse or definite time characteristics.

The voltage control OCR protection gives permanent sensitivity by building the set of overcurrent operating value relative to the applied input voltage. The OCR keeps the voltage under a desired value and controls it by under voltage factor. There are certain coordination difficulties caused by the magnitude of faults that continue for a long time because the sensitivity of the OCR has a preset value. The voltage restraint OCRs are adjusted at different sensitivity levels depending on the voltage sag. Normally, at 25% voltage the sensitivity is four times the one at the rated voltage. It provides better coordination and works as backup for differential relay [11].

For the induction generator operating with DER the fault current contribution depends on the declaration and the required time of fault current for relay to operate. The declaration of fault current in an induction generator depends upon the type, severity, and the means of reactive power compensation. The proper selection of these parameters makes the protection system efficient and the operation becomes smooth.

The DER with inverters behaves as intelligent active elements and, when connected to the main electric grid, cannot afford significant currents under outdoor fault conditions. These currents are usually not more than one and half times their rated load current. The generally applied fault-finding methods using overcurrent standards are not helpful in this case. Inverter-connected DERs rely on other techniques such as voltage deviation or frequency sensing to identify faults in the supply area of electrical power system. Whereas in case of external faults the behavior of the DER can be affected mostly because it is connected to the energy source that is dependent on different types of generator technology, which can supply current to a fault in their neighboring circuit or some other point of area electric power system. Such complications rise after DER integration to the main electric grid because of the dynamic behavior of different technological devices and load uncertainties.

7.2.4 Fault currents change

The DER sometimes works as microgrid, and the change in fault currents depend on the operation means. In DER integrated mode, the utilities contribute to microgrid fault currents and when operating in an islanded mode, the fault currents are less. With the power electronic devices the fault current adding capacity is limited to twice their rated currents. These lower fault currents do not trip the OCRs. Usually under normal conditions, the fault current decreases as the feeder impedance increases when fault point shifts downstream to feeder path. DER equipment, including synchronous and induction generators, contribute more to fault currents compared with DER with power electronic devices. This is because of the dynamic behavior of the machines, so the fault current paths could be different for the DER operation with electric grid or as an islanded mode. Such an arrangement will require different relay settings; the fixed relay settings under such condition might become unfeasible [12].

7.2.5 Smart protection

Voltage stability and coordination among protective devices with high DER penetration, flashing of renewable power output impacts on voltage regulation, and the controlling capability of DER with active and reactive components are the big challenges to cope with. Smart protection refers to the term that the DER integration to the main grid network should have its own protection scheme at the point of link, in order to transfer electrical power in a safe mode. The modern and projected advancements in power system apparatus, processors, and communication networks provide opportunities for pioneering utilities to develop power system organization, competence, and consistency.

Fault location isolation and service restoration (FLISR) application is one of the important developments in smart protection, which enables distributed utilities to implement advanced distribution applications. Fig. 7.3 shows the time line for fault investigation without FLISR, and the delay can be seen clearly. FLISR generates an immediate exchange plan for restarting sections of a distribution circuit that has been shut down as

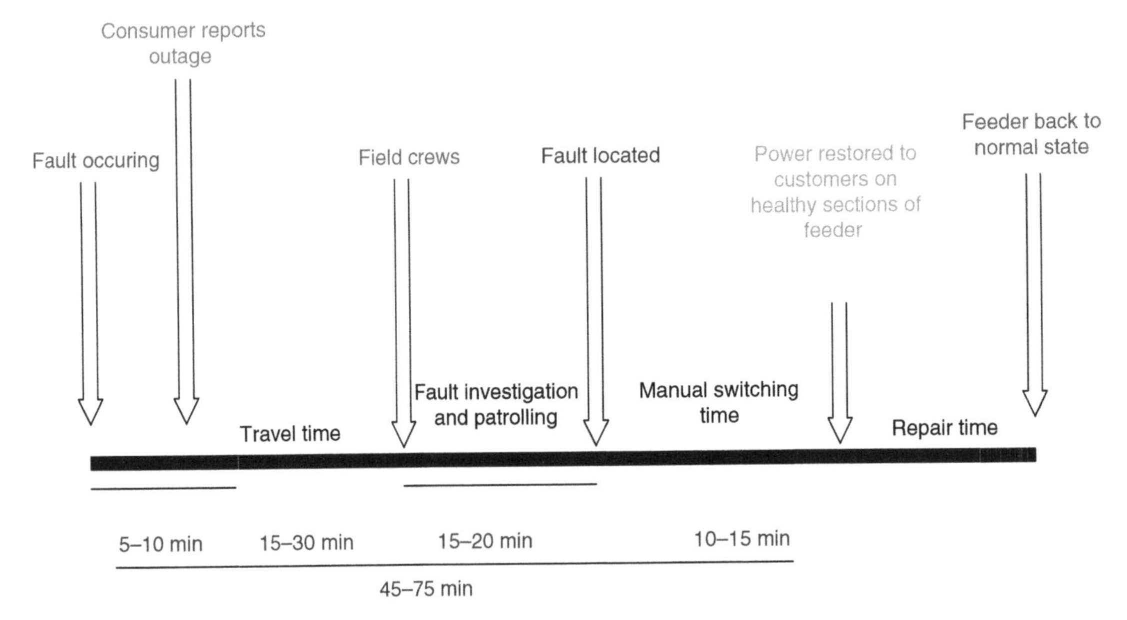

FIGURE 7.3 Time line for fault investigation without FLISR. *(Source: Courtesy of PAC World.)*

a result of an undying feeder fault. Once the exchange plan is generated, FLISR can mechanically implement the plan to reinstate service where feasible, generally in less than 1 min following the original fault incidence.

The FLISR application automatically senses that a fault has arisen and locates the fault between two intermediate switches. And then subject control commands to open the switches where likely to reinstate service to healthy sections of the feeder. This modern technology allows all of these actions to be completed without manual intervention [13].

In smart protection there are several control functions that include regulated power flow on feeders, regulated voltage at the interface of each DER, and load sharing when the system is islanded. It works as a chain reaction and reduces the additional complexity in operation, control, and protection system of the closed integrated loop. The general overview and operating conditions of FLISR are given further.

7.2.5.1 Fault detection

The operating control of FLISR is such set that it operates when a short circuit fault on upstream or downstream occurs. If the feeder becomes de-energized due to manual switching activities, the FLISR will not operate. Such an arrangement can be obtained by putting in more than one fault detectors to trigger FLISR function. Commonly a microprocessor-based control relay or intelligent electronic device (IED) is installed in substations to provide a signal to FLISR for operation.

7.2.5.2 Fault location

After fault detection, the fault segment on the feeder needs to be located. There are different sections on FLISR, which are enclosed by distantly controlled switches. These switches include the faulted circuit indicator (FCI) that determines if fault current has recently passed through the switch. This indicates if a fault is located further from the substation and uses the FCI status and feeder topology to determine the faulty zone.

7.2.5.3 Fault isolation

After fault location, FLISR issues control commands to open the switches desired to completely cut off the damaged segment of the feeder based on the fault location investigation. FLISR suspend all control actions until the routine reclosing sequence is completed and makes sure that reconfiguration of feeder by FLISR is performed. After performing all of the aforementioned checks the FLISR follows the permanent fault to operate.

7.2.5.4 *Service restoration*

This is the last action performed by FLISR, to isolate the damaged segment of the feeder. It attempts to reinstate the service to as many healthy segments of the feeder as possible using the normal source of supply to the feeder and offered backup that have standby capacity to carry additional load.

7.2.5.5 *Power system stability*

With the advent of DER, electrical power systems have been primarily impacted by stability and quality problems. The maloperations and change in protective device settings has dramatically reduced the reliability of DER. Voltage sags are the most significant power system stability problems since the application of the power electronics equipment. The controlling and monitoring devices like programmable logic controller, supervisory control and data acquisition, or automatic voltage regulator are becoming very much sensitive to voltage sags as the complexity of the equipment enhancing [14]. A high share of DER is supporting new technologies to wipe away conventional boundaries and demand in a more practical complexity. Every new operational element adding to the distribution system will lead to associated fluctuation. Like many, DERs are dependent on weather conditions; under normal weather, power generation will be up to the required level. However, any change in weather will at once create a large jerk in the area electric power system and the supply into a given line will drop. Such changing weather conditions may cause voltage variations and create stability problems. Even the change in demand response or real-time charging interacts with programmed home controls to deal with an oversupply of energy on the grid. The fall in prices boosts up energy consumption at domestic and industrial levels and, as a result, a sudden surge in demand is developed [15]. These kinds of issues greatly influence the distribution system in general and the other interconnected systems in particular. Thus the distribution system has to be more imperatively situational based that should consider the load and control dynamics.

7.3 POWER SYSTEM DISTURBANCES

A broad classification of power system disturbances that might occur and cause potential impact on the system behaviors and the equipments is described in this section. The normal operating conditions and the smooth flow of electrical power makes the system stable.

7.3.1 Power quality issues

There are certain power quality issues that need to be addressed and remedies that have to be taken to strengthen the overall power system and increase the life of equipments. Any disturbance in the electrical power system is caused due to the change in the system parameters that affects basic electrical quantities. Some of the quality issues are mentioned further.

7.3.1.1 Voltage behavioral changes

The voltage is an important electrical quantity that generates pressure to make a smooth flow of the current. Any change in this quantity rather than the desired value leads to some unfavorable conditions like voltage dips, voltage surges, overvoltages, undervoltages, voltage fluctuation, voltage imbalance, or short and long voltage interruptions. The voltage dips are exercised due to the local and remote faults in the area of electrical power system with maximum inductive loading and the continuous switching operation of large loads. This impact is found on the sensitive protective equipment in the form of tripping.

Voltage surges are produced due to the capacitive switching, phase faults, and sudden switch off of heavy loads. The surges form high voltage penetration, thus the sensitive equipment operates at once and damages the insulation and windings of machines. The change in voltage is due to the system voltage regulation or heavy network loading, this might impact the equipment dependent on constant steady state voltage. The imbalance in voltage is due to unbalance loads and impedances, which create overheating in motors and generators and mostly interrupt three-phase power operations.

7.3.1.2 Power frequency variations

Power and frequency variations in an electric supply system are mainly due to the extreme loading conditions and loss of generation. The frequency of power system depends on the active power balance and needs to remain constant at different operating conditions. Any change in load affects frequencies in the entire region and tie-line power exchange among regions. Such variations impact on motoring operations and power electronic equipments.

7.3.1.3 Harmonics and transients

The harmonics in power system equipment due to the nonlinearity of transformer core have been always the point of research and interest has increased over past years with the rapid change and modernization in power system devices. Heavy inductive loading in the industries, rectifier equipment, and the source of conventional power generation made this a main power quality problem. Harmonics create maloperation of sensitive power system equipments. The transients are aperiodic current waveforms that flow in a circuit for a very short duration following an electromagnetic disturbance due to various reasons.

7.3.2 Faults

The faults refer to the unwanted or stern variation in the state of voltages or currents under standard working conditions. At normal stages, the electrical power system equipment carries standard voltages and currents that make power system operation secure. However, under abnormal conditions the unwanted signals, known or unknown, damage the power system equipment and stop supply.

Thus to have safe and secure operation, different and technically sound protection schemes are employed that comprise relays, reclosers, circuit breakers, isolators, and other switchgear equipment. In general, electrical power system disturbances are classified into two broad categories, symmetrical and the unsymmetrical faults.

7.3.2.1 *Symmetrical faults*

A fault in an electrical system is a failure that impedes the standard values of voltage and current. Symmetrical three-phase faults are roughly 8–10 %, and cause severe power system disturbances. The currents which flow in unlike parts of the power system instantly after the event of the fault fluctuate from those flowing a few cycles afterward just prior to the circuit breakers operate and open the both sides of faulty line.

The current at this stage is totally different from the currents that flowed under steady state conditions. This can be continued if the faulty line is not disengaged from the line by the efficient use of circuit breakers. Circuit breakers are normally dependent upon the value of the current flowing after the occurrence of fault and the interruption. The analysis of the three-phase fault is carried out on a per phase basis and the information regarding is taken from the switchgear equipment.

7.3.2.2 *Unsymmetrical faults*

The faults that cause imbalance between any two phases are known as unsymmetrical faults. In electrical power system, 80–90% faults are of unsymmetrical type and cause severe unbalancing in the power system operation. Before the occurrence of an unsymmetrical fault, the system is balanced and the positive sequence network is dynamic. As the fault occurs, the sequential networks automatically connect through the faulty line. Fig. 7.4 shows the symmetrical and unsymmetrical fault waveforms.

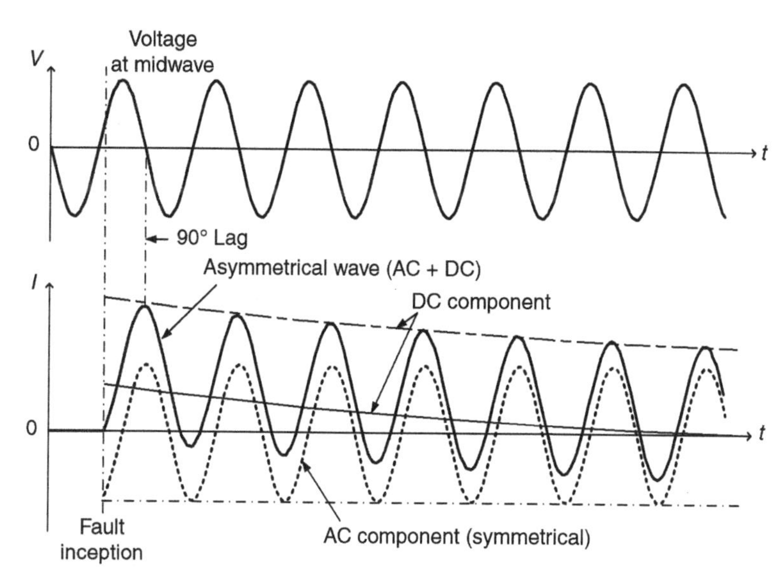

FIGURE 7.4 Symmetrical and asymmetrical waveforms.

These faults can be calculated by applying a bus impedance matrix composed of positive or negative sequence impedances. The method of positive sequence network is applied to determine voltage and current under symmetrical fault conditions. For unsymmetrical faults, negative and zero sequence networks are applied.

Faults in the power system may occur due to certain internal and external conditions that influence directly on power system operations. The reasons behind some of the fault origin conditions are discussed further.

7.3.2.3 Surrounding weather

The fault mostly depends on the condition of the atmospheric weather. Wind, heavy rain, humidity, coastal wind effects on bare conductor transmission lines, ice topping, etc, interrupt the power system continuity and reliability.

7.3.2.4 Equipment

The electrical equipment has to be protected for the all mishaps. As most severe faults in electrical systems are short circuits, the cause could be improper handling, insulation breakdown, or poor cabling. Heavy currents flow through electrical equipments during short circuit fault that could damage or weaken the supply equipment.

7.3.2.5 Mishandling

Mishandling also leads to electrical faults. Such human error could put the lives of the personnel handling the system in danger. The improper rating of the device, loose connections, or poor servicing leads to such problems.

7.3.2.6 Flashover

Flashover in a power line is caused due to the stress on insulators. The corona effect in the transmission line occurs because of ionization in the air. As air is not a perfect insulator, the ionization makes surrounding air conducting and, as a result, the flashover or spark causes the insulators breakdown.

7.3.3 Consequences of electric faults

7.3.3.1 Flow of overcurrent

In an electrical circuit the overcurrent flow produces very low impedance path resulting heavy current being drawn. The current sensing devices trip the circuit at once, any delay could cause loss of insulation or apparatus.

7.3.3.2 Threat for operator

The fault in a system could cause danger for the personnel handling and it depends upon the magnitude of the fault current. Therefore, safety measures and advanced training are essential for the operator handling electrical power system apparatus.

7.3.3.3 Loss of apparatus

The electrical system apparatus can burn completely if the protection equipment could not operate well timed. It is mandatory to keep overall protection coordination smooth to save the devices and the individuals.

7.3.3.4 Effect on healthy systems

Electrical faults may affect a healthy system connection or interconnected circuits of a network to the faulty point.

7.3.3.5 Fire risk

Severe short circuits in a transmission or distribution system due to lightning, heavy winds, or snowfall could cause flashover and sparks that lead extreme fires.

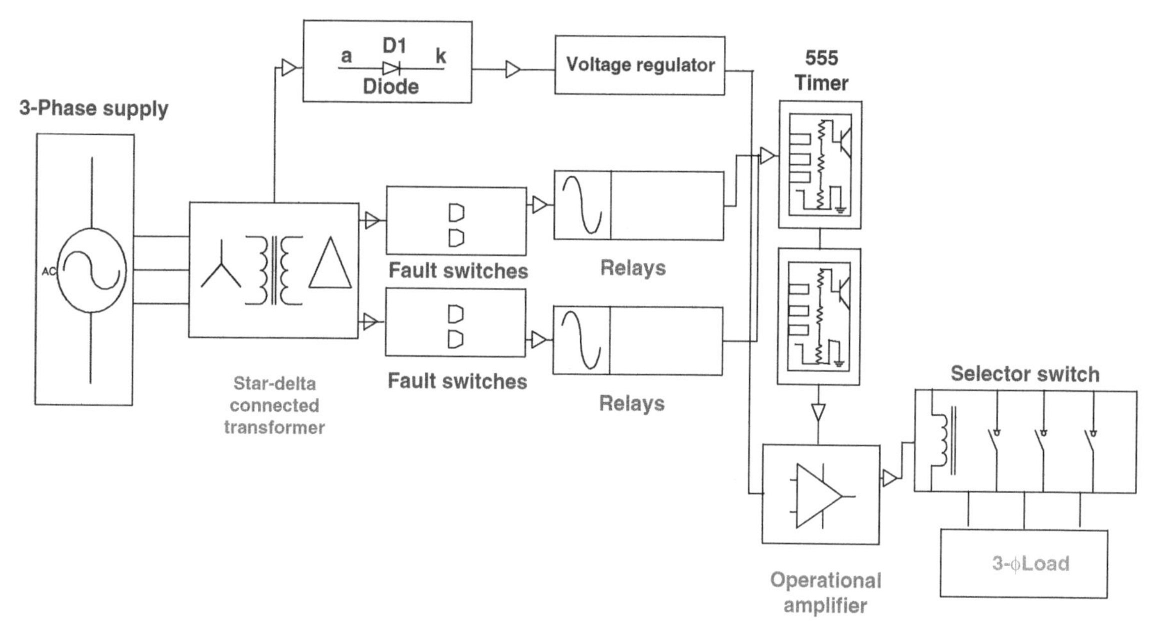

■ **FIGURE 7.5** Three-phase fault analysis [16].

7.3.4 Application based three-phase fault analysis

The circuit shown in Fig. 7.5 has three-phase supply, star-delta connected transformer, and the other fault analyzing devices carrying a three-phase load. The temporary switches connected create temporary and permanent faults. By pressing the switching button, one creates a temporary fault and the 555 timer arrangement trips the line and restores electric supply back to load. By pressing the switch button more than one time, the circuit will create permanent faults thus disconnecting the circuit from the supply line, and the relay arrangement will shut down the system. The distribution protection system also works in the same pattern [16].

7.4 IMPACT OF DER ON PROTECTION SYSTEM

A DER can be considered as small generating station ready to be integrated with the regular electric grid. Generally the DER can be considered as renewable energy or a conventional small-scale generation system, in order to mix share in central power system. There are two options, connect the DER to electric grid or use it in an islanded mode. During any such operation, some diversity is observed that the energy can flow in either track when protection system is sensing and the commonly used radial system of distribution has no bidirectional flows. The low capacity units used in a DER system may affect the distribution system in a number of ways. The overall performance, reliability, and stability may be put in danger if additional protection devices are not installed with proper coordination [17].

DER installation changes the conventional practice of distribution scheduling and engineering by escalating the range and density. Even many conventional rules of thumb and guidelines might not remain valid.

In a conventional electrical network, the electrical power flows in one direction only, that is, from the feeding transformers to the load end. But the DER installed at distribution substations may lead the electrical power flow back and forth. This is because the conventional distribution system protection is not planned to work with DER and different power generator capabilities with changed voltage profile. In total, the DER deals with all power generation voltages including renewable and nonrenewable, which contribute differently during a fault.

All generators give rise to the fault currents during any power system disturbance. But synchronous generators are the main cause of such disturbances because of the field excitation and stability problems. Another challenge for the offered protection of distribution networks is DER integration. Since DER is generally coupled with distribution system, the new connection of a generating source can redistribute the source fault current on the feeder circuits causing loss of relay coordination and possible over voltage [18].

In many countries, most loads are designed to supply radial loads, and a DER running on an islanding mode will make the system more complicated. For the uninterruptable utilities the fault clearing time is of worth importance and they need the fastest fault recovery time, which is impossible if the DER is running in islanding mode.

7.4.1 Protection failure

The trouble when integrating DER with the presented electrical power network is that distribution systems are intended as a passive network, that is, carrying the power unidirectional from the central generation at high voltage and then downstream to the load centers at low voltage.

There are two different ways the protection system might fail to operate correctly. One is the mal-trip (without cause eliminates a nonfaulted element) and the other is fail-to-trip (unable to eliminate a faulted element) as shown in Fig. 7.6.

A mal-trip can occur when a DER unit feeds an upstream fault, which is initiated from any of the DER units. Such a condition may disobey the edge

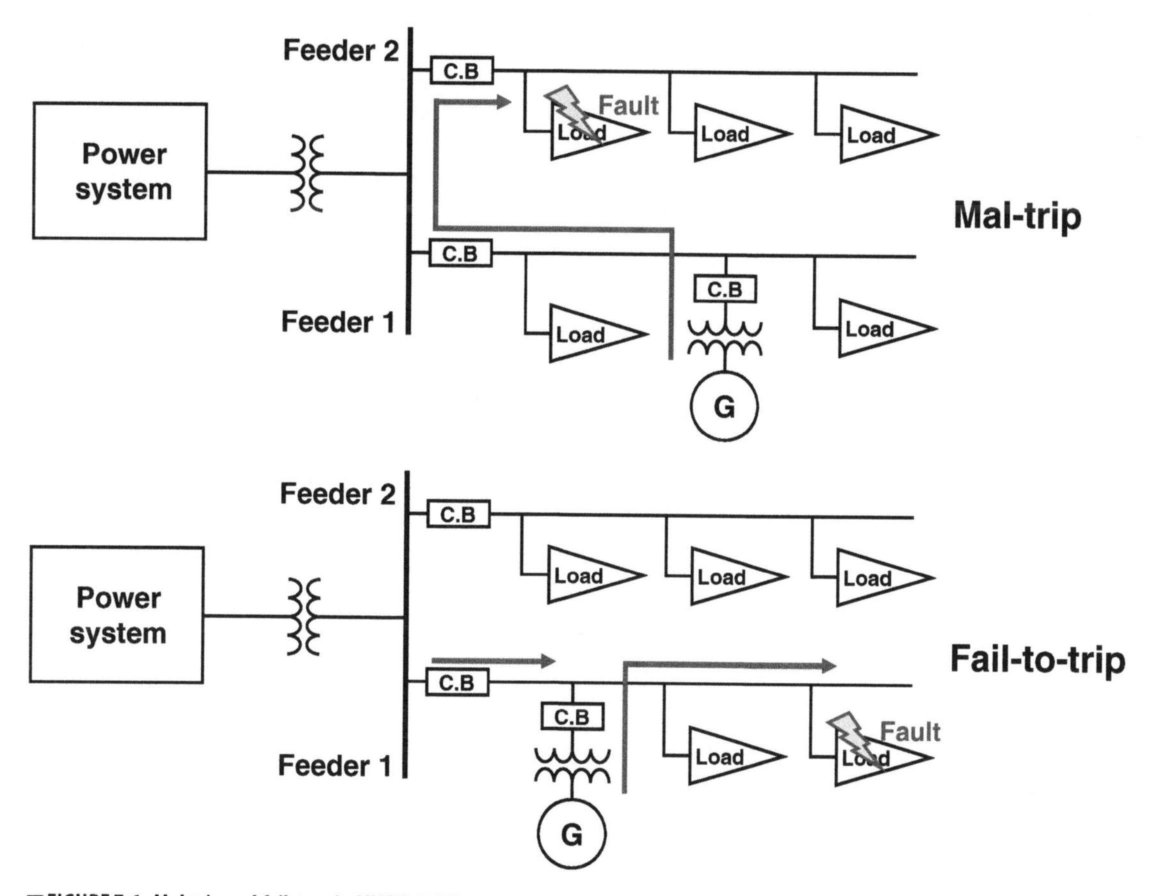

■ **FIGURE 7.6 Mal-trip and fail-to-trip (CIRED 2007).**

of overcurrent protection at the definite feeder, and will disconnect it with the system unnecessarily.

Fail-to-trip occurs for downstream faults and in this case the fault current is mainly composed of the current originated from the DER unit. Therefore, the fault current through the overcurrent protection remains passive and the faulty feeder will not be disconnected from the system [19].

7.4.2 **Hosting capacity**

Let us consider a network without integration of DER that can be modified to consider integration with DER. In order to establish the maximum allowed penetration of DER into an existing network, different indices can be formulated, which are the base for determining hosting capacity. In addition, corresponding to varying degrees of modification, the varying amount of DER can be accepted. The hosting capacity recognizes the degree of DER in the electrical power grid that can be received without putting the reliability or quality of power systems in danger [20]. The performance of the system can be seen from the hosting capacity definition. The estimation of the hosting capacity is repeated for each different phenomenon in power system operation and design and it depends on system parameters including structure of network or type of DER.

One more significant performance condition is that the DER unit must have a selective protection system with the other protections in the network. The selectivity constraint means that the DER should be linked in with the system as long as possible and if selectivity is not a prerequisite, the system and its adjustment will likely be kept simpler. This will allow the DER to be detached without taking into account other protection devices in the network [20].

7.4.3 Loss of coordination

In normal operating conditions, the protective devices correspond in a way that the main protection works before the backup can take some action. The DER integration to the ongoing power system increases the chance of short circuit fault. Depending on the original protection coordination settings along with the size, location, and type of DER, an uncoordinated condition may be acceptable. In such conditions, the backup protection operates before the main, and this results tripping of some loads.

In order to make operation stable, some modifications in coordination between the protective devices and DER should be made. Again it depends on the source of generation being protected because the grid interfacing technology differs for every source, like power electronic converters, induction generators, and synchronous machines.

Here we have developed a generalized method that sets the penetration limit of DER in terms of range, position, and equipment from loss of coordination viewpoint [21]. The procedures involved are summarized in following steps and a flow chart is given in Fig. 7.7.

1. In a given system, first define the different coordination paths.
   A coordination path can be defined as a set of protective devices positioned along a circuit path starting from the main feeder breaker to the most downstream protective device. Most of the fuses selected, which are lateral or sublateral, should be similar and lead to a limited number of different coordination paths. A single coordination path might represent many laterals utilizing the same types of fuses.
2. Construction of coordination charts and organizational study of different protection paths may lead towards better operation.

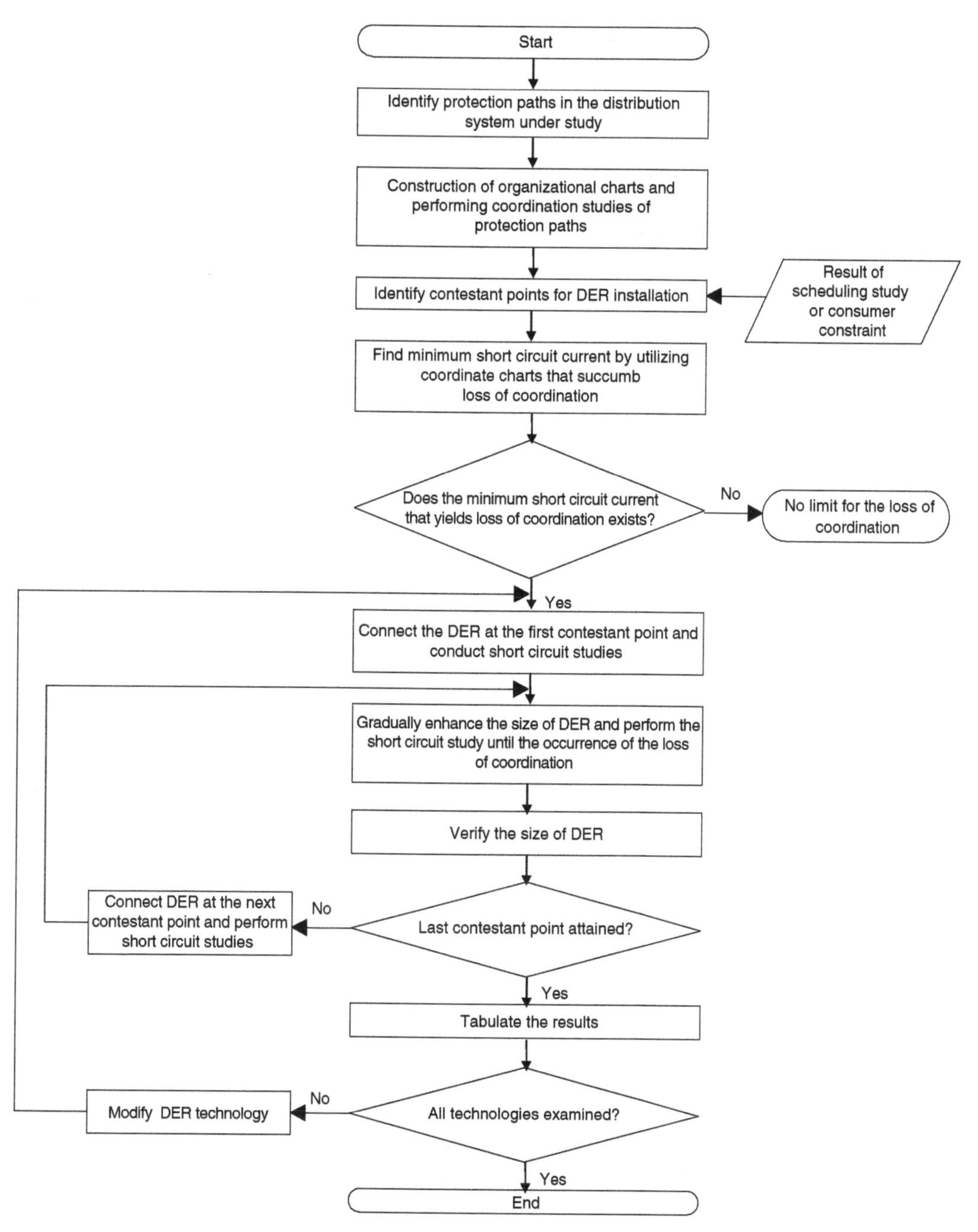

■ FIGURE 7.7 A generalized method to assess the loss of coordination. *(Source: CETC-Varennes 2007.)*

3. The least short circuit current at which the loss of coordination may occur among all protective coordination paths is to be examined. This current may be the junction between coordination curves of two successive protection devices. In a case where there is no junction between the coordination curves, the minimum current may not continue living. Such a condition will not limit the installed DER to breach the system coordination and work efficiently.
4. Describe the contestant points at which the DER may be installed. The penetration limit will be calculated for these specific points. These contestant points may be obtained from a planning study to resolve the finest location of DER to reduce system losses and improve the voltage profile, or it might be stated by the customer.
5. Here the simulation is applied at first contestant point and the capacity of any one of the DER and its interfacing transformer is increased gradually. It will be continued till reaching the minimum short circuit current for the loss of coordination and the specific DER size will be recorded. It should be noted that increasing the size of DER and its interfacing transformer inherently increases the short circuit mega volt-ampere (MVA) capacity of the combined DG/transformer set.
6. Keep on doing the same practice as described in step 5 for other contestant points. This will help in estimation DG impedance estimation using its rating.
7. Tabulate the results.
8. The same procedures could be repeated by altering the DER technology. Consider the cases of synchronous, inverter-based or induction generators, and then repeat steps 5 to 7 by tabulating individual results.

7.4.4 Protection issues of DER

Low voltage distribution systems including DER are divided into limited protective regions which are enclosed by a network or the equipment consisting of transformers, generators, buses, or loads. The well-known criteria 3S provides the requirements for basic design of a distribution protection system and holds good for directional, distance, and differential protection schemes. The abbreviation 3S stands for selectivity, sensitivity, and speed of protection system. Sensitivity refers to the ability to recognize an anomalous state that exceeds a small threshold value. The selectivity makes the protection system target oriented that disconnects only the faulted part of the system at any abnormal condition in order to minimize fault implications. Speed is important for any protection scheme, and the devices should respond to power system disturbances in the fastest possible time in order to avoid risk and keep stability check [22].

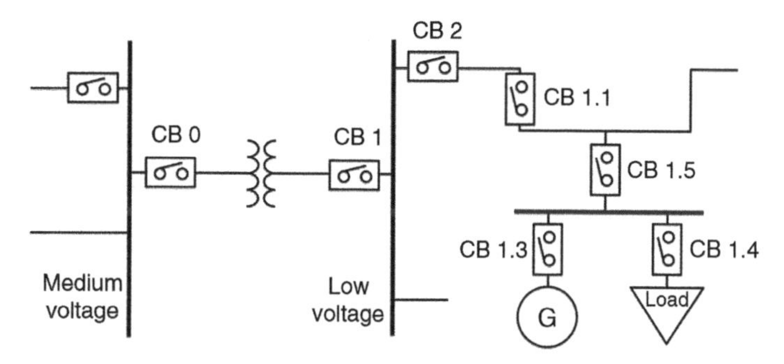

■ FIGURE 7.8 Protection zones of MV and LV circuit breakers.

Fig. 7.8 shows different protection zones of medium voltage (MV) and low voltage (LV) line connected with circuit breakers (CB). CB 0 and CB 1 are protecting the MV and LV lines. The DER generation (labeled G) and the load are connected to the system through switch board.

If a fault occurs between the line and the main electric grid, the MV protection system clears the fault. However, in case of sensitive loads the DER has to be isolated by the system circuit breaker (CB2) as fast as possible. The required time could be 70–80 μs. Also the DER has to be isolated from the main electric grid by CB1 if the MV protection is not operated in the required time. A fault detected by an OCR can be challenging in the case of a DER consisting of power electronics equipment because they have fixed fault current restrictions.

Power electronics devices are capable of supplying a smaller rated current to a fault but the rotary machines specifically designed to supply high fault currents. The fault current depends on the sources of short circuit electrical power. The grid electrical power has generally higher short circuit strikes compared to small DER integrated to normal distribution system. As a result, the fault current sensed by protection relays is less than the fault current sensed when the distribution system is connected to transmission line electric grid in an islanded mode [23]. But still such numbers are lower than a short-circuit current supplied by the main electric grid. In this case a directional OCR with circuit breaker is only practical solution if current is used for the fault finding. Therefore, the setting has to be constantly supervised and modified when DER generation experience changes depending upon the source of generation.

Fig. 7.9 shows a DER monitoring and control system, which is functionally composed of the components with different operational modes. The boxes shown within the interconnection system are associated with power equipment functions whereas the circles represent monitoring and control

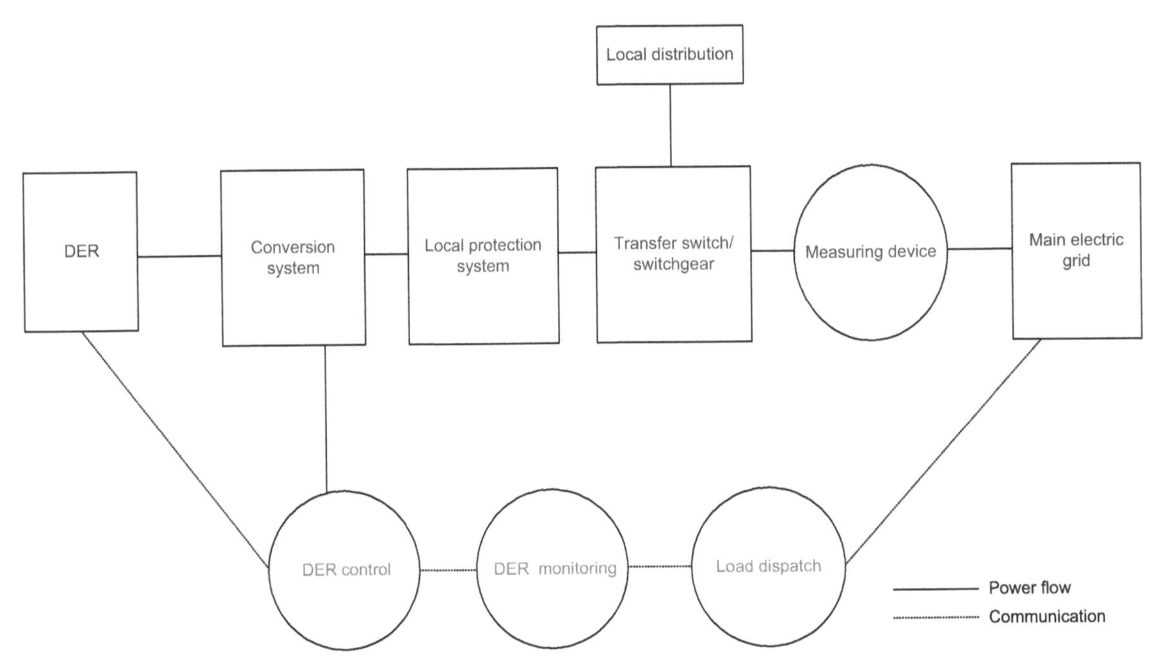

■ **FIGURE 7.9 DER monitoring and control.**

functions. The power equipment functions change the electrical energy from one form to another in order to make it compatible to use. Like the storage devices provide a direct current (DC) source or the microturbines give power at high frequency. The quality issues are also checked by the conditioning devices connected with power equipment to supply a clean AC power [24].

The protection and control monitors the common point of coupling with the overall flow of electrical power supplied by DER. The main function of these blocks is to disconnect the DER connected with main electric grid under abnormalities. By following the IEEE standards, which provides an opportunity for developing procedures and maintenance support, risk can be minimized.

7.5 PROTECTION SCHEMES FOR DISTRIBUTION SYSTEMS WITH DER

A distribution system with DER integration requires a modified protection system because of the many challenges caused by small scale generation in the form of DER. Unusual changes occur during transition from interconnected to islanded network operation. DER applications are reliant on variable primary energy resource, which modify the system working conditions and put down the efficiency of the conventional protection system all through faults occurring.

Presently, many electrical power research institutes, companies, and energy managers are working together for the reliable deployment of DER technology with grid integration. According to recent reports, US military bases are presently looking for consistent and protected energy resources, as these most popular bases are powered by public electrical grids, which in some occasion lead to as many as 300 power outages per year [25].

These disruptions deteriorate the military and other executive institution motivation and create certain protection issues. Reliance on conventional energy supplies may be inadequate to perform important function by these institutions, as well as to other emergency support service agencies, hospitals, and networking data centers. The control and protection hardware are being advanced to permit the supplying of the consumers in island mode as well.

7.5.1 Islanded operation

Islanding is a state in which a part of power system consisting of one or more power sources and load is separated from the rest of the system. Such a state will remain active as long as the DER is capable to meet the load demand. Because of the sudden disconnection from the main supply, the coordination among the protective devices may be lost and lead to vulnerability. Due to the aforementioned facts and associated problems the DER disconnection is preferred if an islanding state is observed.

Islanded operation (IO) can be intentional or unintentional depending upon the condition of the system. For unintentional IO it is planned in advance whereas the system and equipment is designed to deal with the situation. The DER is then capable of controlling the voltage and frequency in IO. An intentional IO regularly exists in manufacturing plants where the procedure has additional energy that can be used to produce electrical power [26].

There are two approaches to deal with islanding operation. The first one deals with the inverter control modes and the other is the synchronous generator control modes [27]. In inverter control mode the DERs have an inverter as a generator interface and they can operate either as a current source or as a voltage source. Normally the voltage source inverter (VSI), with added control loops, can make current control source if required. Due to the voltage and frequency references all the inverters can be operated in active and reactive power (known as PQ) mode.

The VSI controls the magnitude as well the phase of output voltage and the inductor circuit determines the flow of active and reactive power from

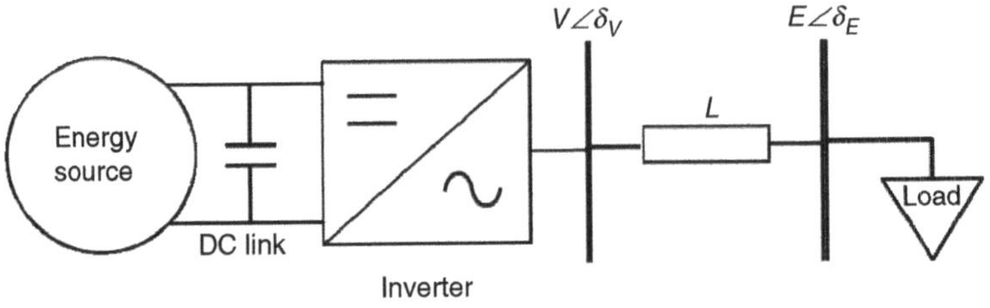

■ **FIGURE 7.10 Inverter mode.**

the DER. Fig. 7.10 shows the basic inverter interfaced generation system. Here the vector connection between the voltage of the inverter, *V*, and the generation voltage, *E*, along with the reactor, *L*, determines the flow of active and reactive power form the DER to the main electrical grid [28].

In synchronous generator control mode the frequency of the DER is maintained tightly; however, following a disturbance the frequency may change quickly due to the low inertia present in DER. Therefore, the control of the main electric grid is very important in order to protect the frequency of the DER during islanded operation [29].

7.5.2 The protection equipment for DER networks

There is a large selection of equipment used to protect the DER networks. The rigorous type of protection used depends upon the scheme of the system and the level of the voltage. Fault clearance is a critical job in power system network. If we interrupt the circuit directly when fault occurs, it reduces the substantial smash up to the power system apparatus and also assets. The DER protection is not dependent on a precise standard but the resource availability, the pattern of scheme, and the utility infrastructure indicates the state of action. The most commonly used devices for DER protection are:

1. OCRs;
2. Reclosers;
3. Sectionalizers; and
4. Fuses.

7.5.2.1 Overcurrent relays

The relays give excellent protection for generators, transformers, and buses. These are not really suitable for feeders and transmission line protection. This is because the current transformer (CT) would need to be located at either end of the line and the secondary leads conducted over the relatively long distances. This is expensive and, more importantly, the impedance of the secondary conductor could give rise to serious inaccuracies. In a relaying system, the differential principle is used for different applications. The very common form of feeder protection is to protect it from overcurrent. For example, if a fault occurred in a feeder, it would give rise to the overcurrent in the line. An OCR connected closer to the breaker would detect the fault and could be set to open the breaker. The standard device number for an instantaneous OCR is 50.

The principle of operation of this type OCR is simple, the CT may use the current in the primary line and the CT secondary passes this current through the coil of electromagnet. The resulting magnetic force pulls the hunched armature that is a clapper against a restraining spring. If the current input to the relay is above the preset pick up level, then the relay contacts will close and energize the tripping circuit. The pickup level can be adjusted by taps on the coil and also by adjusting spring tension.

Another type of instantaneous OCR is the plunging type. In this case the electromagnet pulls the plunger up against the force of gravity. Again the pickup level can be preset by adjusting the taps and also by adjusting the position of the core. When current accedes the pickup level, the instantaneous relay will operate within about 50 ms, that is, about 3 cycles, and energize the tripping circuit to open its associated breakers.

In order to avoid fault consequences the operating pickup must be set to a very high level. This would be adequate for preprotection of severe faults that could affect the power system stability.

One of the recent advancements in OCR for power distribution includes the protection, automation, and control (PAC) system with more than 25 elements added for quick response and advanced-level protection. Fig. 7.11 shows the PAC system model, which provides feeder protection as well bay control along with eight setting profiles. With microprocessor based control and power quality monitoring system, certain effects of sag, swell, transients, and harmonics are detected and recorded.

The protection against low-level faults is achieved by time OCR of standard No. 51. This is a typical type time OCR, and its main components are electromagnet, the operating coil, the rotating disc, pin contacts, and time dial. The secondary current from the CT is passed through the operating coil, which is wound around the central leg of the electromagnet. It sets up this magnetic circuit, the flux passes through the magnetic disc and then returns through the disc again to the outer legs of the electromagnet. In this condition the disc will not rotate as these two fluxes are in phase. However, by placing a shorting coil at one outer leg, a phase displacement

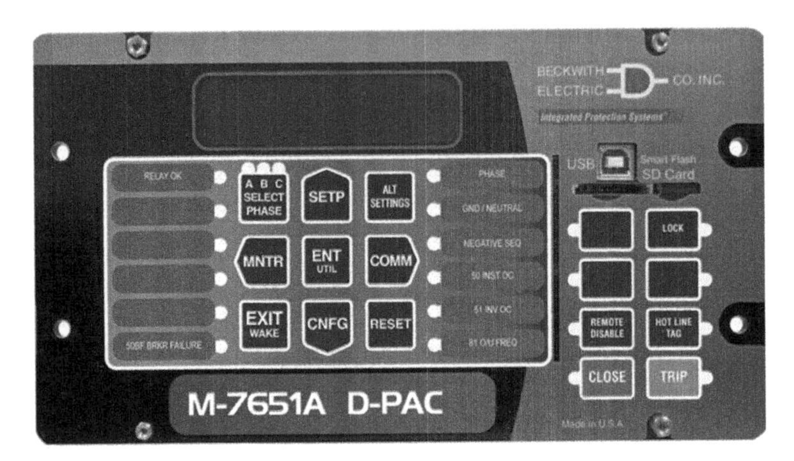

■ **FIGURE 7.11 PAC for distribution protection.** *(Source: Courtesy of beckwithelectric.com.)*

occurs and the flux here will now lag the inner one and cause the disc to rotate.

The disc is normally held stationary by a retaining spring. Only when the sufficient current passing through the operating coils, the disc will start to move. That is the pickup level, and the magnitude of the current in the operating coil is proportional to the primary current along the feeder, which is being protected. Therefore, the greater the primary current, the greater the operating current will be, and the greater the flux, the faster the disc will rotate. Similarly, the higher the level of the fault current, the quicker the operation of the relay will be. These conditions will finally force the circuit breaker to operate.

7.5.2.2 ***Reclosers***

Reclosers are protective devices used to recognize phase to phase, and phase to ground-faults under overcurrent condition. It breaks off the line if the state of overcurrent persists after a programmed instance, and recloses mechanically as the fault is removed. It will remain open after a fixed number of actions, if the fault is still in the line, and will close upon complete isolation of the fault. In overhead system distribution lines 70–80 % of the faults are of momentary nature and last a few seconds or cycles.

Hence, the reclosers, with its opening/closing characteristic, protect a supply circuit being left out of service for momentary faults. Normally, the reclosers are intended to have three open/close operations at maximum and, after these, a last open operation to lock out the succession. One added ultimate operation by instruction manual means is generally permitted. The counting means record operations of the phase or ground-fault units which can also be commenced by externally controlled devices when suitable contact ways are available.

There is a variety of solid dielectric reclosers designed to provide overcurrent protection for distribution system. The structure is composed of epoxy summarized vacuum disrupter attached through a impel rod to a high-speed magnetic activator. Such an arrangement provides excellent insulation properties and secured void-free structure.

The operating time versus current curve of reclosers normally includes three curves, one is the fastest and the other two are delayed. The new recloser devices have a microprocessor-based control, which helps an operator to produce any curve that suits the coordination requirements for both ground and phase faults. This permits encoding of the individuality to understand the consumer's precise needs without changing the apparatus.

Synchronization with other safety devices is significant in order to guarantee that, when a fault arises, a minimal part of the circuit is detached to reduce interruption of supplies to consumers. Normally, the time feature and the progression of action of the recloser are chosen to organize the instrument upstream towards the supply.

At present the new generation recloser device, known as Recloser Control M-7679, is compatible and independent pole with more than 30 elements for best protection, maximal automation, and communication technologies are inserted. Fig. 7.12 shows the recloser and its control mechanism. The device operates with all other protective components in a coordinated way and functions promptly with any abnormality.

7.5.2.3 **Sectionalizers**

Sectionalizes automatically cut off faulted segments of distributed lines prior to an upstream circuit breaker or a recloser interruption. Such a device is normally installed downstream of a recloser. While sectionalizes have no capacity to rupture fault current, so they are used in distribution protection with a supporting device that has fault current breaking capacity. The sectionalizers tally the number of events of the reclosers throughout electrical disturbances, and the number of recloser openings is preselected. Upon reaching the preselected value, the sectionalizer opens and isolates the faulty section of a line. This allows the recloser to shut and reinstate provisions to those regions, which are free of faults. If the fault is momentary, the operating instrument of the sectionalizer is reset.

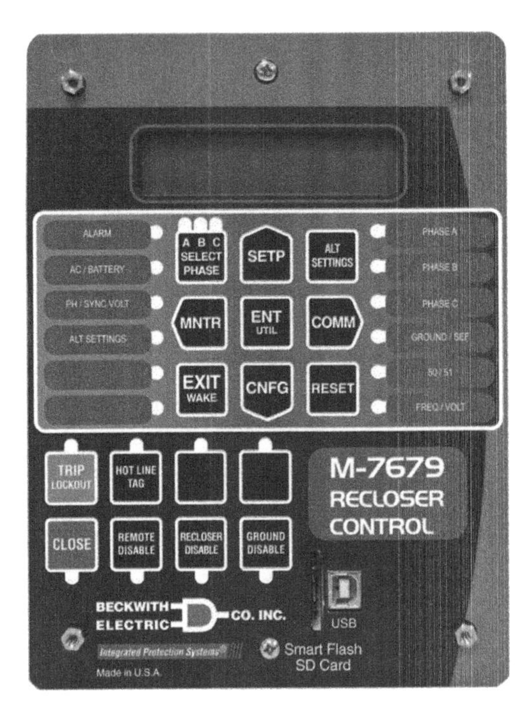

FIGURE 7.12 Recloser and control.
(Source: Courtesy of beckwithelectric.com.)

Sectionalizers operate with hydraulic or electronic mechanism and are constructed to operate for both single and three phase systems. The sectionalizers are coordinated easily with other protective devices as they do not have time–current characteristics. It also brings an added step of protection without any coordination to the protective scheme. Fig. 7.13 shows a three-phase sectionalizer with control mechanism and sectioning rods.

7.5.2.4 Fuses

Fuses are the simplest and most commonly used overcurrent protection device. This is a primary device with slim wire enclosed in a covering or glass which joins two metal parts. It holds a fusing element subjected to the flow of current and blows when the current exceeds a preset value. The physical substitute of a wire is necessary if it blows out due to any abnormal state.

The correct function of a fuse usually depends upon the opening of circuit immediately after the high current interventions. This happens due to the heating effect of high current on the fusing element. However, the majority of fuses used in distribution systems function on the discharge code explicitly having a tube to detain the arc by the inside enclosed with

■ **FIGURE 7.13 Sectionalizer.** *(Source: Courtesy of http://www.entecene.co.kr/eng.)*

deionizing fiber. In the occurrence of a fault, the interior fiber is excited when the fusible element melts and produces deionizing gases which amass in the tube.

When the fault occurs, the high current flow produces an arc which is compressed and expelled out of the tube. The escape of gas from the nail clippings of the tube causes the particles that maintain the arc to be debarred. Similalry, the arc is quenched when current reaches zero.

This is a primary protecting device with slim wire enclosed in a covering or glass which joins two metal parts. The wire melts while extreme current flows in a circuit. There are different types of fuses that depend on the voltage at which it is to work. The occurrence of deionizing gases and the commotion within the tube makes sure that the fault current is not established again after the current passes through zero point. The function zone is limited by two factors, the lower limit is based on the minimum time required for fusing of the element and the upper limit is determined by the maximum total time that the fuse takes to clear the fault. There are different international standards to categorize fuses according to their electrical characteristics. The commonly known categories are voltage and current ratings, time–current relationships, developed features, and some other considerations.

7.5.3 Recent technological trends in DER protection

The organization and control of distribution networks is becoming advanced and changing due the use of new digital protective relay technologies. Broadly, deployment of supervisory control systems that gather grid data within a few seconds is too periodic to expose many of the power system disturbances that cause tripping of relaying circuits.

7.5.3.1 DER protection characteristics

The electricity infrastructure is mainly divided into three zones, generation, transmission, and distribution, whereas the protection is mandatory for all the three zones. The goal of all utility companies is to deliver electricity in a protected, consistent, and efficient way. The relaying system in a protection zone makes the overall system of electricity safe and secure. The protective relaying system comprises current transformers, circuit breakers, and the relays.

The job of the DER protection system is to sense and react to all emerging faults with minimum or no loss to the consumers and equipment. The relaying system does not operate at normal power system operations and also does not limit the capability of a system to carry load currents. However, in electrical systems no device works under ideal conditions, and if the equipment is trying to accomplish some of the objectives described previously it may compromise on others. The limit of this concession is the criterion used to determine positions for fault-interrupting devices, and the sensitivity and working speed of the fault sensing devices. Such different relaying systems are applied in distribution system for overcurrent protection.

7.5.3.2 Overcurrent protection scheme

The overcurrent protection scheme is used to protect the distribution lines of electric grids integrated with DER. This protection scheme is further classified into two categories, the phase overcurrent protection and the ground overcurrent protection. The relays used in such schemes could be directional (operating for in-front events) and nondirectional (will operate for all) depending upon the mode of distribution. The nondirectional relays are mostly used for radial distribution systems.

In all the cases, the feeder protection starts at the electric grid with feeder control mechanism (a breaker or recloser). Reclosers are usually prepared with the inverse time overcurrent device that senses the faults and sends signal to the breaker, and then after a predetermined setback, it operates by reclosing the breaker. All the protective devices should be in sound coordination with

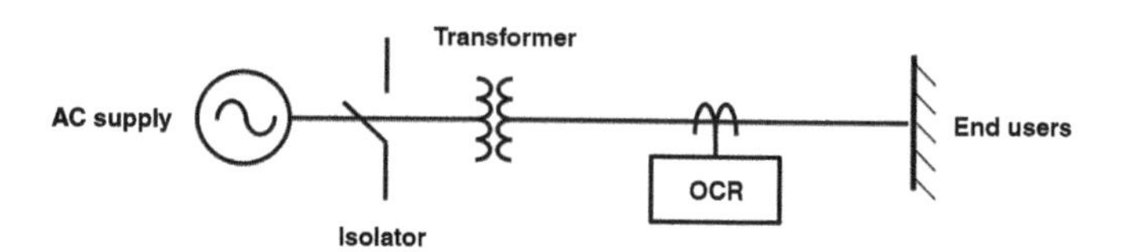

■ **FIGURE 7.14 Overcurrent protection scheme for DER.**

the upstream and downstream links. The upstream link is the high voltage transformers, and the downstream are the protective equipments.

Fig. 7.14 shows the schematic arrangement for overcurrent protection of DER with coordinated protection. The generation is fed to the transformer via isolator switch, and the output power from transformer passing the OCR is supplied to end-users.

7.5.3.3 ***Fuse operating scheme***

Utility companies generally apply two strategies in distribution lines for fuse operating schemes, which are known as fuse-saving and fuse-blowing. The fuse-saving strategy under operation benefits both the company and the consumers. In this scheme, the service is restored automatically and the technical handling is no more required for replacing the fuse. However, one of the drawbacks of fuse-saving schemes is the limited coordination at high currents, as it just coordinates with the next device to trip and functions at the same time. This results in temporary electrical power failure, thus the utility companies have now replaced this scheme with fuse-blowing.

Fuse-blowing schemes are used to limit the total reclosing cycle for the main feeder. This is employed for sensitive loads in distribution feeders, where considerable interruption occurs if the line is momentarily de-energized. One more application of this scheme is commercial and urban load centers where the number of reclosing cycles could result damage to equipments or individuals [30].

7.5.3.4 ***Voltage control scheme***

A direct voltage control method without the inner current control loop can provide a faster and a more robust performance; however, it is unable to limit the current in the DER unit during abnormal conditions. Thus, a fault condition can either trip out the DER unit or damage its power electronic components. A voltage-controlled–distributed energy resource (VC–DER) unit is also prone to dynamic overload conditions since it lacks the capacity to rapidly limit the output power. This is due to the inherent characteristics of the controlling devices.

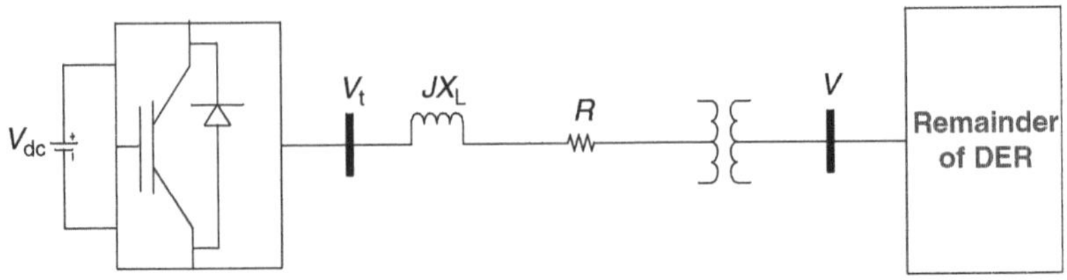

■ FIGURE 7.15 VC–DER unit.

The overload protection scheme limits the output power of DER. This scheme identifies overload conditions based on voltage measurements and limits the output power by assigning suitable voltages to the terminals of the interface voltage-sourced converter (VSC) of the DER unit. Fig. 7.15 shows a VC–DER unit, characterized by a DC voltage source, a VSC, and a series RL filter. The unit is interfaced to the DER microgrids at the point of connection through a step-up transformer. The block denoted by "Remainder of Microgrid" comprises voltage and power-controlled DER units and other components, for example, distribution lines, loads, and capacitor banks [30].

7.6 CONCLUSIONS

This chapter focuses on various protection schemes for distribution systems with DER integration. Since the advancement and rapid utilization of DERs, protection systems have started to change from normal to smart operations. The protection engineers are now working on different control methodologies that can dynamically be applied to promote the reliability and security of electricity supply. Nowadays the capabilities of the advance protective devices have evolved with the addition of more powerful and intelligent components. Electricity is a basic necessity of life and a symptom of modernization, therefore, it is necessary to distribute it efficiently using smart grid or DER integration. Also the protective schemes used in transmission line may be used in the protection schemes for distribution line in future.

This chapter covers a range of DER protection scenarios along with different power system disturbances that weaken the system and cause breakdowns. Also, the cause and effect of each disturbance, advanced protection schemes, and the sensitive equipments used in DER protection is mentioned. The DER integration to the main electric grid creates various overcurrent problems, these problems are discussed in detail by referring the previous work. It is definite that DER is going to be fully operational in the future and it will be the best alternative to meet the growing electricity demands. The electrical energy generation, transmission, and distribution always need coordinated protection, which ensures the stability of power system, and with the integration of DER, the protection system needs to be modified accordingly.

REFERENCES

[1] Basak P, Chowdhury S, HalderneeDey S, Chowdhury SP. A literature review on integration of distributed energy resources in the perspective of control, protection and stability of microgrid. Renew Sust Energ Rev 2012;16:5545–56.

[2] Jenkins N, Ekanayake JB, Strbac G. Distributed generation. London, UK: Institution of Engineering and Technology; 2010.

[3] Tanaka K, Yoza A, Ogimi K, Yona A, Senjyu T, Funabashi T, Kim C-H. Optimal operation of DC smart house system by controllable loads based on smart grid topology. Renewable Energy 2012;39(1): 132–139.

[4] Goya T, Senjyu T, Yona A, Urasaki N, Funabashi T, Kim C-H. Optimal operation of controllable load and battery considering transmission constraint in smart grid. The 9th International Power and Energy Conference, no. P0378, Suntec, Singapore, 27–29 October 2010, pp. 734–739.

[5] Che L, Khodayar ME, Shahidehpour M. Adaptive protection system for microgrids: protection practices of a functional microgrid system. IEEE Electrific Mag 2014;2(1):66–80.

[6] Dugan RC, McGranaghan MF, Santoso S, Beaty HW. Electrical power system quality. 2nd ed. New York: McGraw-Hill, 2004.

[7] Comech MP, Gracia MG, Borroy S, Villen MT. Protection in distributed generation. In: Gaonkar DN, editor, Distributed generation, ISBN:978-953-307-046 9; 2010.

[8] Javadian SAM, Haghifam M-R, Bathaee SMT, Fotuhi Firoozabad M. Adaptive centralized protection scheme for distribution systems with DG using risk analysis for protective devices placement. Int J Elec Power 2013;44(1):337–45.

[9] Jiayi H, Chuanwen J, Rong X. A review on distributed energy resources and microgrid. Renew Sust Energ Rev 2008;12:2472–83.

[10] Institute of Electrical and Electronics Engineers, Inc. IEEE Standard 1547.7. Guide to conducting distribution impact studies for distributed resource interconnection; 2013. Available from: http://www.techstreet.com/ieee.

[11] COMBIFLEX Generator Protection – Application Guide, 1MRK 502 003-AEN. issued on June, 2004.

[12] Shahidehpour M, Khodayar ME, Barati M, Campus microgrid: high reliability for active distribution systems. Proceedings of IEEE Power Energy Society, General Meeting, July 2012, pp. 1–2.

[13] Ulushki RW. Creating smart distribution through automation. EPRI, USA, PAC, March, 2012.

[14] Yeo S-M, Kim C-H. Analysis of system impact of the distributed generation using EMTP with particular reference to voltage sag. KIEE Int Transact PE 2004;4-A(3):122–8.

[15] Aggarwal S, Gimon E, et al. Trending topics in electricity today: getting ready for distributed energy resources. Energy Innovation, Policy &Technology LLC; January 20, 2015.

[16] Gamit V, Karode V, Mistry K, Parmar P, Chaudhari A. Fault analysis on three phase system by auto reclosing mechanism. Int J Res Eng Technol May-2015;04(05):292–8.
[17] Bollen MHJ, Häger M. Power quality: interactions between distributed energy resources, the grid, and other customers. Electric Power Qual Utiliz Mag 2005;1(1).
[18] Antonova G, Nardi M, et al. Distributed generation and its impact on power grids and microgrids protection. 978-1-4673-1842-6/12/$31.00 ©2012 IEEE.
[19] Häger M, Sollerkvist F, Bollen MHJ. The impact of distributed energy resources on distribution system protection. EU-DEEP integrated project, STRI AB, Ludvika; 2006.
[20] Deuse J, Grenard S, et al. Effective impact of DER on distribution system protection. CIRED 19th International Conference on Electricity Distribution, Vienna, May 21–24, 2007.
[21] Abdel-Galil TK, Ahmed EB, Elanien A, El-Saadany EF, Yasser AG, Mohamed Magdy A-RI, Salama MA, Zeineldin HHM. Protection coordination planning with distributed generation. Presented to Scientific Authority: Chad Abbey Natural Resources Canada (NRCan); June 2007.
[22] Oudalov1 A, et al. Novel protection systems for microgrids. TC2 Technical requirements for network protection; 2009.
[23] Mahat P, Chen Z, et al. A simple adaptive overcurrent protection of distribution systems with distributed generation. IEEE Trans Smart Grid 2011;2(3):428–37.
[24] Friedman NR. Distributed energy resources interconnection systems: technology review and research needs; September 2002, NREL/SR-560-32459.
[25] Mariah Energy Development Corp. Ellsworth, KS 67439, sighted April 2015.
[26] Strath N. Islanding detection in Power Systems. Licentiate Thesis, Department of Industrial Electrical Engineering and Automation, Lund University; 2005.
[27] Pecas Lopes JA, Moreira CL, Madureira AG. Defining control strategies for microgrids islanded operation. IEEE Trans Power Syst 2006;21(2):916–24.
[28] Kamel RM, Chaouachi A, Nagasaka K. Three proposed control techniques applied upon inverters which interfacing micro sources with the islanded micro grid. ISESCO J Sci Technol Vis 2011;7(11):76–84.
[29] Giri V, Illindala M. Microgrids and sensitive loads. Proceedings of IEEE Power Engineering Society Winter Meeting, IEEE Press; 2002, pp. 315–322.
[30] Girgis A, Brahma S. Effect of distributed generation on protective device coordination in distribution system. Proceedings of Large Engineering System Conference on Power Engineering, LESCOPE'01, July 2001, pp. 115–119.
[31] Etemadi AH, Iravani R. Overcurrent and overload protection of directly voltage-controlled distributed resources in a microgrid. IEEE Trans Ind Electron 2013;60(12).

Chapter 8

Lightning protections of renewable energy generation systems

Shozo Sekioka

Department of Electrical & Electronic Engineering, Shonan Institute of Technology, Japan

CHAPTER OUTLINE

8.1 Introduction 193

8.2 Lightning protection principle 196

8.2.1 Reduction 196

8.2.2 Suppression 197

8.2.3 Shielding 199

8.2.4 Lightning characteristics for lightning protection design 201

8.2.5 IEC international standard 206

8.3 Lightning protection for wind power generation systems 209

8.3.1 Lightning damage in wind power generation system 209

8.3.2 Lightning protection methods for wind turbine 211

8.3.3 Grounding resistance 212

8.3.4 Lightning protection of blade using receptor 215

8.3.5 Energy coordination of surge arrester/surge protective device 216

8.4 Lightning protection of wind farms 218

8.5 Lightning protection for photovoltaic power generation systems 222

8.5.1 Lightning damage in photovoltaic system 223

8.5.2 Lightning protection against lightning overvoltages in photovoltaic systems 223

8.5.3 Direct lightning flash to photovoltaic system 224

References 226

8.1 INTRODUCTION

The top of a blade of a wind turbine, with a couple of MW capacity, is more than 100 m high. In general, lightning tends to strike higher structures [1]. Accordingly, lightning often strikes a wind tower or a blade. Most lightning

Integration of Distributed Energy Resources in Power Systems

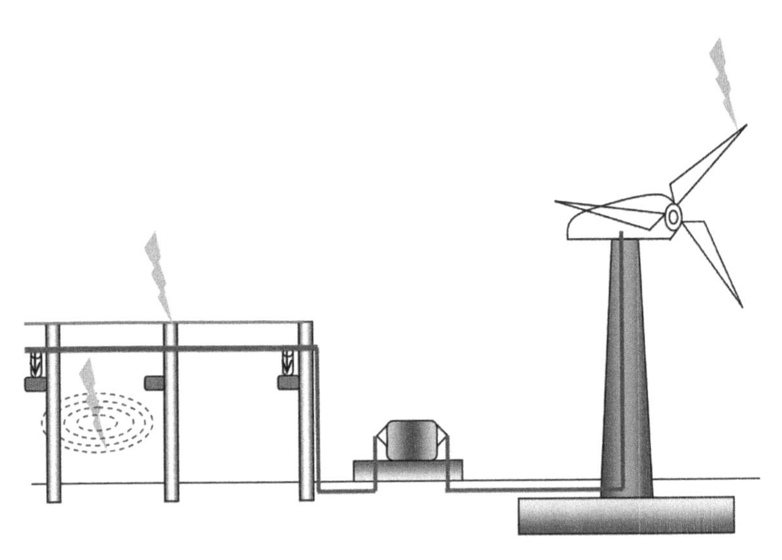

■ **FIGURE 8.1 Lightning events in a wind turbine generation system.**

currents have high amplitude and high frequency, and sometimes have large energy. The current generates a high voltage in the power apparatus and measurement and control systems in the wind turbine generation system, and causes damage to the apparatus, causing the system to malfunction. Distributed energy resources outside a house or building have a risk of lightning damage.

Lightning surges come into a wind turbine generation system consisting of single wind turbine and an overhead line as illustrated in Fig. 8.1. Lightning surges comes from an overhead line such as a distribution/transmission line or a telecommunication line, direct lightning strikes a wind tower or a blade, and ground potential rise caused by lightning hit to the ground or the tower.

The mechanisms that cause lightning damage are classified as follows.

1. Lightning overvoltage: high current with high frequency generates high voltage and causes breakdown.
2. Energy: lightning current with long duration has large energy and causes meltdown or burnout of conductors and a blade.
3. Lightning electromagnetic impulse: lightning current with high frequency generates impulse voltages. The impulse voltage on an overhead line caused by a nearby lightning is called lightning-induced voltage. The crest voltage and energy of the lightning-induced voltage are relatively low. Thus, the lightning-induced voltage is not main cause of serious lightning damage. The lightning electromagnetic impulses frequently appear in various networks, and sometimes cause

malfunctions. The insulation level of a power distribution line is relatively low. Therefore, power distribution lines have been targeted for the lightning-induced voltage in lightning protection design [2]. The use of surge arrester (Ar) is effective for the protection of the distribution lines against the lightning-induced voltages, and lightning outages have been decreased. Thus, the lightning protection for distribution lines mainly considers a direct lightning hit to the line [3].

Wind turbine blades sometimes explode and scatter due to the lightning flash. Serious damage to a wind turbine blade causes serious economic losses due to long-term loss of service and its cost to repair the blade [4]. Abnormal lightning such as winter lightning, which is frequently observed along the coast of the Sea of Japan [5–8], causes serious damage to wind power generation [4,9]. Many utilities in Japan gave up maintaining wind power generation because some lightning damage in winter seasons needed long time and much money to repair the blade during the winter. Therefore, the rational lightning protection design for wind turbine is very important to stable generation.

Home photovoltaic panels are set on the roof of a house or top of a building. Considering lightning sometimes strikes an antenna on a roof, lightning might strike a photovoltaic panel on a roof. Air termination systems such as a lightning rod or a shielding wire are used to prevent photovoltaic panels from damage caused by direct lightning flashes. The insulation level of photovoltaic system is very low in comparison with that of distribution line. Therefore, the lightning protection design for photovoltaic system is targeted for lightning-induced voltages [10,11]. However, lightning flashes with low lightning current might strike a photovoltaic panel on a roof of a house based on an electrogeometric model (EGM) [12] even if an antenna stands besides the panel. Direct lightning flash causes serious damage to photovoltaic systems. If low lightning current cause no damage in the system, direct lightning hit should be considered in lightning protection design of photovoltaic system. Photovoltaic generation needs large area to generate MW of electric power. The height of the system is low, but the probability of the lightning striking a photovoltaic panel is not small.

Wind turbine and photovoltaic systems related to the lightning protection should be designed on the basis of the IEC international standards 62305 series [13–16] and 61400-24 [17]. Lightning current parameters exceeding the values used in the IEC standards are sometimes observed. Thus, lightning phenomena are not well known. Moreover utilities want to make the cost for lightning protection as low as possible. Therefore, it is very important to know lightning protection methods for distributed energy resources, and to improve them. This chapter describes the lightning protection methods of wind turbine and photovoltaic systems, which are frequently damaged by lightning.

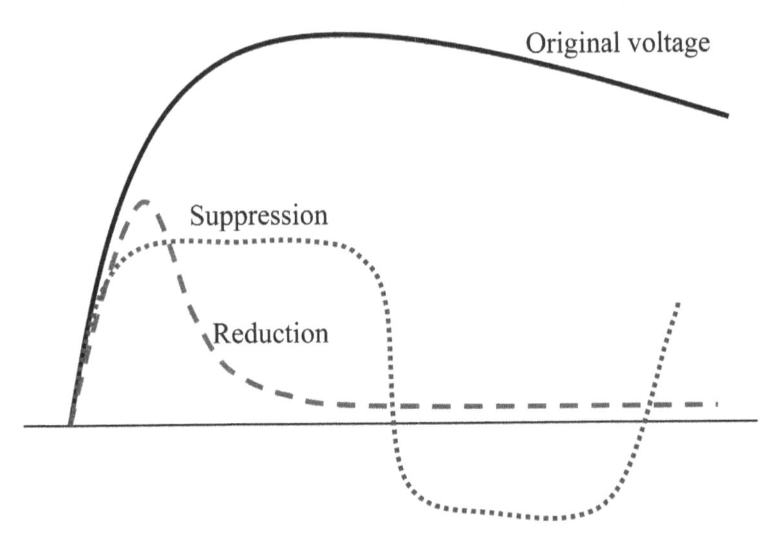

■ FIGURE 8.2 Reduction and suppression for lightning overvoltage.

8.2 LIGHTNING PROTECTION PRINCIPLE

The lightning protection of electrical equipments and signal transmission systems should be carried out using the following methods.

1. Shielding: air termination system such as a shielding wire or a lightning rod catches a lightning discharge. Overhead lines, buildings, and structures can be protected from a direct lightning strike using the air termination system.
2. Reduction: lightning overvoltage becomes lower by reducing impedance. Equipotential bonding method is a kind of reduction method because this bonding reduces voltage differences in grounding system.
3. Suppression: surge arrester and surge protective device (SPD) suppress lightning overvoltage.

Fig. 8.2 illustrates the reduction and suppression against lightning overvoltage.

8.2.1 Reduction

The reduction method makes lightning overvoltages lower by reducing impedances. The overvoltage is proportional to impedance. Therefore, breakdown occurrences are decreased by the reduction of the overvoltage. However, if lightning current is much high, the voltage in a system exceeds the breakdown voltage, and lightning damage occur. The reduction of grounding resistance is a typical reduction method. The reduction of the grounding resistance at a lightning striking point is very effective. The grounding resistance is approximately inversely proportional to the surface area of a grounding electrode. A large grounding electrode must be driven in the ground to obtain low grounding resistance. Terminal voltage V (V) is approximately given by

$$V = L\frac{di}{dt} + Ri \tag{8.1}$$

where L is the inductance of the down conductor (μH), R is the grounding resistance (Ω), and i is the injected current into the grounding electrode (A). L is approximately 1 μH/m.

Even if the grounding resistance is very low, high inductance of the down conductor raises the terminal voltage for high frequency current. Thus, the down conductor should be as short as possible.

8.2.2 Suppression

Zinc oxide varistor has an excellent nonlinear characteristic. Fig. 8.3 shows an example of voltage–current characteristics of an SPD using the ZnO element. The horizontal axis in Fig. 8.3 is logarithmic scale. Therefore, the SPD or Ar terminal voltage is almost constant for the variation of the injected current. The Ar and SPD are modeled by a cell in convenience. Thus, the terminal voltage with an Ar or SPD is suppressed under the value, and equipments can be protected from lightning overvoltages. The voltage is effectively suppressed at which an Ar or SPD is installed. The Ar or SPD

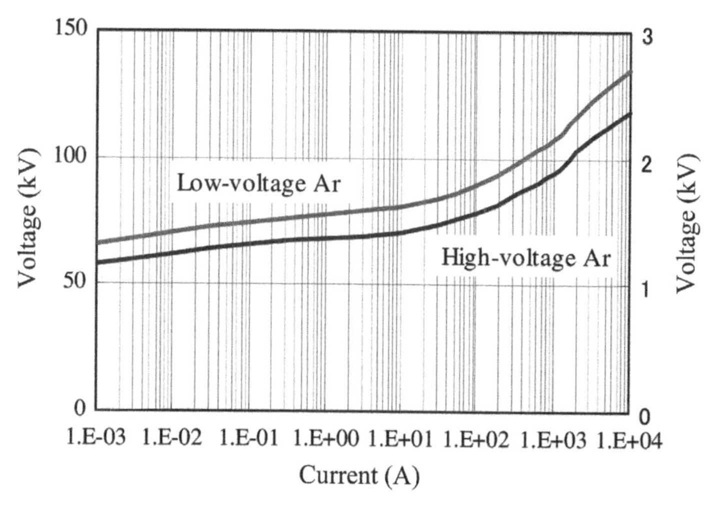

■ **FIGURE 8.3 V–I characteristic of SPD.**

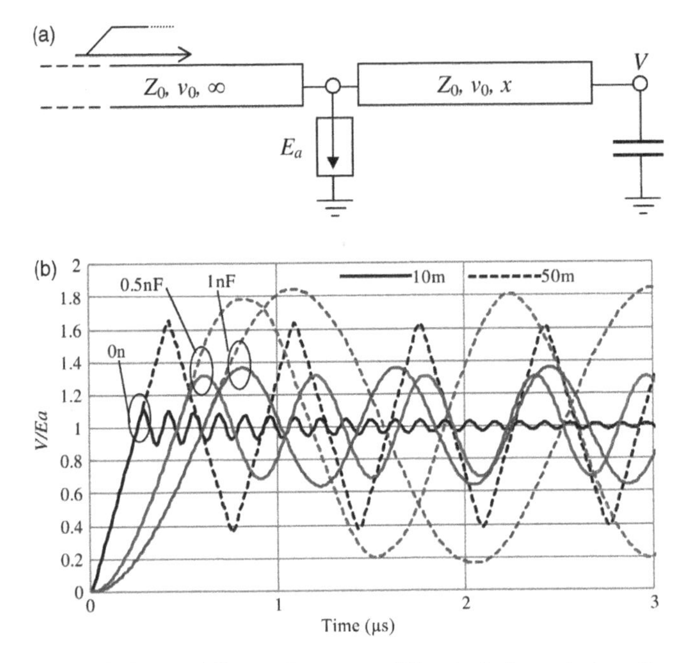

FIGURE 8.4 Influence of distance between an SPD and a device on device voltage. (a) Simulation circuit for an SPD and an apparatus. (b) Lightning overvoltage at line end.

cannot cover all the system, and many Ars or SPDs must be installed in the system. Fig. 8.4a illustrates a calculation circuit including an SPD and a device. Devices are often represented by capacitance in lightning surge analysis. The voltage of the SPD is constant E_a, the line characteristics are the surge impedance Z_0 of 500 Ω, the surge velocity v_0 of 300 m/μs, which is equal to the speed of light. Forward traveling voltage $e = at$ for $t < T_f$, where T_f is the rise time of the forward traveling voltage, propagates on a line from far end of the line, which is semi-infinite length of line. The line is terminated at the end of the line through a capacitance C, which represents a device. An SPD is connected to the line at the distance x from the capacitance. Fig. 8.4b shows simulation results of the terminal voltage V in cases of x = 10 m and 50 m. aT_f in the simulation is $2E_a$, and T_f = 1 μs. The electromagnetic transients program is used. The capacitance of the device is an important factor. The higher the capacitance, the terminal voltage becomes higher [18]. The amplitude of resonance voltage is dependent on the distance from the SPD. This phenomenon can be explained by the oscillation of a circuit consisting of line inductance and device capacitance [19]. The SPD/Ar must be installed as close as to the device to be protected.

The grounding method affects the effectiveness of SPD on the lightning protection level. The circuit for discussion consists of an impulse voltage generator, an SPD including a series gap, and a device for testing. Experimental results are shown in Fig. 8.5 [20]. Upper curve in the measurement results are applied currents. The following three cases are discussed on the influence of grounding system on overvoltage on a device.

1. Down conductor of the device is connected to the SPD, and is grounded at the SPD.
   SPD terminal voltage, V_1, and device terminal voltage, V_2, are effectively suppressed by the SPD, and these voltages become almost same.
2. Down conductor of the device is connected to the SPD, and is grounded at the instrument.
   V_2 is much higher than V_1. This difference is caused by the inductance voltage in the down conductor of the SPD due to the SPD discharge current; $V = L \times di/dt$.
3. Grounding of the SPD is isolated from that of the device.
   V_2 is much higher than V_1. The voltage difference in this case is higher than that in case (b). Thus, an SPD and a device to be protected should be commonly grounded at the SPD.
   The equipotential bonding method is useful technique as is clear from the experimental results. The grounding system related to down conductors must be carefully selected to suppress lightning overvoltages using SPDs effectively.

8.2.3 Shielding

For simplicity, the EGM proposed by Armstrong and Whitehead [12], which is used in lightning protection design of electric power system, is adopted in this chapter. Fig. 8.6 illustrates an example of the application of the EGM to a transmission line. The lightning striking distance, r_c (m), is dependent on the crest value of lightning current, and is given by:

$$r_c = kI_m^{\alpha} \tag{8.2}$$

where I_m is the crest value of lightning current (kA). k and α are empirically determined based on experimental or observation results, and many values are proposed [1].

Lightning flashes with higher currents generate longer lightning striking distance, and a shielding wire catches the lightning strike. There are three conductors: a shielding wire, a phase conductor, and the ground. Lightning strikes LS_{1f} with low current goes down outside of the exposure arc SB, and

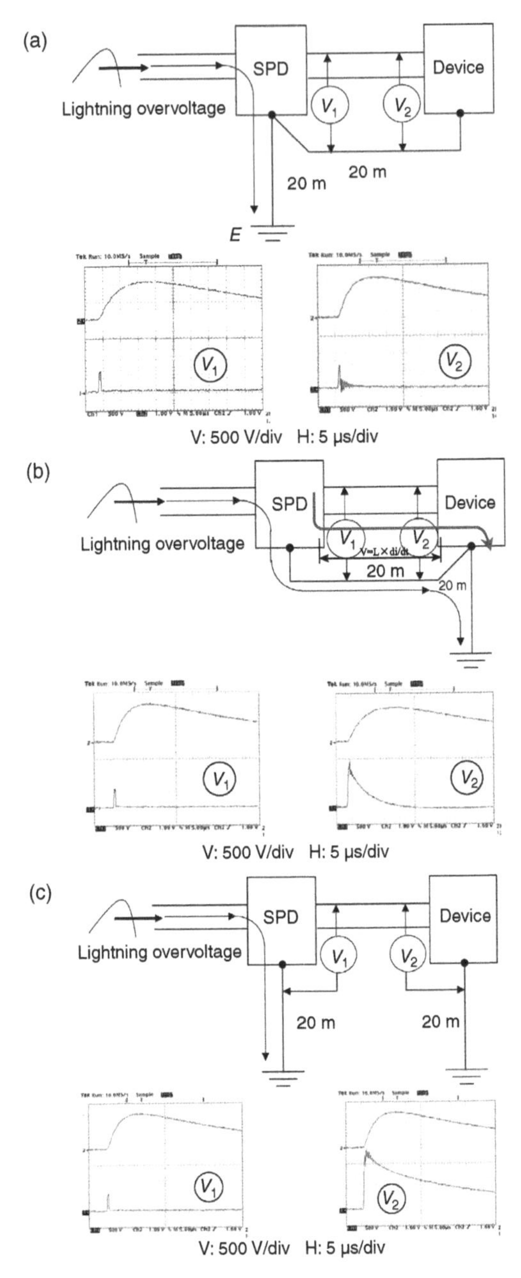

■ **FIGURE 8.5 Experimental study of SPD grounding methods [20].** (a) Down conductor of the instrument is connected to the SPD, and is grounded at the SPD. (b) Down conductor of the instrument is connected to the SPD, and is grounded at the instrument. (c) Grounding of the SPD is isolated from that of the instrument.

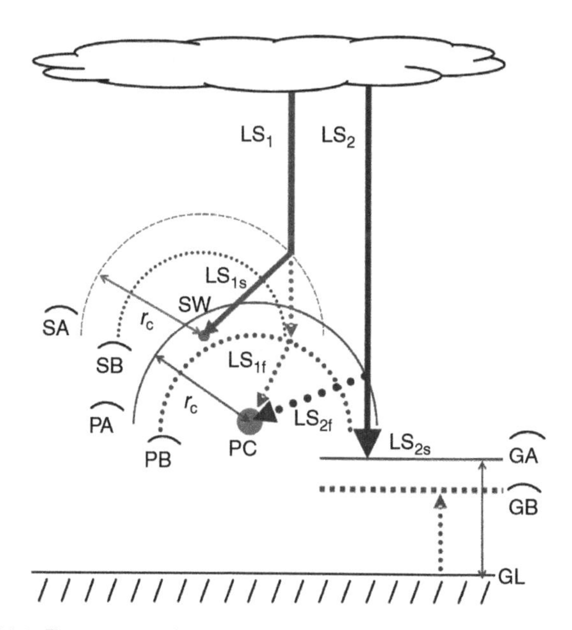

■ **FIGURE 8.6 Electrogeometric model.**

reaches the exposure arc PB. This means the lightning strikes the power conductor, namely shielding failure. In case of higher lightning current, the exposure arc SA catches the lightning strike LS_{1s}, and the shielding wire succeeds in shielding. Thus, lower currents tend to cause shielding failure. When lightning strikes are located outside an exposure arc of the power conductor, the lightning hits the ground (LS_{2s}). Line and structures can be prevented from direct lightning hit by properly arranging the air-termination system such as shielding wires or lightning rods.

8.2.4 Lightning characteristics for lightning protection design

Applied voltage waveform for lightning impulse withstand voltage test is a double exponential function $k(e^{-\alpha t} - e^{-\beta t})$. A typical lightning current waveform is shown in Fig. 8.7 [21]. Most lightning currents show the negative polarity. Two time scales, A and B, in Fig. 8.8 are the average current waveforms for negative first and subsequent strikes, respectively. The waveform is normalized by the crest value. The lightning current waveforms are similar to the test voltage waveform. Most of lightning currents have high amplitude, but wave duration is less than 1 ms. This type of lightning current mainly causes breakdown or electromagnetic interference.

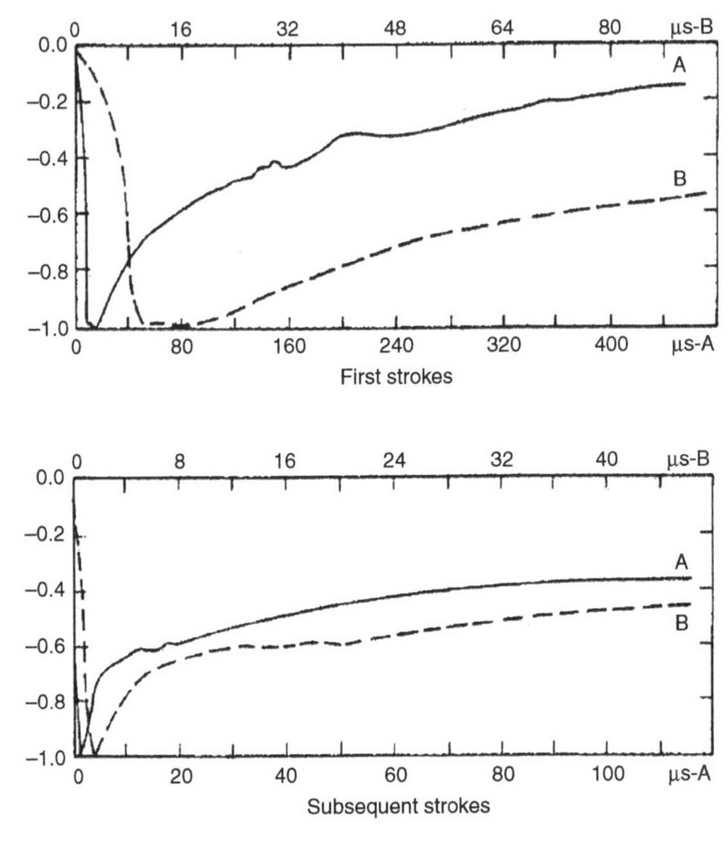

FIGURE 8.7 Average negative first and subsequent stroke waveform [21].

The Heidler function as a lightning current waveform model is frequently used in lightning surge analysis [22]. This model is given by:

$$I(t) = \frac{I_0}{\eta} \cdot \frac{(t/\tau_1)^n}{(t/\tau_1)^n + 1} \exp(-t/\tau_2) \tag{8.3}$$

A triangle wave shape is also often used in lightning protection design and surge analysis because this model is very simple and convenient [23].

Fig. 8.8 shows the cumulative statistical distributions (solid line curves) of return-strike peak currents for (1) negative first strikes, (2) negative subsequent strikes, and (3) positive strikes (each was the only strike in a flash) [21,24]. Subsequent return strikes are characterized by 3–4 times higher current maximum steepness (current maximum rate of rise).

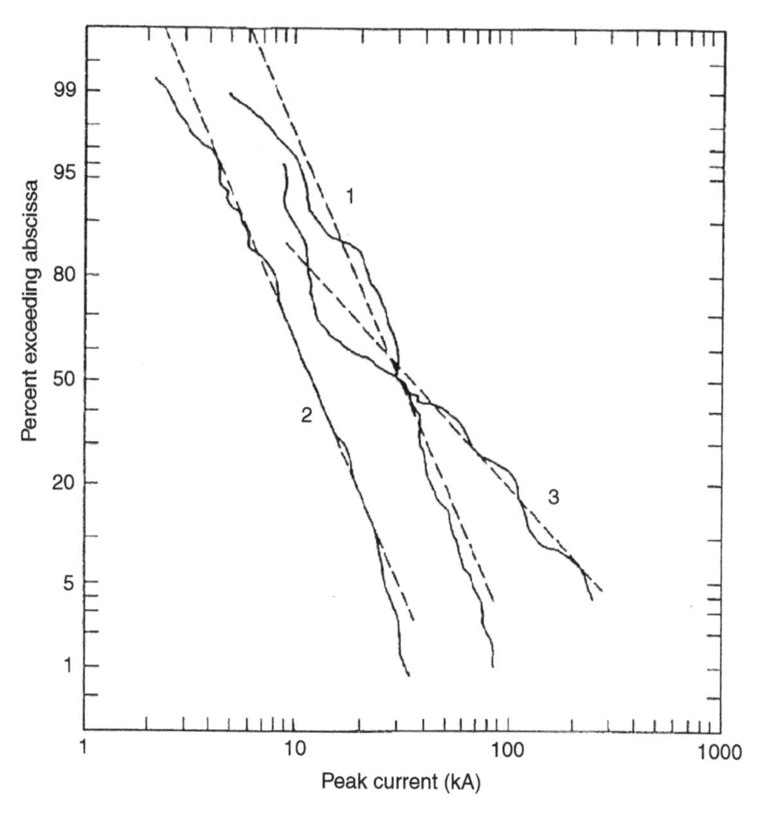

■ FIGURE 8.8 Cumulative statistical distributions of peak values of directly observed lightning currents (solid lines) and their log-normal approximations (dashed lines). (1) Negative first strokes, (2) negative subsequent strokes, and (3) positive first strokes [21].

The winter lightning current shows a variety of and complex waveforms such as long-duration waveform plus many pulses and having both polarities. The average peak value of the winter lightning current is approximately equal to that of the summer lightning. The winter lightning current has much longer wave tail duration. Consequently, the winter lightning sometimes shows large electric charge. This is equivalent to large energy, and causes serious damage in wind turbines. Fig. 8.9a illustrates a wind power station in Japan, and Fig. 8.9b shows an observation result of lightning current injected into the lightning tower in the power station [25]. The peak value of the lightning current is estimated to be about 240 kA. This value much exceeds that used in the IEC standard. The lightning current shown in Fig. 8.9b is one of giant lightning flashes. The current waveform is quite different from the summer

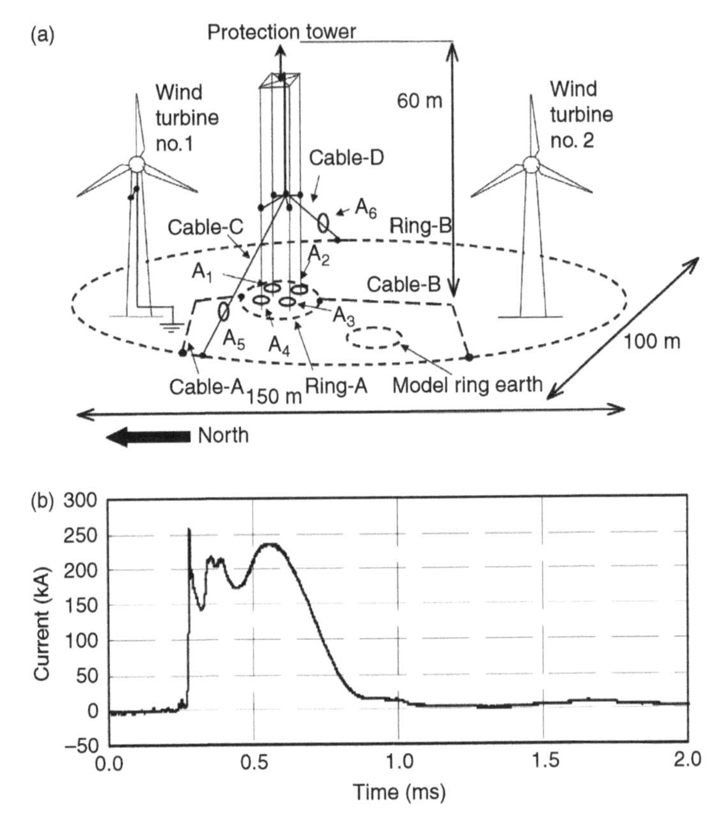

FIGURE 8.9 Observation of lightning current in a wind power station [25].
(a) Observation point at Nadachi Wind Power Station. (b) Observation result of lightning current.

lightning. Observation results of the lightning currents in the wind turbines suggest the following remarks [26]:

- Some of the winter lightning currents have shown a time duration of more than 200 ms.
- Some of the winter lightning currents with the electric charge of more than 300 C, which is the protection level recommended by the IEC international standard, have been observed, and the maximum value is approximately 1270 C.

Thus, the IEC international standard is not sufficient for the Japanese wind turbine generation systems against the winter lightning in Japan.

The stepped leader of summer lightning develops from a cloud to the ground, and the return strike with high current starts from the ground, a structure or a tree to the cloud. A downward lightning is observed in summer as shown in Fig. 8.10a. On the other hand, the winter lightning frequently has an upward

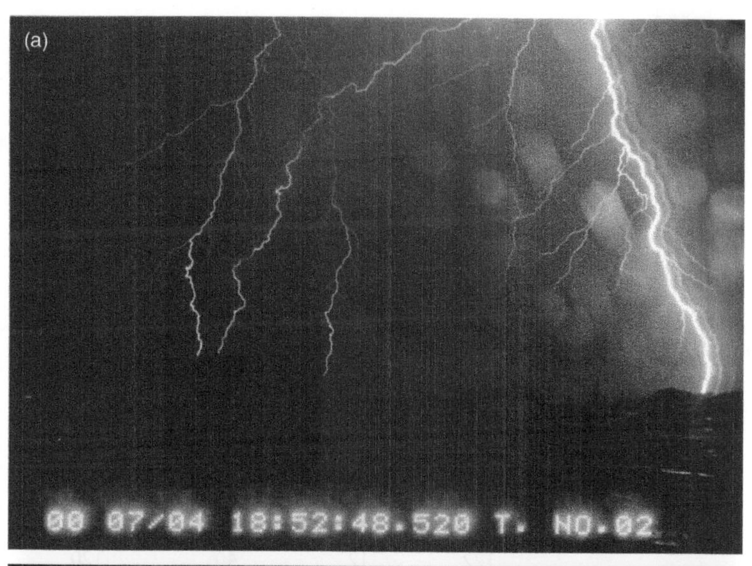

■ FIGURE 8.10 Example of lightning strokes in summer and winter in Japan [27].
(a) Summer lightning. (b) Winter lightning.

lightning strike as in Fig. 8.10b [27]. The difference of lightning strikes affects the lightning striking point and shielding failure.

The other parameters related to the lightning such as ground flash density, lightning channel impedance, and return strike velocity should be known. Updated data on the lightning parameters for engineering including theses characteristics are available [28] (Table 8.1).

Table 8.1 Parameters of Downward Negative Lightning [24]

Parameters	Units	Sample Size	Exceeding Tabulated Value (%)		
			95%	50%	5%
Peak current (minimum 2 kA)	kA				
First strokes		101	14	30	80
Subsequent strokes		135	4.6	12	30
Charge (total charge)	C				
First strokes		93	1.1	5.2	24
Subsequent strokes		122	0.2	1.4	11
Complete flash		94	1.3	7.5	40
Impulse charge (excluding continuing current)	C				
First strokes		90	1.1	4.5	20
Subsequent strokes		117	0.22	0.95	4
Front duration (2 kA to peak)	μs				
First strokes		89	1.8	5.5	18
Subsequent strokes		118	0.22	1.1	4.5
Maximum di/dt	kA/μs				
First strokes		92	5.5	12	32
Subsequent strokes		122	12	40	120
Stroke duration (2 kA to half peak value on the tail)	μs				
First strokes		90	30	75	200
Subsequent strokes		115	6.5	32	140
Action integral ($\int i2dt$)	A^2s				
First strokes		91	6.0×10^3	5.5×10^4	5.5×10^5
Subsequent strokes		88	5.5×10^2	6.0×10^3	5.2×10^4
Time interval between strokes	ms	133	7	33	150
Flash duration	ms				
All flashes		94	0.15	13	1100
Excluding single-stroke flashes		39	31	180	900

where I_0, η, τ_1, n, and τ_2 are constants, and are empirically determined.

8.2.5 IEC international standard [13–17]

The lightning protection zone (LPZ) as shown in Fig. 8.11 and Table 8.2 was introduced in the IEC 62305 for lightning protection design. Protection measures such as the lightning protection system (LPS), shielding wires, magnetic shields, and SPD determine the LPZs. IEC 62305 also introduces four lightning protection levels (LPL) I–IV. For each LPL, a set of maximum and minimum lightning current parameters is fixed. The maximum values of lightning current parameters for the different LPLs are given in Table 8.3 [13], and are used to design lightning protection components. The minimum amplitudes of lightning current for the different LPLs are used to determine the rolling sphere radius to define the LPZ 0B, which cannot be reached by direct lightning strike. The minimum values of lightning current parameters together with the related rolling sphere radius are given in

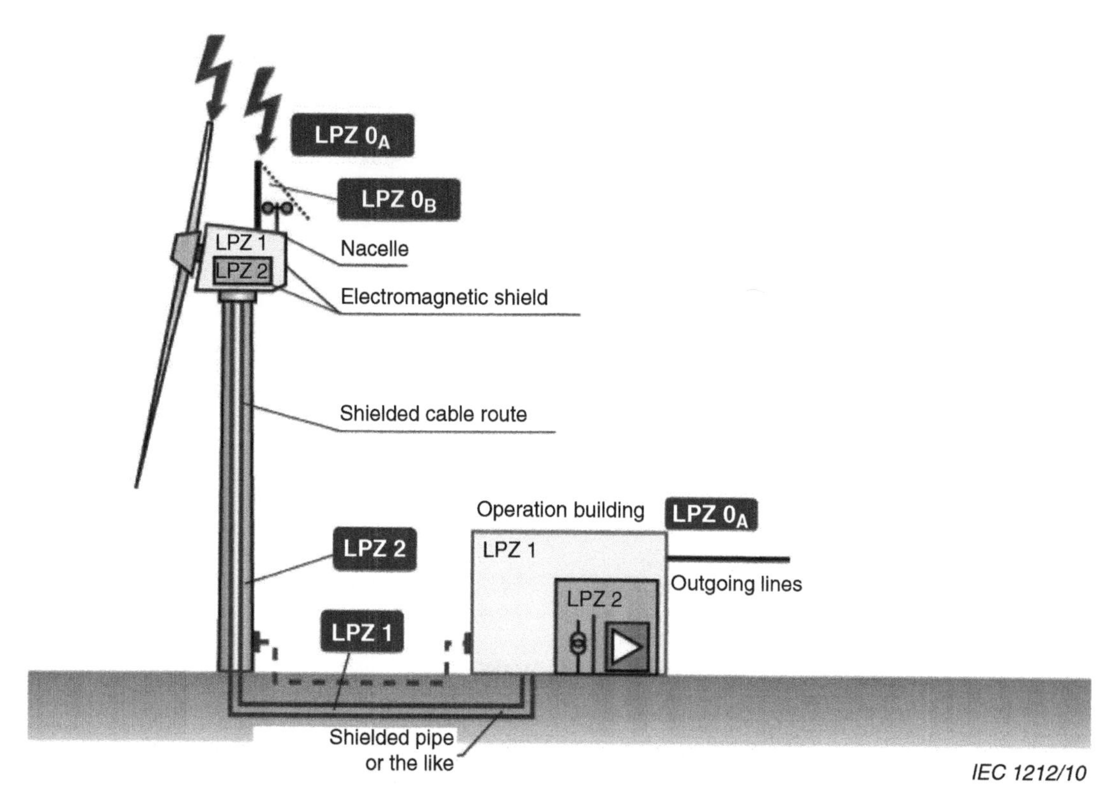

■ **FIGURE 8.11 Division of wind turbine into different lightning protection zones [17].**

Table 8.2 Definition of Lightning Protection Zones

Outer Zones	
LPZ 0	Zone where the threat is due to the unattenuated lightning electromagnetic field and where the internal systems may be subjected to full or partial lightning surge current.
LPZ 0_A	Zone where the threat is due to the direct lightning flash and the full lightning electromagnetic field. The internal systems may be subjected to full or partial lightning surge current.
LPZ 0_B	Zone protected against direct lightning flashes but where the threat is the full lightning electromagnetic field. The internal systems may be subjected to partial lightning surge currents.
Inner Zones	
LPZ 1	Zone where the surge current is limited by current sharing and by SPDs at the boundary. Spatial shielding may attenuate the lightning electromagnetic field.
LPZ 2,...*n*	Zone where the surge current may be further limited by current sharing and by additional SPDs at the boundary. Additional spatial shielding may be used to further attenuate the lightning electromagnetic field.

Table 8.3 Maximum Values of Lightning Parameters According to LPL

First Short Positive Stroke			LPL			
Current Parameters	**Symbol**	**Unit**	**I**	**II**	**III**	**IV**
Peak current	I	kA	200	150		100
Short stroke charge	Q_{short}	C	100	75		50
Specific energy	W/R	MJ/Ω	10	5,6		2,5
Time parameters	T_1/T_2	μs/μs		10/350		
First Short Negative Stroke[a]			**LPL**			
Peak current	I	kA	100	75		50
Average steepness	di/dt	kA/μs	100	75		50
Time parameters	T_1/T_2	μs/μs		1/200		
Subsequent Short Stroke[a]			**LPL**			
Current Parameters	**Symbol**	**Unit**	**I**	**II**	**III**	**IV**
Peak current	I	kA	50	37,5		25
Average steepness	di/dt	kA/μs	200	150		100
Time parameters	T_1/T_2	μs/μs		0,25/100		
Long Stroke			**LPL**			
Current Parameters	**Symbol**	**Unit**	**I**	**II**	**III**	**IV**
Long stroke charge	Q_{long}	C	200	150		100
Time parameter	T_{long}	s		0,5		
Flash			**LPL**			
Current Parameters	**Symbol**	**Unit**	**I**	**II**	**III**	**IV**
Flash charge	Q_{flash}	C	300	225		150

[a] *The use of this wave shape concerns only calculations and not testing.*

Table 8.4 [13]. The rolling sphere model [29] is applied to a wind turbine. This model is an EGM. The lightning striking distance in the IEC 62305 is dependent on the LPL. They are used for positioning of the air-termination system and to define the LPZ 0B.

The nacelle, the tower, and the transformer kiosk are protection zone LPZ 1. The devices inside metal cabinets in LPZ 1 are in protection zone LPZ 2. For instance, control systems inside a cabinet or a metal tower are in LPZ 2, but in a metal cabinet outside the tower they are in LPZ 1 or LPZ 2. Very sensitive equipment may be placed within a still more protected zone LPZ 3.

Table 8.4 Minimum Values of Lightning Parameters and Related Rolling Sphere Radius Corresponding to LPL

Interception Criteria			LPL			
	Symbol	Unit	I	II	III	IV
Minimum peak current	I	kA	3	5	10	16
Rolling sphere radius	r	m	20	30	45	60

The division of the wind turbine into LPZs is a tool to ensure systematic and sufficient protection of all components of the wind turbine. For instance, protection against overvoltages is only necessary for cables passing from one zone into another zone with more sensitive components, whereas internal connections within the zone may be unprotected. Basic protection measures in an LPMS according to IEC 62305-4 include bonding, magnetic and electrical shielding of cables, line routing (system installation), coordinated SPD protection, and grounding.

When the structure is protected by an LPS, a Type 1 SPD needs to be provided at the line entrance to provide equipotential bonding between the line and the grounding system. When there is no LPS, the SPD at the entrance is still needed to provide a border between an LPZ 1 inside the structure and an LPZ 0 outside the structure.

8.3 LIGHTNING PROTECTION FOR WIND POWER GENERATION SYSTEMS

8.3.1 Lightning damage in wind power generation system

The main reason for outage for long periods is damage to blades. Some types of damage of blades are found in [4]. An example of the damage is shown in Fig. 8.12 [4].

Fig. 8.13 shows outage period of blades and control equipments due to lightning as a parameter of season [9]. This period reflects an influence of the damage on power supply. The number of the damaged control equipments is larger than that of the blades. However, damage of a blade caused by winter lightning stops generation for long periods. Thus, the damage of the blades is the most serious problem for wind power generation. It is hard to repair the damaged blades in winter due to the strong snow and wind. Damage to control systems are mainly observed. This is caused by malfunctions due to the lightning electromagnetic impulse.

FIGURE 8.12 A case of lightning damage in a wind turbine (same turbine and same event) [4]. (a) Blade burnout; (b) blade burnout and wire melting.

Observed lightning currents often have high amplitude or large electric charge. However, the large lightning flash does not always cause damage [26]. A rational lightning protection design by properly arranging SPDs and constructing air-termination system is required.

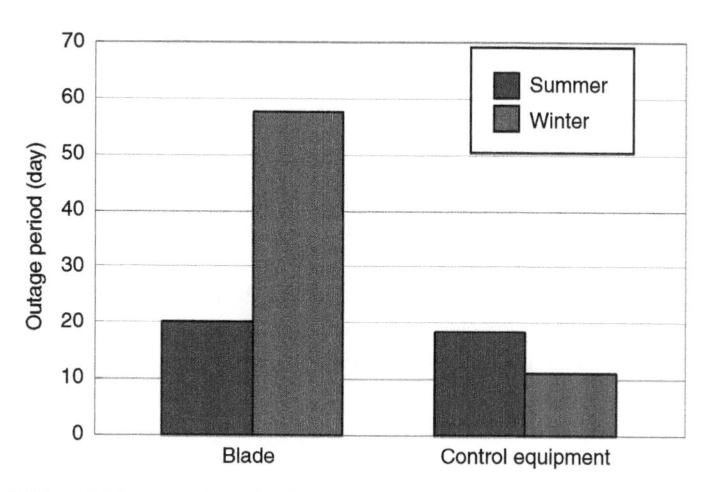

■ **FIGURE 8.13 Damage of wind turbine caused by lightning.**

8.3.2 Lightning protection methods for wind turbine

Fig. 8.14 gives an example of lightning protection measures for a wind turbine generation system [17]. SPDs must be installed at entrances of wind tower to a service line and a telecommunication line. Wind tower and nacelle themselves are regarded to be Faraday cages. Equipments and instruments

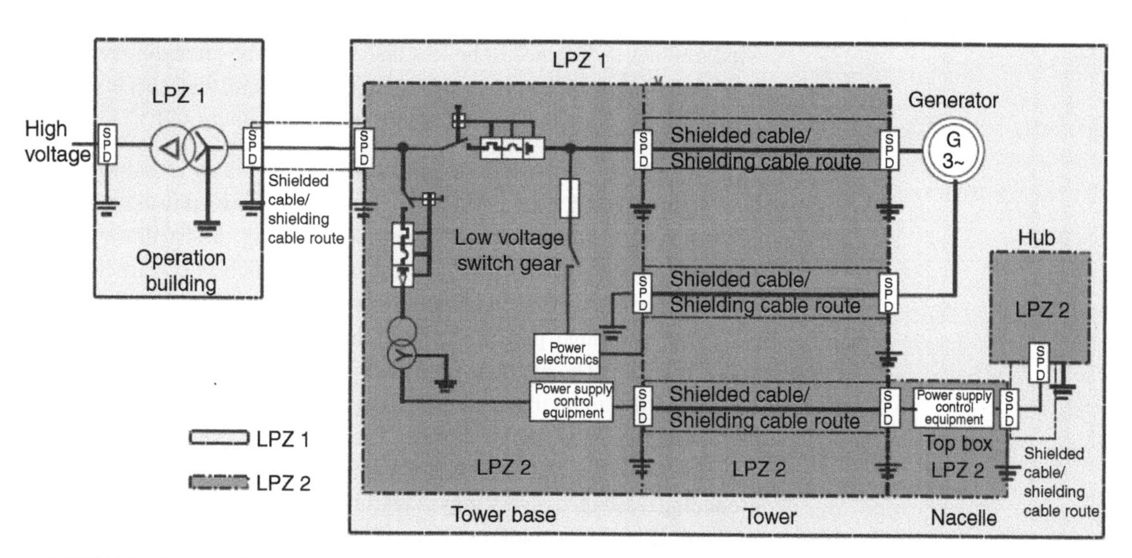

■ **FIGURE 8.14 An example of lightning protection management system considering LPZs [17].**

in the tower and the nacelle can be protected from lightning overvoltages by the equipotential bonding and the installation of the SPDs.

Fig. 8.14 is an example of how to document LPMS division of electrical system into protection zones with indication of where circuits cross LPZ boundaries and showing the long cables between tower foundation and nacelle [17]. It is the best method to install a number of SPDs at every entrances of equipment and boundary of the LPZs. A cable with metallic sheath reduces an influence of external electromagnetic fields by grounding the metallic sheath. The core conductor is located very close to the metallic sheath, and the voltage between the core conductor and the metallic sheath is low because of the coupling factor. Considering the nacelle and the wind tower are a kind of Faraday cage, and grounding the metallic sheath and the installation of SPDs properly on the basis of the equipotential bonding, the voltages on the equipment in the nacelle and the tower can be less than the lightning impulse withstand voltage.

8.3.3 Grounding resistance

Lightning performance of a grounding system must be considered for rational lightning protection design [30]. Transient response of grounding resistance is simply represented by circuits shown in Fig. 8.15. High grounding resistance of more than 100 Ω often shows a capacitive type and a low resistance of less than 10 Ω the inductive type. Thus, the grounding resistance should be treated as impedance. The steady state grounding resistance for a wind turbine is designed to be as low as possible. The grounding resistance of wind turbines is designed to be less than 2–10 Ω. The grounding resistance is lower as the grounding electrode is larger. Therefore, high cost is required to obtain very low grounding resistance in high resistivity soil.

The ground potential rise of a grounding system of an actual wind turbine was measured. The measurement results are verified using the finite-difference time-domain (FDTD) method [31]. Fig. 8.16a illustrates the grounding system consisting of a foundation, which is constructed with reinforced concrete, a grounding mesh and foundation feet. Measurement result of the ground potential rise of the grounding system is shown in Fig. 8.16b [32]. The applied current has stepwise waveform, and the voltage is regarded to be transient grounding resistance. The transient grounding resistance is the inductive type. The maximum value is much higher than the steady state grounding resistance. Thus, the transient response of grounding resistance is very important for estimating lightning overvoltages in wind turbine generation system. Simulation results using the FDTD method agree well with the experimental results. Thus, the FDTD method is a useful tool to predict transient response of wind turbine grounding

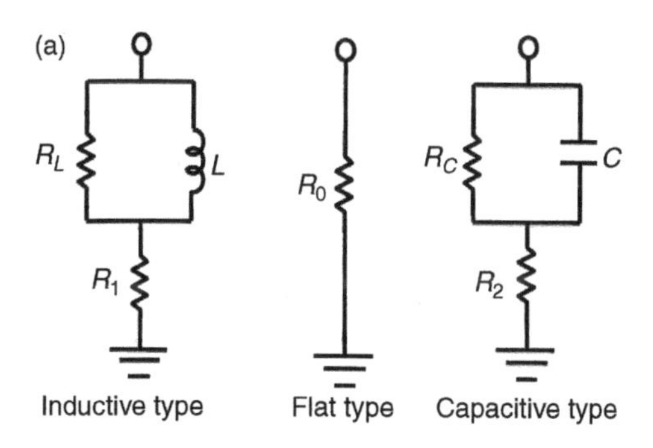

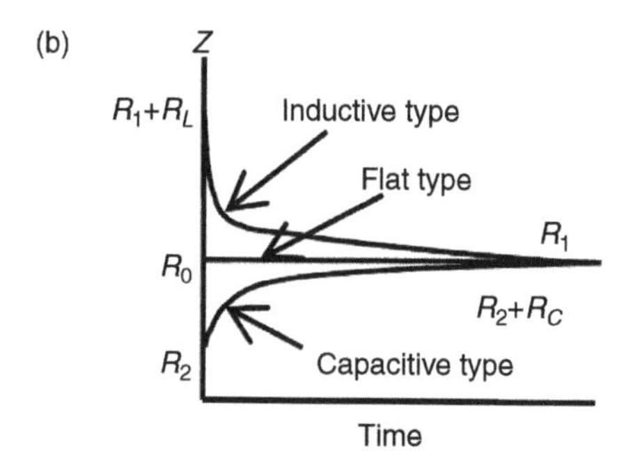

FIGURE 8.15 Grounding resistance for lightning current. (a) Equivalent circuits of grounding resistance. (b) Transient response of grounding resistance.

system. Mesh type grounding electrode, which is frequently used in substations, also shows the same variation [33].

A tower foundation can be regarded to be a large grounding electrode, and is an excellent grounding electrode, and very low grounding resistance can be obtained by the tower foundation alone. However, it is difficult to obtain low grounding resistance, when the wind tower is constructed in high-soil-resistivity yard. Auxiliary grounding electrodes such as horizontal and vertical grounding conductors, and a ring electrode are often used to obtain the low steady–state grounding resistance. The grounding electrodes of the towers in the wind farm are sometimes connected each other using the long grounding conductors to obtain low steady–state grounding resistance [34].

For simplicity, the grounding resistance R_p of two grounding electrodes of a tower foundation and an auxiliary grounding electrode is given by:

$$R_p = \frac{R_1 R_2 - R_m^2}{R_1 + R_2 - 2R_m} \tag{8.4}$$

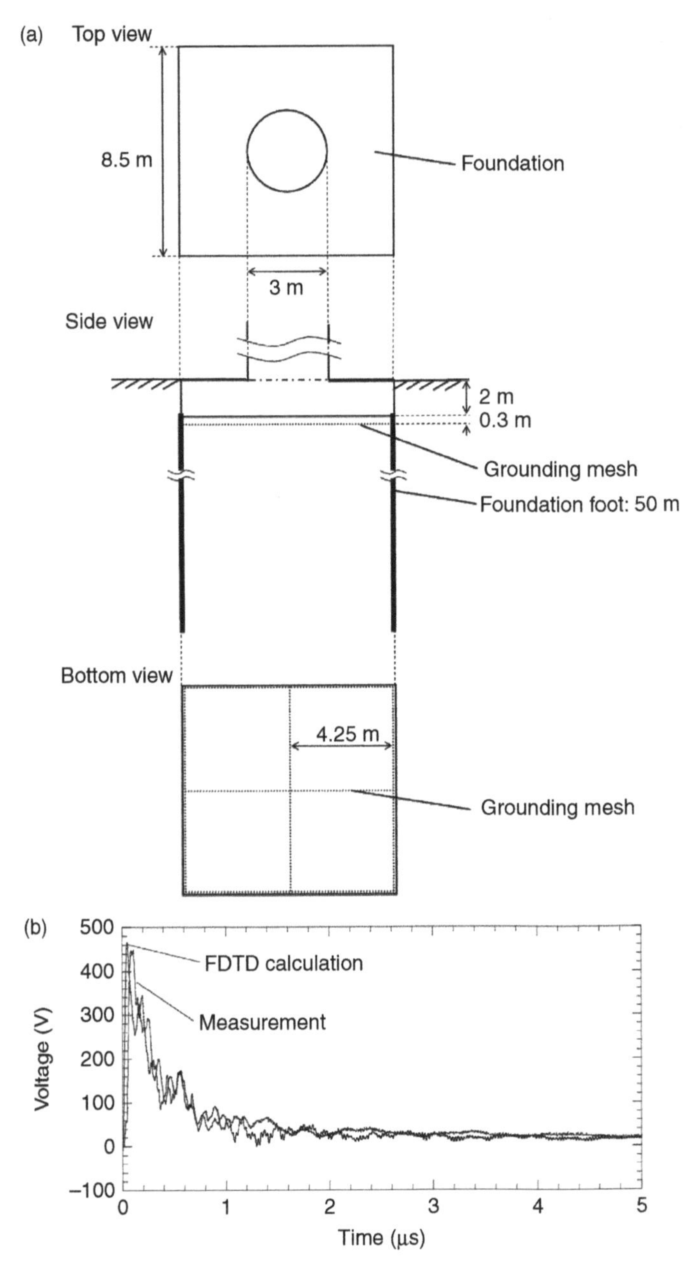

FIGURE 8.16 Measurement results of the grounding system compared with FDTD calculated results [32]. (a) Configuration and dimension of grounding system of an actual wind turbine generator system. (b) Comparison of measured results of ground potential rise with simulation results using FDTD.

where R_1 and R_2 are self-grounding resistances of a tower foundation and an auxiliary grounding electrode respectively, and R_m is the mutual grounding resistance.

The following two cases must be carefully considered in the grounding system design:

1. $R_1 \approx R_m$, a ring type grounding electrode is sometimes set near and around a tower foundation. If these electrodes are closed, R_m is almost equal to R_1. As a result, the total grounding resistance is approximately equal to the tower footing resistance, namely $R_p \approx R_1$. In this case, the ring type electrode is not expected to reduce the grounding resistance. The ring type electrode has an important role to reduce foot voltage around the tower.
2. $R_1 \ll R_2$, in case the grounding resistance of an auxiliary grounding electrode is sufficiently higher compared with R_1, the total grounding resistance is approximately equal to the tower footing resistance. Thus, the reduction effect of the auxiliary grounding electrodes is not expected in this case.

Thus, a tower foundation is very useful grounding electrode, and auxiliary grounding electrodes are carefully selected so that mutual and self-grounding resistances are as low as possible. Simulation tools based on electromagnetic theory such as the FDTD method and the method of moment are useful for designing the grounding system.

Considering the reduction of the grounding resistance for lightning currents due to the soil ionization is small in case of low grounding resistance [33], the soil ionization effect on the grounding resistance of a wind turbine is not expected.

A down conductor for equipments in a nacelle is sometimes set along a wind tower. Breakdown and short circuit occurs between the down conductor and the tower due to lightning overvoltages [35]. Radius of a wind tower is very large, and the tower has low surge impedance. Therefore, the wind tower should be treated as a down conductor.

8.3.4 Lightning protection of blade using receptor [4,17]

Wind turbine blades are large and long hollow structures manufactured of composite materials such as glass fiber reinforced plastic, wood, wood laminate, and carbon fiber reinforced plastic. Wind turbine blades should be considered for lightning protection design as follows:

1. Height of a blade is very tall, and lightning frequently hits the blade.
2. Shielding wire, which is used in electric power systems, is not fundamentally adopted considering wind condition. Thus, shielding effects by the shielding wire are not expected.
3. Blade materials, such as FRP, are easy to burn.
4. Most of blades are composed of two pieces of shells and are hollow. Lightning discharges can intrude into the cavity resulting in a rupture of a blade.

Receptors are installed on a blade surface for lightning protection so that the damage of blade rupture or surface tearing can be reduced more effectively. A down conductor is installed inside the cavity of the blade, and leads the lightning current safely to the root end of the blade and onward to the earthed structure of the wind turbine. Serious blade damage can be mitigated by the receptor. A metal cap receptor on the tip edge of each blade is an improved type of the receptor. The receptors should be installed close to the tip part of a blade. The down conductor should be a thick wire for lightning current with large energy.

8.3.5 Energy coordination of surge arrester/surge protective device [36]

Winter lightning and continuing current sometimes have large energy enough to burn Ars and SPDs. When air-termination system works well, the most of lightning current is injected in to the ground. The ground potential rise due to the lightning current is the main reason to cause the large energy into the Ars and SPDs. A relation between current $I(t)$ and electric charge $Q(t)$ is given by:

$$Q(t) = \int_0^t I(T)\,dT \tag{8.5}$$

Assuming impedance, Z, is constant, energy $En(t)$ is given by:

$$En(t) = \int_0^t Z\{I(T)\}^2\,dT = Z\int_0^t \{I(T)\}^2\,dT \tag{8.6}$$

$\int\{I(t)\}^2$ corresponds to the action integral in Table 8.1. It is difficult to estimate Z and most of impedances are not constant. The electric charge is useful to estimate an influence of lightning current on a lightning damage. As is shown in Fig. 8.3, Ar and SPD voltages can be regarded to be constant, and

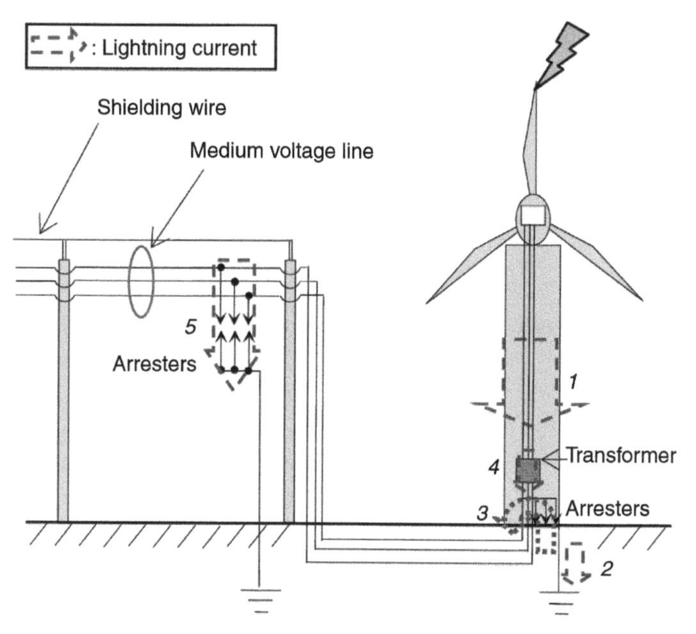

■ **FIGURE 8.17 Routes of lightning currents in case of lightning hit to a receptor.**

the Ar and the SPD are modeled by the direct current (DC) voltage source, E_0. The energy is given by:

$$En(t) = \int_0^t E_0 I(T)\,dT = E_0 \int_0^t I(T)\,dT = E_0 Q(t) \tag{8.7}$$

and is estimated from the electric charge.

Fig. 8.17 illustrates routes of lightning currents when the lightning strikes a receptor. In case of a wind farm, the distribution line can be replaced by neighboring towers.

The lightning current injected into the tower from the receptor (arrow 1) is flown into the ground through grounding electrodes (arrow 2). The current into the ground raises the ground potential rise. When the ground potential rise is higher than the surge arrester operation voltage, the transformer arrester acts, and a part of the injected current into the ground goes through the surge arresters (arrow 3). The lightning surge on a power cable in the tower transfers through the transformer (arrow 4). However, this surge is low, and can be ignored. A pert of the lightning current is flown into the ground through distribution line arresters (arrow 5).

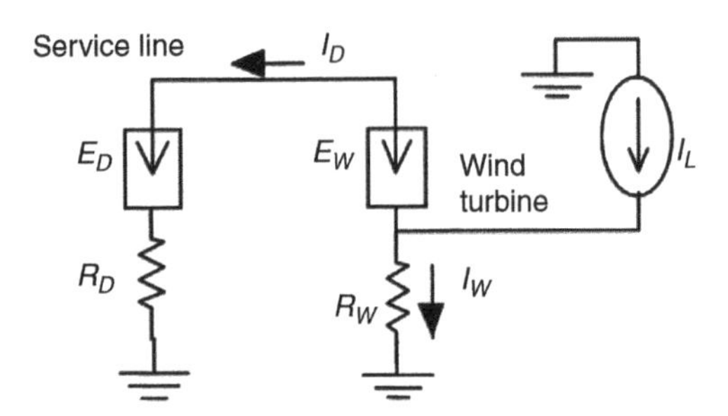

■ **FIGURE 8.18 Equivalent circuit for current distribution in case of lightning hit to a receptor.**

Fig. 8.18 shows an equivalent circuit for direct lightning hit to a receptor. Current into the service line I_D and the service line voltage V_D are given by:

$$I_D = \frac{R_W I_L - (E_W + E_D)}{R_D + R_W} \tag{8.7}$$

$$V_D = \frac{R_D R_W}{R_D + R_W} I_L - \frac{R_D E_W - R_W E_D}{R_D + R_W} \tag{8.8}$$

Calculated results using the equations are shown in Fig. 8.19, where $E_D = E_W = 15$ kV, and dotted line is flashover voltage of the service line. The grounding resistance of the wind turbine should be as low as possible because most of the lightning current is injected into the ground. The line voltage becomes high in comparison with line flashover voltage even if Ars are installed, because the ground potential rise becomes high. Thus, Ar capability is carefully selected, and wind turbines and lines must be coordinated on the insulation and energy.

8.4 LIGHTNING PROTECTION OF WIND FARMS

One of the countermeasures to protect wind turbine blades from the lightning is constructing lightning towers besides the wind turbine towers as shown in Fig. 8.20.

The receptor is an air-termination system for blades. However, the receptor is small, and does not always catch the lightning. The lightning tower is a kind of independent giant lightning rod. When the lightning towers are constructed in a wind farm, those cover some wind towers against the lightning. The lightning tower needs to be as high as possible. However, the dimension and location of the lightning tower is restricted because the lightning tower affects the flow of wind.

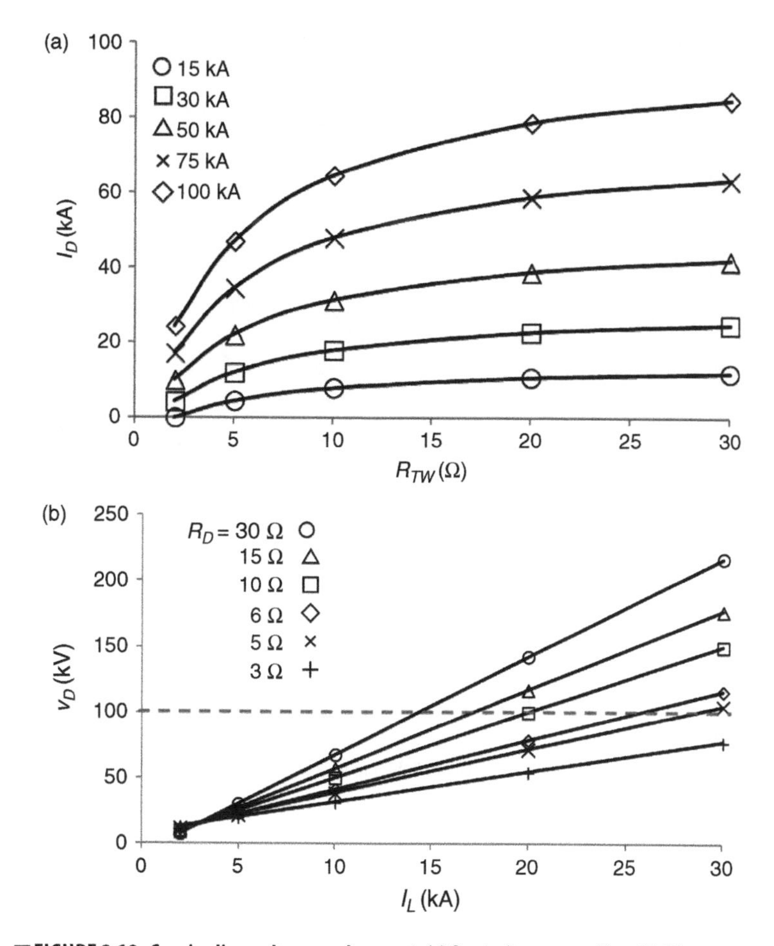

FIGURE 8.19 Service line voltage and current. (a) Service line current ($R_D = 15\ \Omega$); (b) Service line voltage ($R_W = 10\ \Omega$).

There are 15 wind turbines in Nikaho wind park, which is located in Akita prefecture of Japan and facing the coast of the Sea of Japan. Observation of lightning discharges using still cameras got 99 lightning hits to wind turbines. An uneven distribution of the number of lightning hits on the wind turbines is shown in Fig. 8.21. The wind turbines, which are located at the end of a wind turbine line or facing the coast, are more often hit by lightning. Thus, the lightning discharge in a wind farm is dependent on the location of wind towers. This observation result suggests that if a lightning tower is constructed at the location to which the lightning most frequently hits, most lightning flashes can be caught by the lightning tower.

■ **FIGURE 8.20 Shielding by lightning tower.** *(Source: Photo is taken by Uchinada-cho.)*

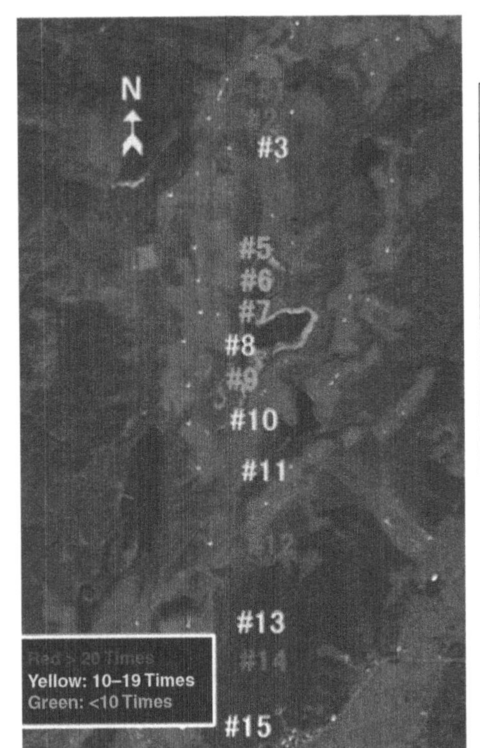

Lightning frequency by photo observation from 2005 to 2011

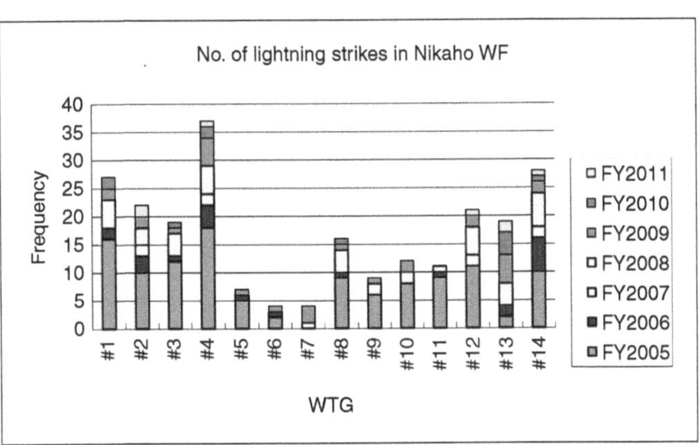

Target of photo observation #1 - #14 WTG
(#15WTG is not covered)

Frequency from 2005 is 118, 22, 9, 41, 28, 11, 7

No. of strikes to #1, #4, #14 WTG is larger

No. of strikes to #5, #6, #7 WTG is less

More flashes strike to both ends and ocean side WTGs

■ **FIGURE 8.21 Lightning discharge observation at Nikaho wind park [4].** WTG, Wind turbine generator.

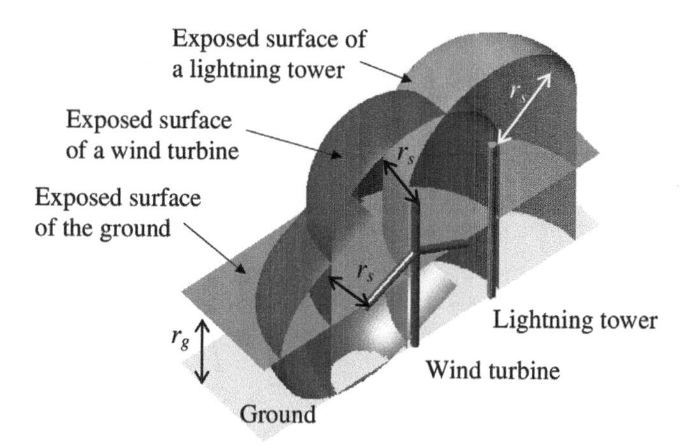

FIGURE 8.22 Exposed surface with a lightning tower and wind turbines [37].

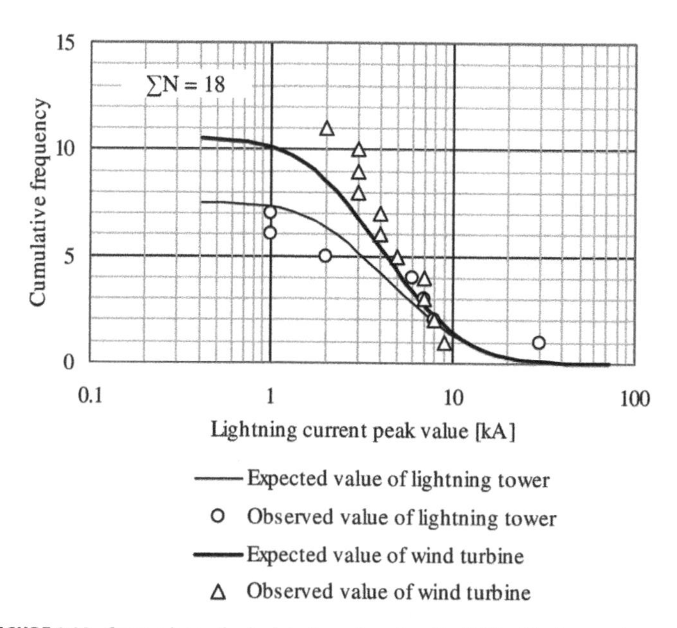

FIGURE 8.23 Comparison of calculated results using the EGM with observation results for a wind tower and a lightning tower [37].

Lightning striking position is dependent on the dimension and location of structures, lightning striking angle and lightning peak current. The EGM is applied to the determination of the lightning striking point as illustrated in Fig. 8.22 [37]. It should be mainly considered to be spherical in the case of wind turbines. According to Fig. 8.22, the exposed surface changes due to not only the lightning current but also the direction and orientation of the blades. Fig. 8.23 shows a comparison of the calculated results using the

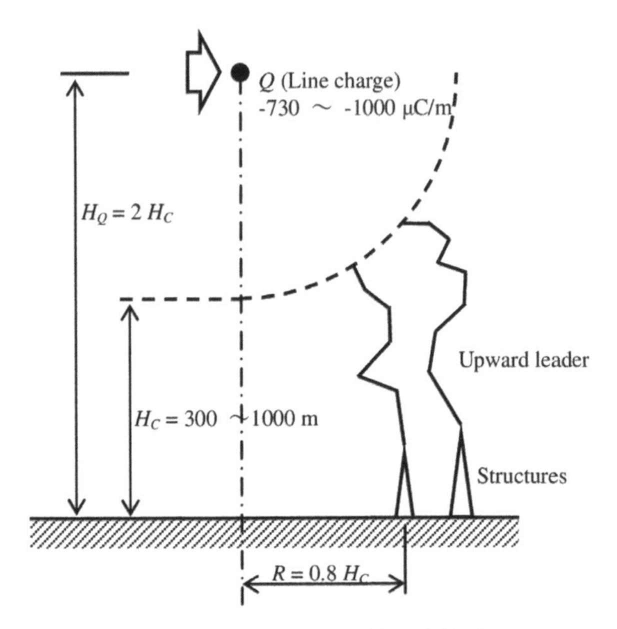

■ **FIGURE 8.24 Charge configuration used in the ULP model [41].**

EGM with the observation results. The comparison of the expected rate of lightning strikes seems to be reasonably good.

Winter lightning observed along the coast of the Sea of Japan, where many wind turbine generators are built, shows upward lightning discharge. The upward leader progression (ULP) model has been proposed to predict lightning protective effect of upward lighting [38]. The EGM was developed on the basis of the downward lightning developing from a cloud to the ground, and is hard to evaluate lightning shielding failure against winter lightning, namely upward lightning. The ULP model is applicable to winter lightning and a shielding effect of a lightning tower [39]. Fig. 8.24 illustrates configuration of cloud, a lightning tower, and wind turbine towers. The relationship between the direction of the cloud movement and location and height of the towers is very important. The estimated ratio of lightning strike to wind turbines using the ULP model is satisfactorily agrees with that of observation result [40] (Fig. 8.25).

8.5 LIGHTNING PROTECTION FOR PHOTOVOLTAIC POWER GENERATION SYSTEMS

The height of photovoltaic systems is low, and the probability of direct lightning hit to the photovoltaic system is low. Moreover, the withstand voltage of the photovoltaic system is much lower compared with the other electric power systems. Therefore, the lightning protection design for photovoltaic

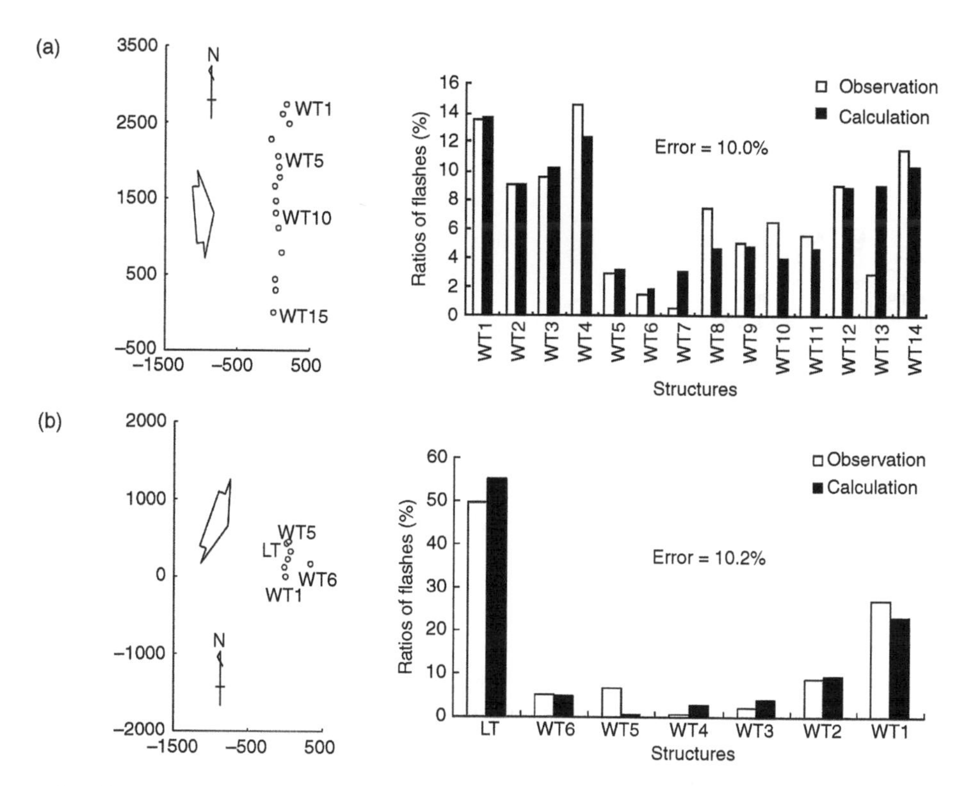

FIGURE 8.25 Comparison of calculated results using the ULP model with observation results for a wind tower and a lightning tower [40]. (a) Nikaho wind park. (b) Taikoyama wind park.

system concerns the air-termination system and lightning electromagnetic impulse such as lightning-induced voltage. Lightning protection for photovoltaic system is fundamentally based on the IEC standard.

8.5.1 Lightning damage in photovoltaic system

Only a few lightning damage in photovoltaic array has been reported. Most of lightning damage in actual photovoltaic systems occur in power conditioner and measurement and indicator equipment [41]. Fig. 8.26 illustrates an example of lightning protection measures for a photovoltaic system. Megasolar system frequently adopts lightning rods [42].

8.5.2 Lightning protection against lightning overvoltages in photovoltaic systems

SPDs must be properly selected and installed according to the LPZ. The distance between the LPS and the metal structure of the photovoltaic array

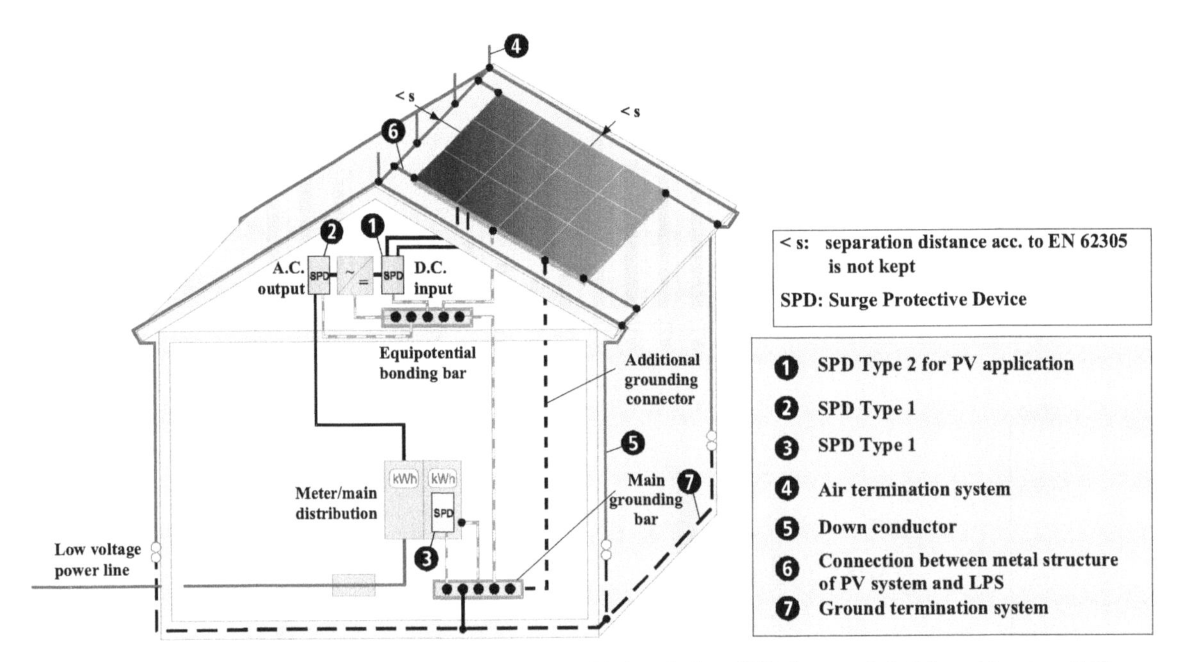

■ FIGURE 8.26 Possible installation of SPDs in case of a building with external LPS when separation distance is not kept [42].

should be kept as short as possible to prevent the photovoltaic array from lightning damage. Insulation level of photovoltaic modules is basically determined by reverse withstand voltage of backward diodes.

LEMP generates high voltage but small energy. However, photovoltaic array continues generating DC voltage, and large energy is flown into SPDs. Therefore, a disconnector for an SPD in the DC side must be provided to disconnect the failed SPD from a line in case the failed SPD has been damaged due to high lightning surge exceeding its capability.

8.5.3 Direct lightning flash to photovoltaic system

Lightning strike with low current might strike a photovoltaic array based on EGM. An experiment of discharge to a photovoltaic array using an impulse voltage generator was carried out. Lightning impulse voltage with 1.2/50 μs is applied to a rod using the impulse voltage generator. Specification of a photovoltaic array is 70 W and 11.9 V. The thickness of the cover glass (tough glass) on photovoltaic cells is 2.5 mm, and the breakdown voltage of the glass is approximately 210 kV. The metal frame of the array is grounded in the experiments, but is not directly connected to the cells. Fig. 8.27 illustrates a photovoltaic array. The cell is polycrystalline silicon type.

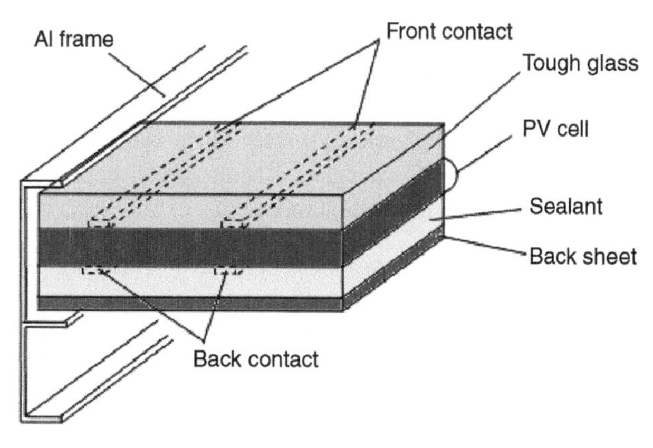

FIGURE 8.27 Configuration of a photovoltaic array.

The 50% sparkover voltage of the rod-plate gap is almost equal to that of the rod array gap. Thus, the photovoltaic cells act a metal plate for lightning impulse voltages. Fig. 8.28 show an example of discharges in case of 0.3 m gap including sparkover and surface discharge [43]. The sparkover is attached to a point on the glass of the array under the rod, and then the surface discharge towards a frame, which is grounded, occurs. Generally speaking,

FIGURE 8.28 Example of observation results of surface discharge on a photovoltaic array.

surface discharge occurs on an insulated material above a conductor. Thus, photovoltaic cells can be regarded as a metallic plate for lightning impulse voltages. The experimental results suggest that a direct lightning strike with small lightning current to a photovoltaic array might not cause serious damage due to surface discharge on the glass. Therefore, it is possible to protect photovoltaic systems from direct lightning strikes.

A large-scale photovoltaic system should be protected from direct flash with lightning current having high amplitude using air-termination system such as tall lightning rods. Lightning flashes with low current might strike the system. SPDs must be installed according to the IEC standard. The grounding system design is very important in both cases [11,44]. The photovoltaic array structure itself is a good down conductor. They should be connected each other using grounding conductors.

REFERENCES

[1] Rakov V, Uman MA. Lightning—Physics and effects. UK: Cambridge University Press; 2003.

[2] Nucci CA, et al. Lightning-induced voltages on overhead power lines, part I: return stroke current models with specified channel-base current for the evaluation of the return stroke electromagnetic fields, part II: coupling models for the evaluation of the induced voltages. Electra 1995;162:74–102.

[3] Yokoyama S. Lightning protection on power distribution line against direct lightning hits. Trans IEE Japan 1994;114-B(6):564–8 (in Japanese).

[4] CIGRE WG C4.409. Lightning protection of wind turbine blades, CIGRE Technical Brochure no. 578, 2014.

[5] Nakahori K, Egawa T, Mitani H. Characteristics of winter lightning in Hokuriku district. IEEE Trans Power Ap Syst 1982;101(11):4407–12.

[6] Miyake K, Suzuki T, Takashima M, Takuma M, Tada T. Winter lightning on Japan Sea coast—lightning striking frequency to tall structures. IEEE Trans Power Deliv 1990;5(3):1370–7.

[7] Miyake K, Suzuki T, Shinjou K. Characteristics of winter lightning on Japan Sea coast. IEEE Trans Power Deliv 1992;7(3):1450–7.

[8] Asakawa A, Miyake K, Yokoyama S, Shindo T, Yokota T, Sakai T. Two types of lightning discharges to a high stack on the coast of the Sea of Japan in winter. IEEE Trans Power Deliv 1997;12(3):1222–31.

[9] New Energy and Industrial Technology Development Organization (NEDO), Activity report on study of lightning protection measures for wind turbine generator. 2007 (in Japanese).

[10] Takahashi M, Kawasaki K, Matsunobu T. Lightning surge characteristics and protection methods of photovoltaic power generation system. Trans IEE Japan 1989;109(10):443–50 (in Japanese).

[11] Hernandez JC, Vidal PG, Jurado F. Lightning and surge protection in photovoltaic installations. IEEE Trans Power Deliv 2008;23(4.):1961–71.

[12] Armstrong HR, Whitehead ER. Field and analytical studies of transmission line shielding. IEEE Trans Power Ap Syst 1968;87(1):270–81.

[13] Protection against lightning electromagnetic impulse—part 1: general principles IEC 62305-1; 2006.
[14] Protection against lightning electromagnetic impulse—part 2: risk management IEC 62305-2; 2006.
[15] Protection against lightning electromagnetic impulse—part 3: physical damage to structures and life hazard IEC 62305-3; 2006.
[16] Protection against lightning electromagnetic impulse—part 4: electrical and electronic systems within structures IEC 62305-4; 2006.
[17] Wind turbines—part 24: lightning protection IEC 61400-24; 2010.
[18] Yamamura Y, Oka S. Distribution of surge voltages in the substations and the new surge analogue computer. J IEE Japan 1958;78:54–60 (in Japanese).
[19] Kawamura T, Kouno T, Kojima S, Ishizaki S, Ishihara H. New estimation methods of lightning overvoltages in substation and their application, IEEE PES Summer Meeting, A78. pp. 502–507.
[20] Wind turbine grounding systems for lightning protection, IEE J Tech Rep, 1270; 2013.
[21] Berger K, Anderson RB, Kroninger H. Parameters of lightning flashes. Electra 1975;80:223–37.
[22] Heidler F. Traveling current source model for LEMP calculation. Proceedings of the sixth international Zurich Symposium on Electromagnetic Compatibility; 1985. pp. 157–162.
[23] Ametani A, Kawamura T. A method of a lightning surge analysis recommended in Japan using EMTP. IEEE Trans Power Deliv 2005;20:867–75.
[24] CIGRE WG C4.404. Cloud-to-ground lightning parameters derived from lightning location systems—the effects of system performance, CIGRE Technical Brochure no. 376; 2009.
[25] Minowa M, Yoda M, Kusama Y, Maeda Y. Observation of winter lightning flashes to Nadachi wind power station on the Cast of Japan. Proceedings of International Conference on Grounding and Earthing; 2002. pp. 53–57.
[26] NEDO. Research and development of next-generation wind power generation technology for technology corresponding to natural environment etc. for measures of lightning protection; 2015.
[27] Lightning damages in wind power plants and mitigation against them, IEEJ Technical Report, No. 1126; 2008.
[28] CIGRE WG C4.407. Lightning parameters for engineering applications, CIGRE Technical Brochure. No. 549; 2013.
[29] Horváth T. Interception of lightning air termination systems constructed with rolling sphere method. Proceedings of the International Conference on Lightning Protection. pp. 555–560; 2006.
[30] Lorentzou MI, Hatziargyriou ND, Cotton: I. Key issues in lightning protection of wind turbines. WSEAS Trans Circuits Syst 2004;3(5):1408.
[31] Noda T, Yokoyama S. Thin wire representation in finite difference time domain surge simulation. IEEE Trans PWRD 2002;17(3):840–7.
[32] Yamamoto K, Yanagawa S, Yamabuki K, Sekioka S, Yokoyama S. Analytical surveys of transient and frequency-dependent grounding characteristics of a wind turbine generator system on the basis of field tests. IEEE Trans Power Deliv 2010;25(4):3035–43.
[33] Sekioka S, Hayashida H, Hara T, Ametani A. Measurements of grounding resistances for high impulse currents. IEEE Proc Gener Transm Distrib 1998;145(6):693–9.

[34] Cotton I. Windfarm earthing. Proceedings of International Symposium on High Voltage Engineering, vol. 2. London, UK; 1999. pp. 288–291.

[35] Sekioka S. Lightning surge analysis model of reinforced concrete pole and grounding lead conductor in distribution line. IEEJ T Electr Electr 2008;3:432–40.

[36] Sekioka S. A study of lightning surge in a medium voltage line caused by direct lightning hit to wind turbine generator system, High voltage meeting, IEEJ; 2015. HV-15-043.

[37] Sakata T, Yamamoto K, Sekioka S, Yokoyama S. Risk evaluation of lightning damage to wind turbines with the electro geometrical method. IEEJ Trans 2007;127(12):1320–4.

[38] Shindo T, Aihara Y. A shielding theory for upward lightning. IEEE Trans Power Del 1993;8(1):318–24.

[39] Sakata T, Yamamoto K, Sekioka S, Yokoyama S. Lightning protection effect of lightning tower set up in wind power station based on upward Leader progression model. IEEJ Trans 2011;131(2):215–21 (in Japanese).

[40] Sakata T, Yamamoto K, Sekioka S, Yokoyama S. An examination of upward leader progression model simulating thundercloud charge with a point charge, High Voltage Meeting. IEEJ; 2014. HV-14-081.

[41] NEDO, Analysis and evaluation of lightning damage condition and damage decrease countermeasure technique of lightning damage for PV systems; 2009.

[42] CIGRE WG C4.408, Lightning protection of low-voltage networks. CIGRE Technical Brochure. No. 550; 2013.

[43] Sekioka S. An experimental study of sparkover between a rod and a photovoltaic panel. Proceedings of International Conference on Lightning Protection; 2012.

[44] Charalambous CA, Kokkinos ND, Christofides N. External lightning protection and grounding in large-scale photovoltaic applications. IEEE Trans Electromagn Compat 2014;56(2):427–34.

Chapter 9

Distributed energy resources and power electronics

Masahide Hojo

Department of Electrical and Electronic Engineering, Faculty of Engineering, Tokushima University, Japan

CHAPTER OUTLINE

9.1 Power electronics in PV power generation systems 231
9.2 Power electronics in wind power generation systems 232
9.3 Power electronics in battery energy storage systems 233
9.4 Power quality problems with related to DERs 234

Most of the distributed energy resources (DERs) generate DC or AC power with variable frequency but the power system is operated with an almost constant frequency. In order to inject electric power from DERs to a power system, a suitable power conversion has to be required. From the DC power source, the power conversion is provided by an inverter, which converts the electric power from DC to AC. On the other hand, the AC power with variable frequency is often converted to DC first, and then it is converted to the AC power by an inverter, synchronizing with the power system. Therefore, most of DERs have an inverter at the coupling point with the power system. The inverter usually configured by a sinusoidal voltage source, which is emulated by a self-commutated static power converter. In addition, the converter normally equips some functions not only to protect itself from a power system disturbance but also to provide some ancillary services. Therefore, the inverter is called a utility interactive inverter.

This chapter discusses popular power conversion technology for several DERs. At first, a basic technology is discussed in order to inject the electric power generated by the DERs to the power system with the constant frequency. In the following sections, three types of popular DERs, a

Integration of Distributed Energy Resources in Power Systems

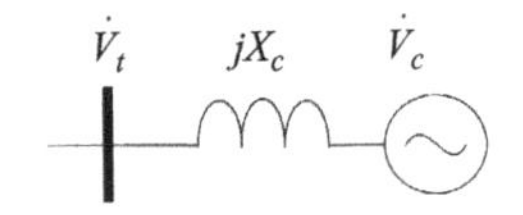

■ **FIGURE 9.1 The simplest model of the DER.**

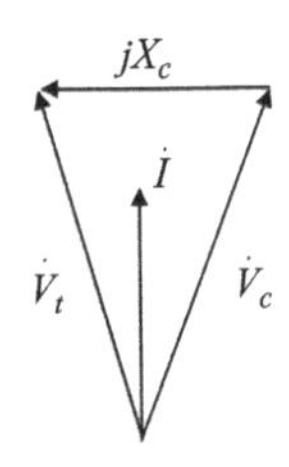

■ **FIGURE 9.2 A phasor diagram around the DER.**

photovoltaic (PV) generation system, a wind turbine, and an energy storage system by a battery, are explained separately. And finally, models for considerations of interferences among the DERs are considered.

In a stable power system, a node voltage, which the DER is connected, can be considered as a sinusoidal voltage source with a nominal frequency. Although the amplitude and frequency of the node voltage may actually fluctuate by some disturbance, they stay within a small range in general stable power system. Therefore, the node voltage used to be considered as an ideal sinusoidal voltage source in the research field of the utility interactive inverter.

Fig. 9.1 introduces a one-line diagram with the simplest model of the DER, which can trade an active and a reactive power between the DER and the power system. V_c is the output voltage phasor of the DER, V_s is the node voltage at the coupling point of the DER, and jX_c is a reactance, which represents an interconnecting inductor placed between the utility interactive inverter and the power system. The percent reactance of the inductor may be 3–5% of the DER.

The relations between V_c and V_s can be displayed as shown in Fig. 9.2. In this diagram, the active and reactive power of the DER is determined by the relation between V_c and the current phasor I. On the other hand, practical restrictions are given by the generated power of the primary energy resources such as solar panels or wind turbine, and the rated apparent power of the utility interactive inverter. The former defines the active power uniquely, and the latter restricts the maximum value of the reactive power. Therefore, the controller of the DER has to define the output current phasor of I depending on the voltage phasor of V_s at the coupling point of the power system by a phase locked–loop controller. In general, the current phasor is defined by the two-axis theory based on the reference phase angle given by the phase locked–loop controller. The inverter controller detects its three-phase output current as two-axis components, compares them with their reference values, and defines the appropriate modification of the output voltage. When the reference output voltage phasor is decided, it should be converted to three-phase sinusoidal reference voltage waveforms by decomposition of the two-axis theory.

Generally, the sinusoidal voltage V_c is realized by a self-commutated inverter from an equivalent ideal DC voltage source, which is provided by the DER. Various circuit topologies and converter control techniques can be found, but the most popular one is a full bridge inverter and a pulse width modulation control. When the frequency of its carrier wave is sufficiently higher than the nominal frequency, the utility interactive inverter does not basically cause harmonic problem. It should be high frequency under the restriction by the semiconductor devices because the inverter may cause audible noise depending on the switching frequency. In addition, the utility interactive inverter has some necessary functions for electric safety and protection of the apparatus.

9.1 POWER ELECTRONICS IN PV POWER GENERATION SYSTEMS

Fig. 9.3 shows a basic configuration of the PV generation system for the three-phase power system. It consists of three components; a solar panel, a DC–DC converter, and a utility interactive inverter. The DC–DC converter is the first stage of power conversion. It regulates the DC voltage across the solar panel so as to derive its maximum power as well as keeps a constant DC voltage at the DC side of the utility interactive inverter. And then, the DC power is converted to AC power by the utility interactive inverter. When the active power output of the utility interactive inverter is controlled to keep the DC voltage constant, this power conversion can work successfully.

Fig. 9.4 represents a simple model of the PV generation unit, which consists of solar panels. It has a typical relationship between output current and voltages shown in Fig. 9.5. By this relationship, the power curve by the open-circuit voltage can be derived as shown in Fig. 9.6. As can be seen in Fig. 9.6, the PV generation unit has an optimal operating point at which the unit can generate its maximum power. However, these curves shown in Figs. 9.5 and 9.6 depend on the conditions of solar irradiation and the temperature. In the circumstances, the input voltage of the DC–DC converter in Fig. 9.3 has to be considered as a variable voltage, depending on the conditions around the solar panels. Therefore, the DC–DC converter

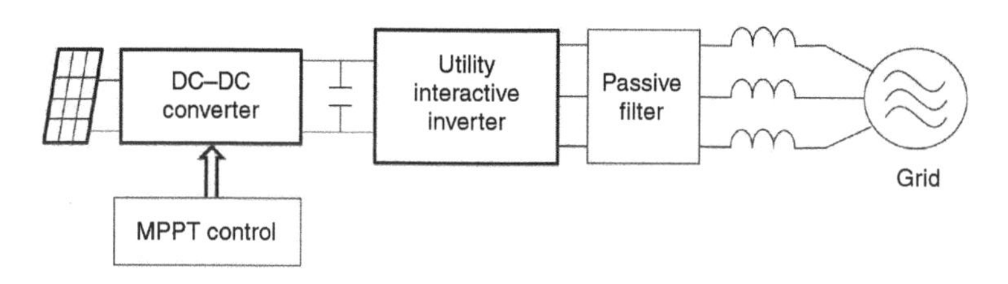

FIGURE 9.3 A basic configuration of the PV generation system.

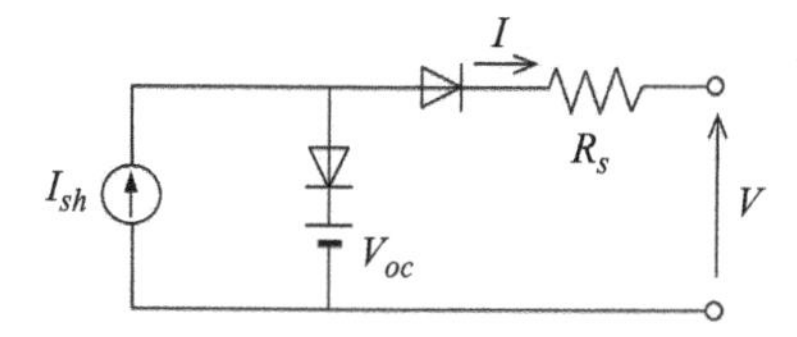

FIGURE 9.4 A simple model of the PV generation unit.

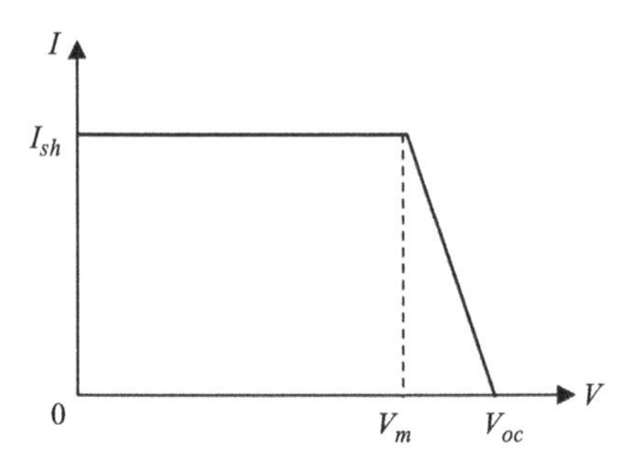

FIGURE 9.5 The relationship between output current and voltages of the model.

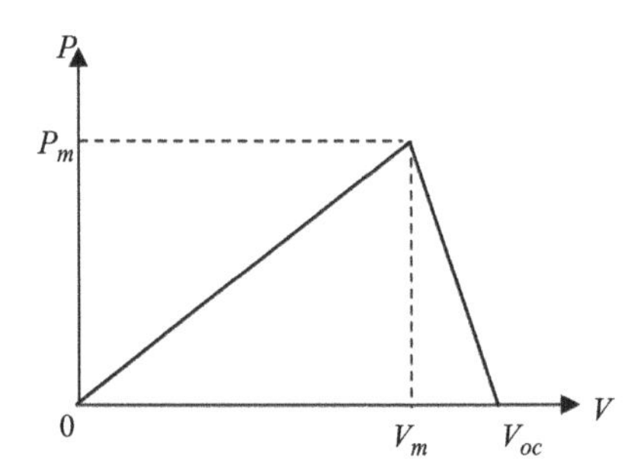

FIGURE 9.6 Characteristics of the output power of the model.

is required to connect the variable input voltage and the constant output voltage for the utility interactive inverter.

Normally, the voltage across the PV generation unit is designed as smaller than the DC voltage of the utility interactive inverter. The power flow through the DC–DC converter is only one direction from the solar panels to the grid. Therefore, the DC–DC converter is realized by a boost chopper in many cases. There are some alternative circuit topologies such as a high-frequency DC–DC converter with an isolating transformer for a small rated power unit.

9.2 POWER ELECTRONICS IN WIND POWER GENERATION SYSTEMS

The most simple wind power generation unit simply consists of an induction motor. If a wind turbine is accelerated by the wind over the nominal rotating frequency of the grid, the induction motor becomes to generate the electric power.

Today, a lot of high power wind turbines are installed. The mechanical torque from the wind turbine is converted to the electric power by a synchronous generator or a variable speed induction generator.

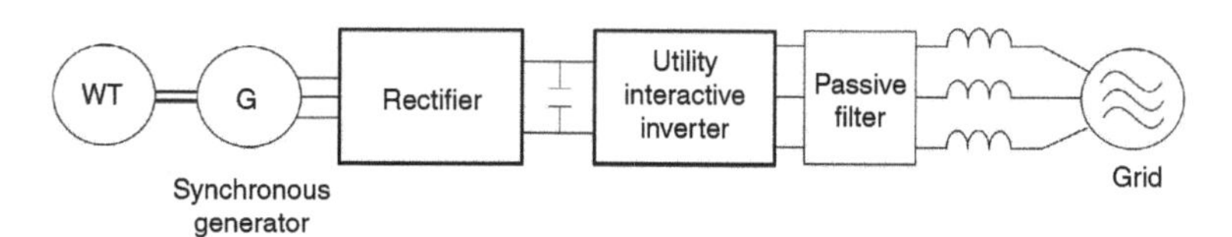

■ **FIGURE 9.7 A basic configuration of the wind power system based on a synchronous generator.**

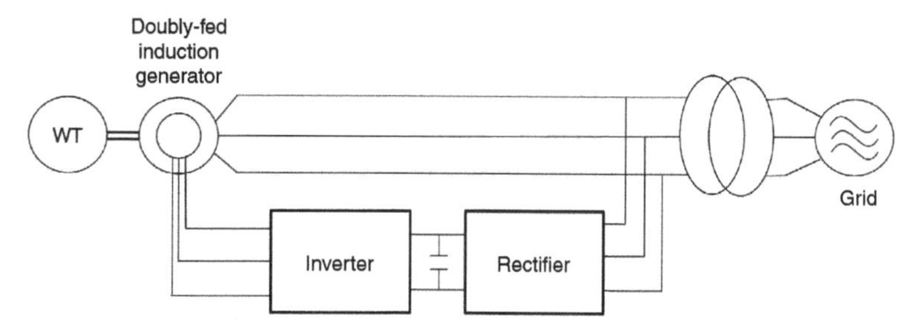

■ **FIGURE 9.8 A basic configuration of high power variable speed wind power system.**

The former needs a utility interactive inverter because the synchronous generator supplies AC power with variable frequency depending on the wind speed. Fig. 9.7 shows a basic configuration of the system based on the synchronous generator. The power conversion consists of two steps. First, the generated AC power with variable frequency is converted to DC power by the rectifier, and then it is sent to a utility interactive inverter.

Fig. 9.8 shows a typical wind power system by variable speed induction generator. The generated main power does not flow through a utility interactive inverter. The back-to-back converter applies magnetizing current to the rotor of the induction generator. As a result, the wind turbine can rotate with frequency different from the nominal frequency of the grid.

In addition, the wind power generation system is connected or combined with high- or medium-voltage DC transmission system in such as offshore windfarms.

9.3 POWER ELECTRONICS IN BATTERY ENERGY STORAGE SYSTEMS

Fig. 9.9 shows a basic configuration of the battery energy storage systems. As the battery unit trades DC power, it is constructed similarly to the PV generation system. It consists of three components; a battery unit, a DC–DC

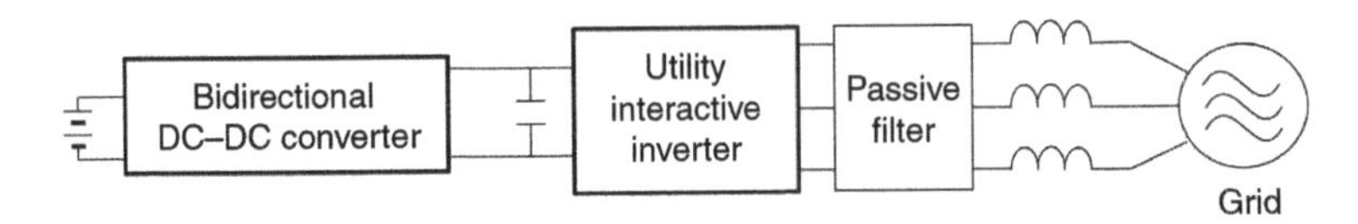

■ FIGURE 9.9 A basic configuration of the battery energy storage system.

converter, and a utility interactive inverter. By contrast to the PV generation system, the DC–DC converter has to be a bidirectional-type converter. In addition, the battery voltage slightly varies along its state of charge. Moreover, voltage imbalance can be caused among the battery cells in some cases. In this case, the DC–DC converter has to provide a function of balancing control.

9.4 POWER QUALITY PROBLEMS WITH RELATED TO DERs

For power system analysis, it is important to build a model suitable for analysis. In many cases, the power system is analyzed based on effective values. All the voltages and currents are represented by effective values and the other variables such as active and reactive power will be calculated by them. In such analysis, system behaviors in the range over tens of cycles are considered. In this case, the DER unit can be modeled by current source shown in Fig. 9.10 represented by effective value of its output because the current controller of the DERs works well and the utility interactive inverters can regulate its output voltage at sufficiently quick speed. In this case, various DERs can be modeled by a common style as shown in Fig. 9.10 because engineers need not to consider what the background energy source is but to know how their output are regulated by the DERs controller. This model is effective for study on energy management among the DERs.

On the other hand, some engineers consider more detail phenomena, which vary around the several cycles. For example, transient phenomena after a system disturbance should be considered by more detail model. In such cases, the voltage source model shown in Fig. 9.1 may be useful. By applying a current controller to the voltage source model, the transient phenomena can be analyzed with varying node voltage.

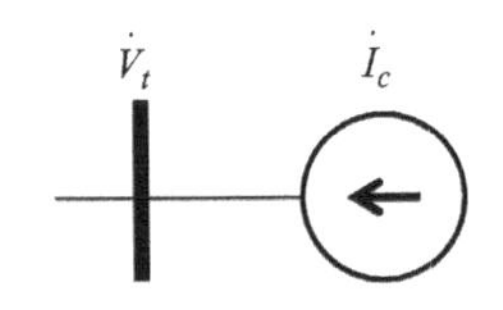

■ FIGURE 9.10 The simplest current source model of the DER.

Moreover, some engineers consider more detail phenomena, which vary shorter than several cycles.

For example, harmonics and high-speed transient behaviors should be considered by instantaneous values because other frequency components must be considered in such case. The converters must be configured by a static power converter including switching devices.

There can be some challenges of power quality improvements such as reduction of harmonics or suppression of three-phase imbalance by the utility interactive inverters. Both can be detected as AC components in two-axis theory. Therefore, if the utility interactive inverter has sufficient surplus capacity, the power quality improvement can be realized by the converter control, which regulates the d- and q-axis components as constant values.

Chapter 10

AC/DC microgrids

Kazuto Yukita

Aichi Institute of Technology, Department of Electrical Engineering, Toyota, Japan

CHAPTER OUTLINE

10.1 Basic concept of AC microgrids 237

10.1.1 Islanding mode 238

10.1.2 Connected mode 239

10.1.3 Backup mode 239

10.1.4 Experiment field and device specification 239

10.1.5 System operation 240

10.1.6 Measurement of power quality 242

10.2 Battery charge pattern and cost 244

10.2.1 Battery charge method 246

10.2.2 Test results 246

10.3 Supply and demand control of microgrids 251

10.3.1 Peak cut/peak shift mode operation 251

10.3.2 Receiving constant power mode operation 253

10.4 Basic concept of DC microgrids 254

10.4.1 System operation of DC microgrid 255

10.5 Examples of microgrids in the world 258

10.6 Conclusions 259

References 259

In recent years, the severity of worldwide environmental problems such as global warming and extreme weather events has intensified, while efforts to reduce adverse environmental impacts and to introduce alternative energy sources to replace fossil fuels have been actively promoted. In the field of electrical energy, distributed power sources that utilize natural and renewable energy sources, such as photovoltaic (PV) and wind power generation (WG), as well as biomass power generation and fuel cell technology, continue to be active areas of research and development.

Power generation systems that use natural energy have been highly anticipated; however, with some distributed power sources, it is difficult to achieve stable power generation owing to the dependence of the energy sources on weather and other natural conditions. Thus, research into power supply systems configured from distributed power sources that have already been introduced within specific areas is being actively pursued with the aim of achieving both environmental compatibility and power supply stability. Furthermore, with the widespread popularization of information and communication technology, consumers have adopted stringent requirements with regard to power quality (frequency, voltage, and harmonics).

In this chapter, AC/DC microgrids that have introduced an electrical generating system supplied with renewable energy are described.

10.1 BASIC CONCEPT OF AC MICROGRIDS [1]

The concept of the typical AC microgrid is shown in Fig. 10.1. It consists of DGs (PV, WG, FC, etc.), a bidirectional converter, an ACSW, and Valve-Regulated Lead-Acid battery (VRLA) batteries. The bidirectional converter and the DGs are connected to an AC grid, and the ACSW is inserted between the utility and the AC grid. The VRLA batteries are connected through the DC bus to the bidirectional converter. The system changes between each operational mode easily and without any interruptions. The

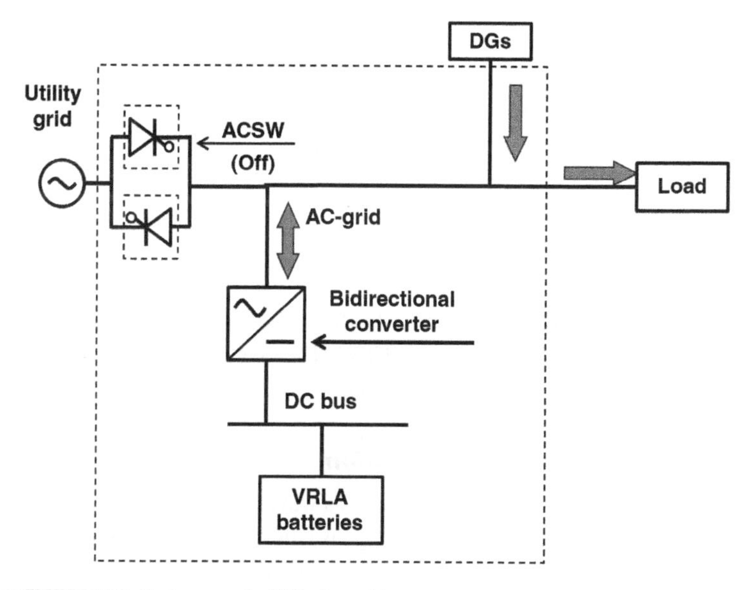

■ FIGURE 10.1 Basic concept of AC microgrid.

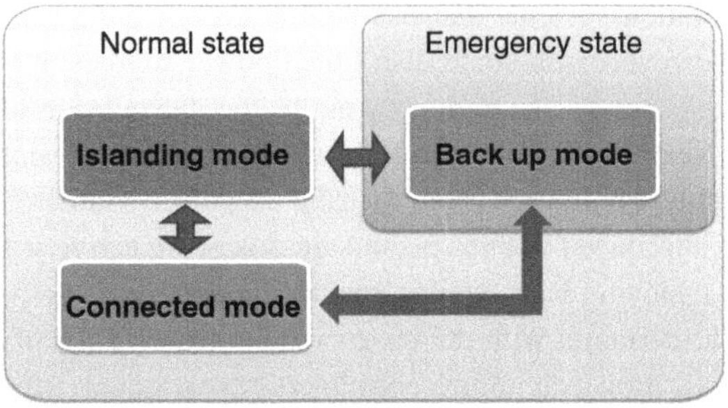

■ FIGURE 10.2 State transition diagram of the system

Table 10.1 System Operation Mode

Mode	ACSW	Bidirectional Converter Operation	AC Voltage/ Frequency	DGs	Harmonics
Islanding	OFF	Asynchronous operation • Asynchronous AC output with utility grid Synchronous operation • Synchronous AC output with utility grid	Constant	Maximum power point tracking (MPPT)	Repression of voltage distortion in AC-grid N/A
Connected	ON	• Active filter for harmonics current • Battery charging	Depends on utility grid		Repression of harmonic current as AC input
Back up	OFF	• Asynchronous AC output with utility grid	Constant		Repression of voltage distortion in AC grid

transition of the system is shown in Fig. 10.2. Table 10.1 shows the operational modes. This system has three modes, and their details are explained subsequently. Then, in all modes, the inverter and the converter for DGs are operated using maximum power point tracking.

10.1.1 Islanding mode

When the system operates in the islanding mode, it is disconnected from the utility grid by turning off the ACSW. The operator then chooses either asynchronous or synchronous operation.

10.1.1.1 *Asynchronous operation*

The output of a bidirectional converter is a waveform that has constant voltage and constant frequency. The voltage is adjusted so that it does not deviate from the nominal voltage (200 V) by more than 2%, and it also does not deviate from the typical frequency (60 Hz) by more than 0.1%. In this mode, even if the bidirectional converter has a problem, the ACSW will be turned on after an interval.

10.1.1.2 *Synchronous operation*

The output of a bidirectional converter is a waveform that is synchronous to the utility grid. Voltage amplitude and voltage phase in the AC grid are synchronized with the utility grid. Thus, if the voltage and the phase of the utility grid vary, so do the voltage and the phase of the system. Even though the bidirectional converter may have a problem, the system continues to supply power to the load when in connected mode. When the ACSW are turned on, there is no phase jump.

10.1.2 Connected mode

When the VRLA batteries discharge, the voltage of the DC bus drops to the lowest threshold, and the system starts to charge the VRLA batteries. In that case, the operation mode is changed to the connected mode. In this mode, the thyristors (SCR) serving as ACSW are turned on alternately every half cycle. In the connected mode, the bidirectional converter operates not only as a rectifier for batteries charging but also as an active filter for rejecting harmonic current from loads. When the bidirectional converter charges the VRLA batteries, the bidirectional converter is initially operated by constant current (CC) control. When the DC bus voltage increases to the highest threshold, the system changes to the islanding mode. Furthermore, we can change each value of the current via CC control.

10.1.3 Backup mode

During the charge operation, any problems in the utility grid, for example, a dip in voltage, can turn the ACSW off. Then, the operational mode of bidirectional converter is changed to asynchronous operation of the islanding mode, the system continues to supply power.

10.1.4 Experiment field and device specification

Fig. 10.3 shows the two sets of microgrid systems used in the study. Both systems are almost the same, except for the capacity of the VRLA batteries. One of the systems was set up in Building 12, a building with five floors, with rooms usually used as classrooms and offices. The electric loads are lights and air conditioners. The other system is installed in a library. The library has four floors with book storerooms and reading rooms. The electric loads are lights and PCs. Because 50 kVA transformers were put into the

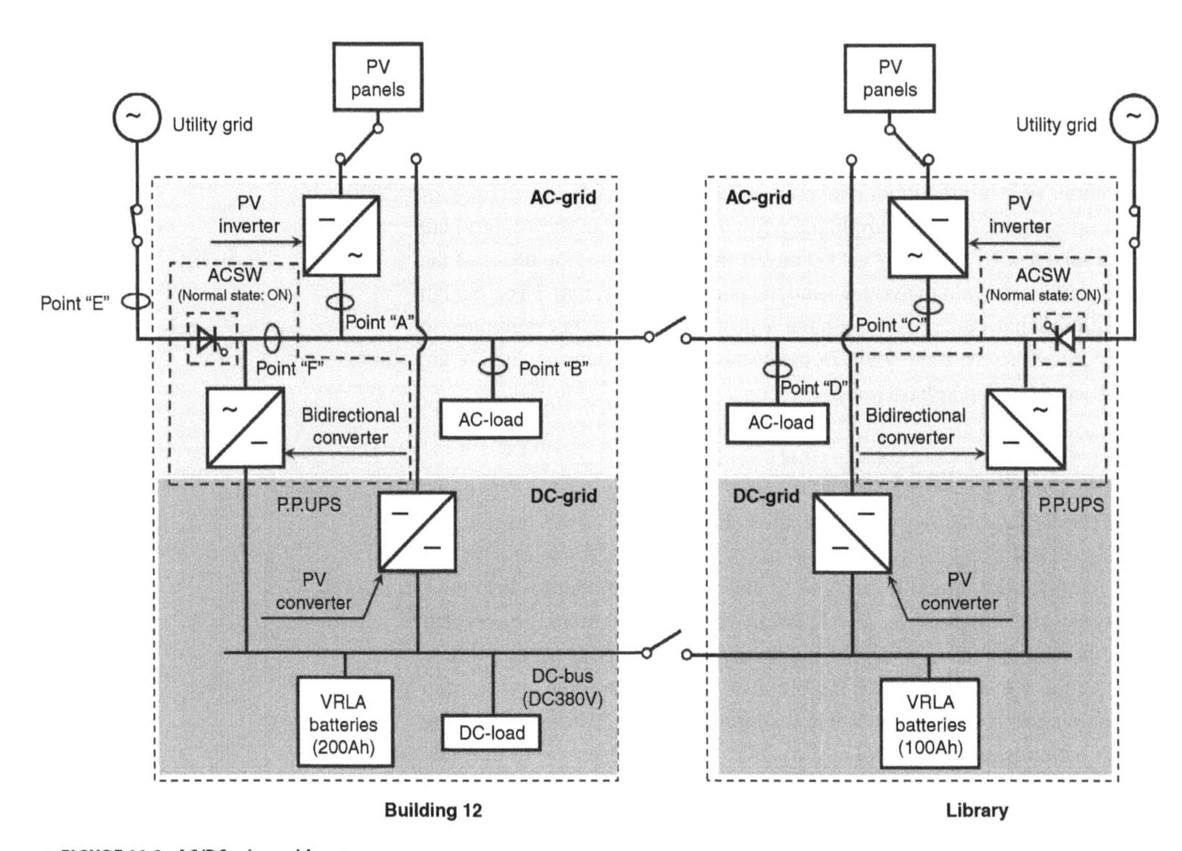

FIGURE 10.3 AC/DC microgrid systems.

buildings, the capacities of the bidirectional converters were set to 50 kW. We also set up two kinds of PV panels. One of them is a polycrystal silicon type that has a typical output power of 10 kW. The other type is monocrystal silicon, with a voltage of 10 kW. We use VRLA batteries because they are a commonly used energy source.

10.1.5 System operation

The load duration curves and frequency distributions of the PVs' output are shown in Fig. 10.4. The data were collected in May 2008. The data were taken every 5 s for 24 h, that is, at 17,280 points. As shown in Fig. 10.3, the measurement points are "A," "B," "C," and "D." Points "A" and "C" are outputs of the PVs. Points "B" and "D" are inputs of the loads. We can see that the load consumption of the library is fixed in both operation and nonoperation modes. At Building 12, the connected loads were controlled. Therefore, the total load consumption was smaller than the amount of electric power generated by PV.

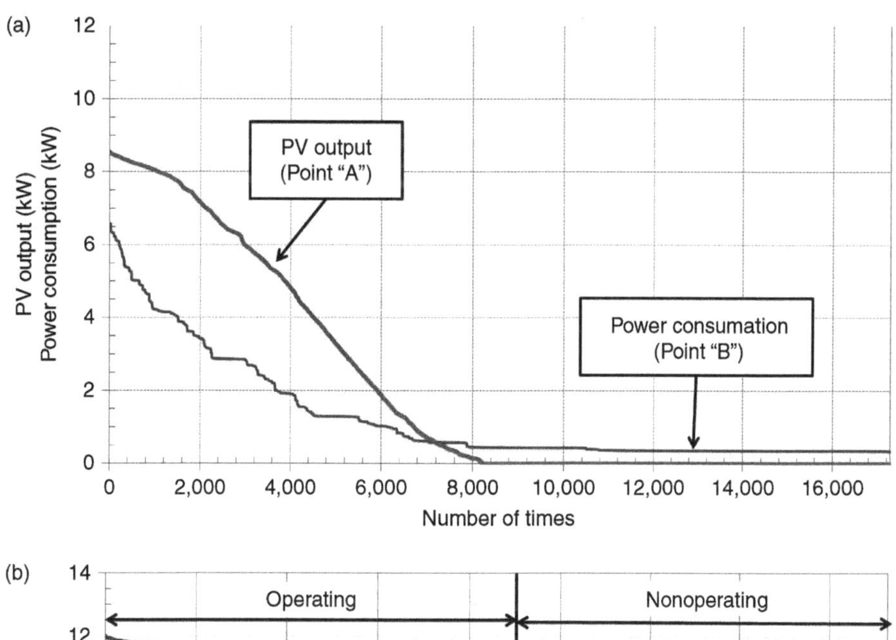

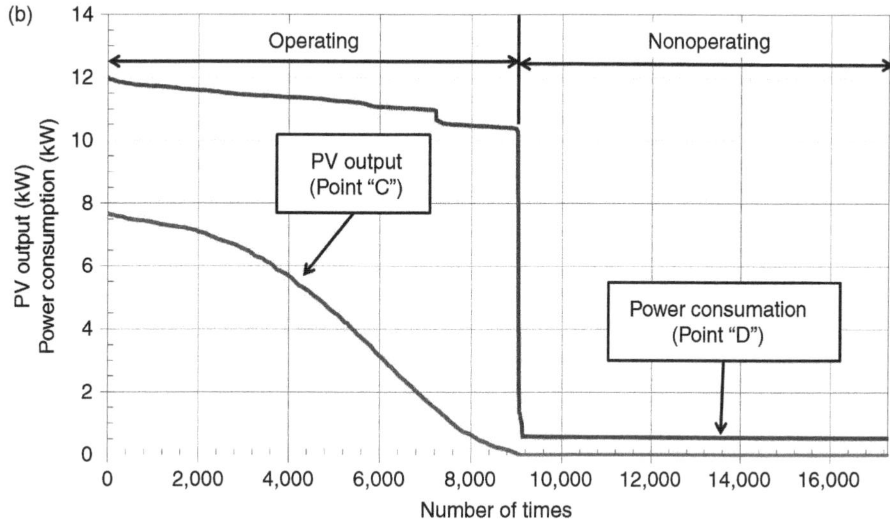

FIGURE 10.4 Load duration curve and PV output. (a) Building 12; (b) library.

However, at the library, the total load consumption is greater than the electric power generated by PV, because the connected loads were not controlled. The power flows in the AC grid of Building 12 (i.e., those of the utility grid), PV output, and load consumption are shown in Fig. 10.5. The utility grid power increased from 5 to 8 am, which indicates the battery charging in the connected mode. Further, the power supplies to loads continued for a whole day, which shows that state migrations were complete.

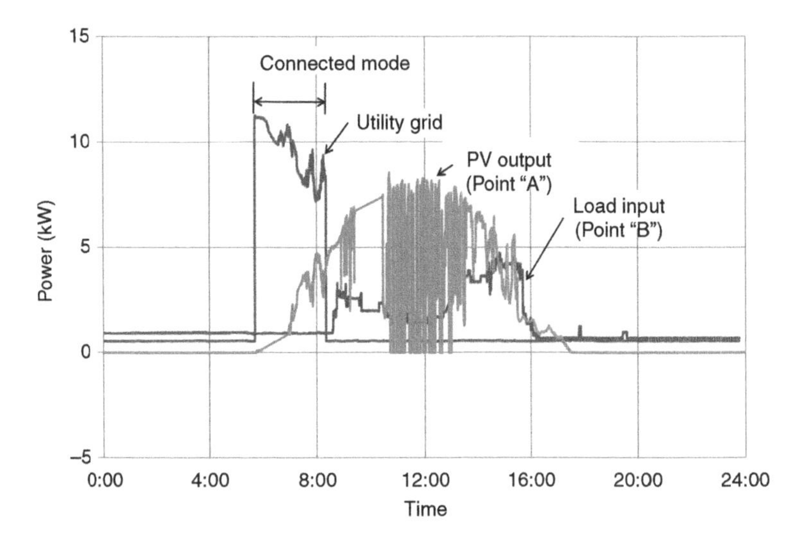

■ FIGURE 10.5 Daily load curve of Building 12.

10.1.6 Measurement of power quality

To measure power quality, the PV power is connected through the PV inverter, and the bidirectional converter and ACSW are connected with the AC grid. Further, the VRLA batteries are connected through the DC bus to the bidirectional converter. The loads are the lights and the air conditioners in Building 12. The load power was set in the range from 5 to 8 kW, and the PV output ranged from 2 to 8 kW. The system operated in the islanding and connected modes.

10.1.6.1 *Voltage and frequency*

The voltage–frequency characteristics are shown in Fig. 10.6 for asynchronous operation and asynchronous operation in the islanding mode. The root mean square (RMS) voltage of the AC grid in Building 12 (Point "F" in Fig. 10.3) is plotted on the vertical axis, and the frequency is plotted on the horizontal axis. In addition, the areas enclosed by the dashed line and the solid line show how much operational voltage is required for the induction motor, written as JEC-2137-2000 in Japanese. The first area shows the short-term operational requirement, and the second shows the long-term operation requirement. Although both modes meet each requirement, in the case of the asynchronous mode, the improved power quality is clear.

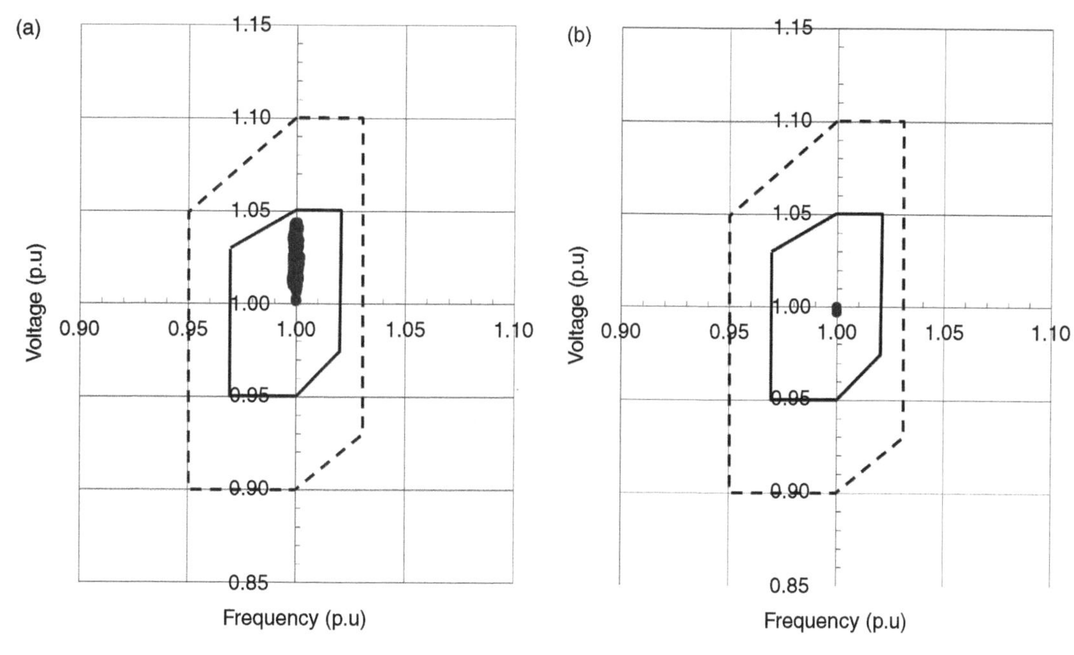

FIGURE 10.6 Voltage–Frequency characteristics. (a) Synchronous; (b) asynchronous. Samples: 14,400 points (for 12 h period in 2008).

10.1.6.1.1 Voltage harmonics

The total harmonic displacement (THD) and the rate of content that is in accordance with the degree of voltage harmonics at the AC grid in Building 12 (Point "F" in Fig. 10.3) is shown in Fig. 10.7. There is a guideline in Japan requiring that the THD of harmonic voltage in a high-voltage

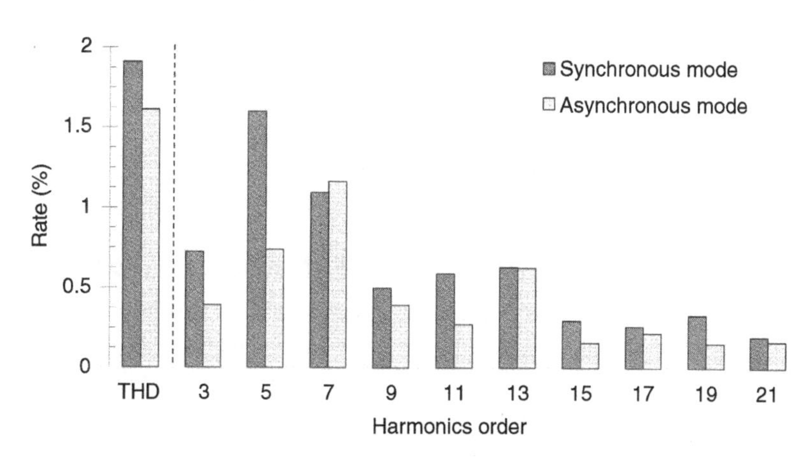

FIGURE 10.7 Distortion rate and THD of harmonic voltage.

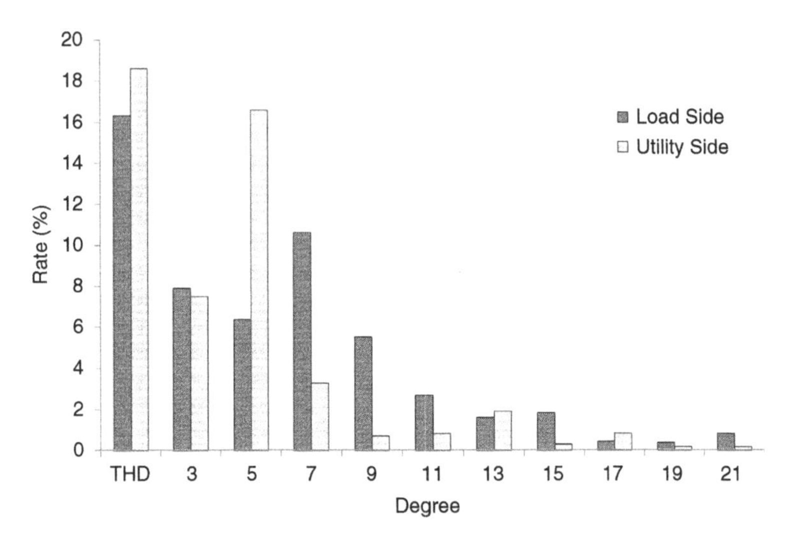

■ FIGURE 10.8 Distortion rate and THD of harmonic current.

line be limited to 5%. Although both operational modes meet this requirement, when the synchronous mode is compared to the asynchronous one, we can see the THD decreases by 0.25% and several higher degrees also decrease. In addition, the cause of the increased 7th harmonics is under investigation.

10.1.6.1.2 Current harmonics

The THD and several degrees of harmonic currents, which are the load sides (Point "F" in Fig. 10.3) and the utility-grid sides (Point "E" in Fig. 10.3), are shown in Fig. 10.8. By comparing both sets of data, we can see the effect of using the active filter with a bidirectional converter. As shown in Fig. 10.8, the THD decreases by 1.27%, and there is also a decrease of several higher degrees.

10.2 BATTERY CHARGE PATTERN AND COST [2]

In this section, we focus on receiving power from the utility grid and study ways to reduce the cost of electricity. When this microgrid system receives power from the utility grid, most of the power is used to charge the battery. Therefore, we tested several patterns for charging the battery and calculated the cost of electricity based on these test results. As a result, we found that when we reduce the peak power and extend the charge times, the cost of electricity is reduced.

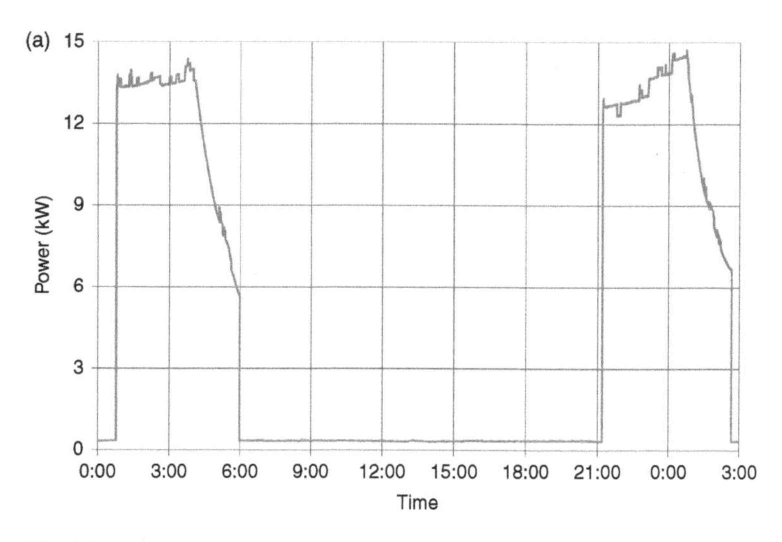

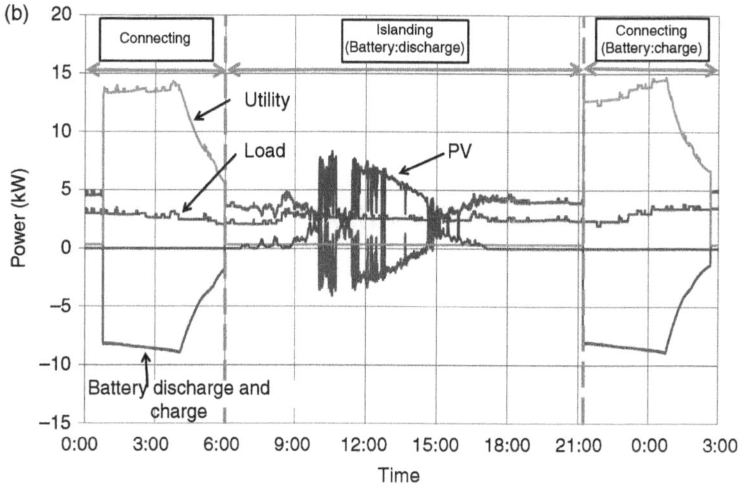

■ **FIGURE 10.9 Daily power load curve.** (a) Utility power in a day; (b) power flow and load curve in a day.

Fig. 10.9a details the utility power that is shown in Fig. 10.9b. The AC microgrid system is based on supplying to the load with the minimum utility power necessary. However, it is difficult for the PV system to supply stable power generation to the load. Therefore, in fact, we have relied on receiving power from the utility grid. Table 10.2 shows the average battery charge times from Jan. to Dec. 2008. This table shows that it is charged more than once per day on average. We can see that it is difficult to supply power only by PV generation and the battery. Considering this, our focus was on the

Table 10.2 Average Battery Charge Times

Operated Days	Charged Times	Average Per Day
169 (days)	269 (times)	1.59 times/days

economical operation of this system and a reduction of the cost of electricity. In particular, when this system receives power from the utility grid, its purpose is to charge the battery. Therefore, we changed the charge pattern for the battery, focusing on the following.

1. The cost of electricity is decided based on the peak power from the utility grid. We studied charge patterns to reduce the use of peak power.
2. The price of electricity is lower at night than at other times of day. We studied the charge pattern to make maximum use of the night rate.

Based on these two points, we supposed that charging the battery is done at the night rate. We also calculated the effect of charge patterns on the cost of electricity used. In practice, because the PV power and load change from day to day, the battery energy does not stay constant.

10.2.1 Battery charge method

The battery is initially charged by constant current (CC) control. Then, the DC bus voltage increases to specified values, and the charge mode automatically changes to constant voltage (CV) control. Taking this into consideration, to reduce the commercial power received, we reduce the CC charge value in CC control. However, if this continues, the amount of battery energy is expected to be insufficient. Therefore, we will make up for the energy shortfall to extend the time of the CV control. To use the night rate more, we set the charge terms of CV control at 8 h. In this test, the charge for the battery is controlled by the bidirectional converter, so we change the setup value. The test results are described further.

10.2.2 Test results

10.2.2.1 Charge pattern

We set the following three cases as charge patterns for the battery. We measured the voltage, current, peak power, and charge energy for the battery.

Pattern (A), CC charge value: 25 A, CV control time: 2 h
Pattern (B), CC charge value: 18 A, CV control time: 2 h
Pattern (C), CC charge value: 25 A, CV control time: 8 h

Fig. 10.10a–c shows the battery voltage and current for the aforementioned patterns.

The test circuit was the same as in Fig. 10.3, and the specification was the same as in Table 10.3. The load curve was almost the same as described in Fig. 10.9. The peak demand was about 5 kW.

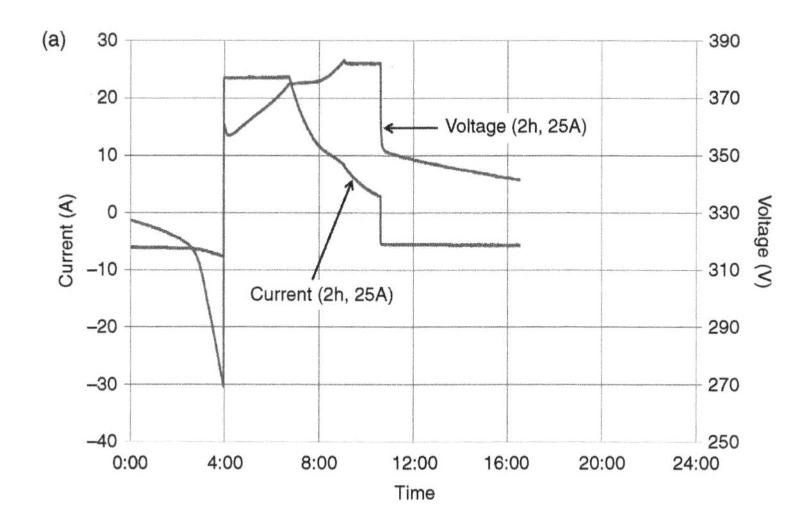

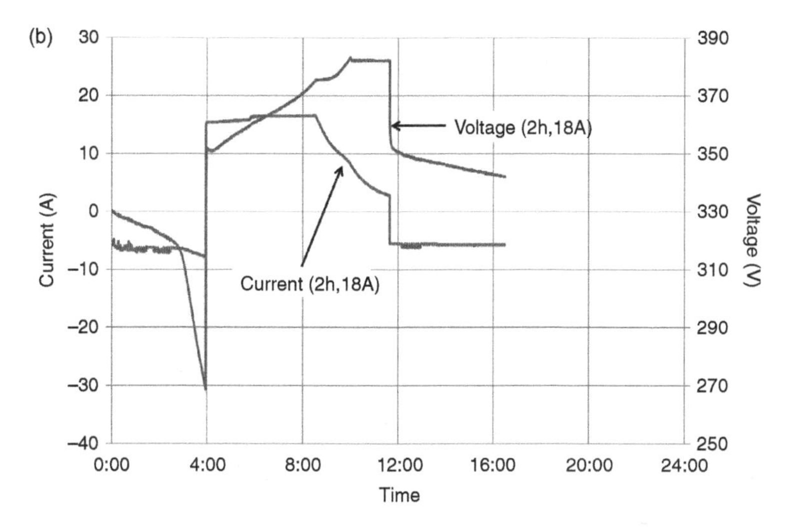

■ FIGURE 10.10 Current and voltage during battery charge. (a) CC charge value 25 A, CV charge time 2 h; (b) CC charge value 18 A, CV charge time 2 h; (c) CC charge value 25 A, CV charge time 8 h.

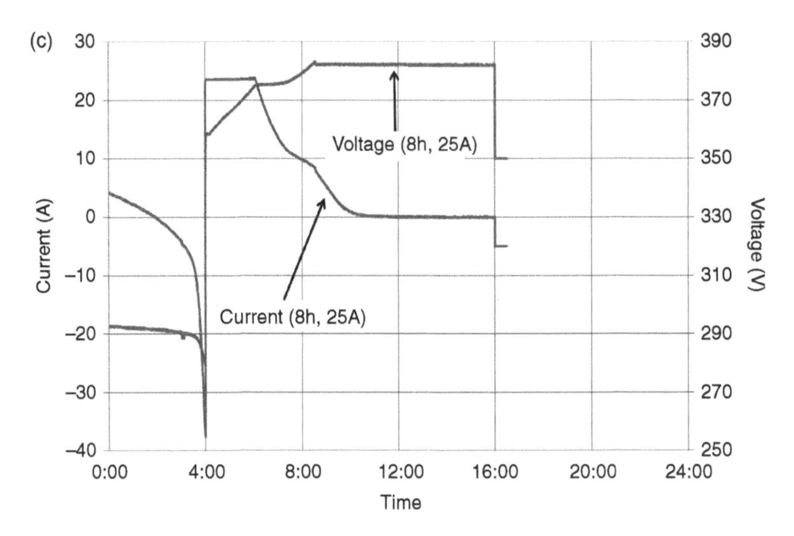

■ **FIGURE 10.10** *(Continued)*

Table 10.3 Device Specifications

Devices	Quantity	Capacity	Remarks
Photovoltaic (PV) panels	2	10 kW	
PV inverter	2	10 kW	
PV converter	2	10 kW	
Bi-directional converter	2	50 kW	
VRLA batteries	2	2V 200Ah 2V 100Ah	168 (cells/set)

10.2.2.2 ***Test result***

Fig. 10.10 shows the period from beginning to measure the data. After 4 hours, the direction of the battery current changes from negative to positive, changing the operating mode of the system from islanding to connected and charging the battery. Immediately after the connected mode starts, the battery begins to be charged by the CC control mode. After the battery voltage approaches a constant value, the charge mode changes to CV control. Furthermore, from the results shown in Fig. 10.10, we could ensure that the system is operated accurately with this CC and CV control mode.

Fig. 10.11 shows a comparison of the charge power of the battery under different circumstances. Fig. 10.11a shows the case of reducing CC value, and Fig. 10.11b shows the case of extending CV terms. Furthermore, the results

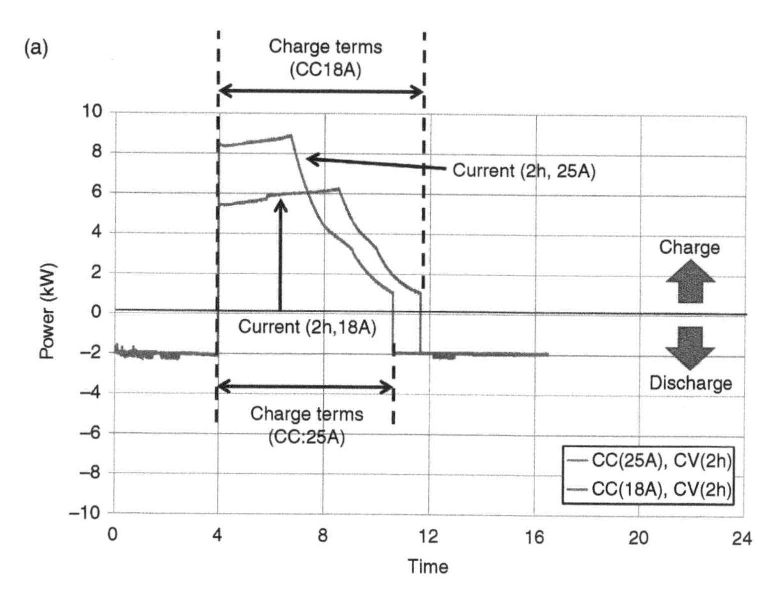

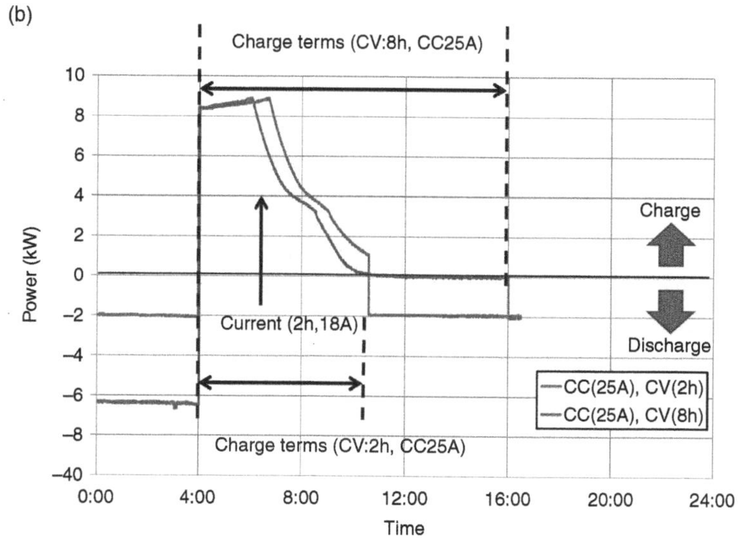

■ **FIGURE 10.11 Comparison of the charge power.** (a) The case of reducing CC value; (b) the case of extending CV terms.

of calculating electric energy are shown in Table 10.4. From Fig. 10.11, we see the effect of peak power reduction on reducing the CC charge value. We also investigated changing the CV control time from 2–8 hours when the CC charge value was the same. It is clear that the charge power of 8 hours was less than that of 2 hours. As Table 10.4 suggests, when we changed the

Table 10.4 Charge Pattern

	CC Charge Value (A)	CV Charge Value (h)	Peak Power (kW)	Electric Energy (kWh)
1	25	2	8.9	38.52
2	18	2	6.27	34.07
3	25	8	9.22	31.74

charge pattern, the charge power required was reduced by 17% from 38.52–31.74 kWh. The reason is thought to be that distinguishing the discharge value before the battery is charged affects the charge power.

10.2.2.3 *Effect on electric rate*

We estimated the cost of electricity for one month based on the test results in the previous paragraph. As a prerequisite for this calculation, the charge power was almost the same the entire time. In this situation, we studied the effect on reducing the cost of electricity used. When we changed the CC charge value arbitrarily, we would also change the CV control time so that the charge power was always kept constant. We used the following conditions to calculate the cost of electricity:

- Basic charge = peak power × basic unit price
- Energy charge = total charge times × energy unit price (total charge times = CC charge time + CV charge time)
- Receiving power is considered to be as follow: 1 day of test data × 30 days
- The battery finishes charging at the night rate.
- The charge power is constant the entire time. (Prices set by Chubu Electric Power.)

The calculation results are shown in Fig. 10.12. We see that, if we assume that the charge power for the battery is always the same, lengthening the total charge times and lowering the peak power is effective for reducing

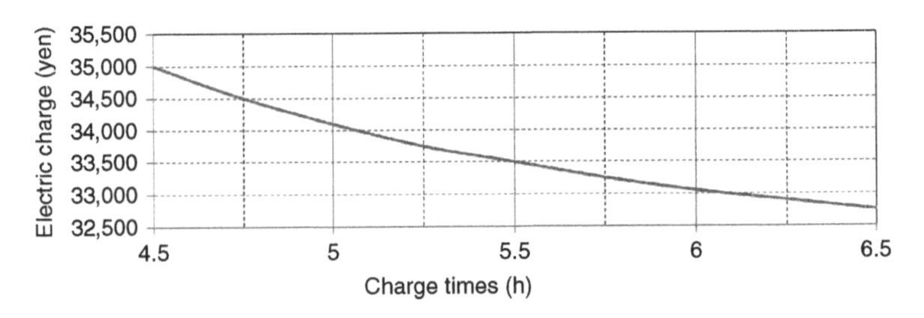

■ **FIGURE 10.12 Effect of charge times on electricity rate.**

the cost of electricity. In this system, we changed the total charge times from 4.5–6 hours, reducing the cost of electricity by 10%. It is preferable to charge the battery at the night rate, and it is much better for it to finish charging before the PV generator starts to generate power.

10.3 SUPPLY AND DEMAND CONTROL OF MICROGRIDS [3,4]

The peak cut/peak shift mode operation and the receiving constant power mode operation are described as an example of the control of a supply and use of a demand control in this section.

10.3.1 Peak cut/peak shift mode operation

Fig. 10.13a shows the electricity demand that can be put in the grid using an energy storage system during a peak of electricity demand. Then, Fig. 10.13b shows the relationship between the electricity rate and the charge pattern for the VRLA batteries. It is important to charge the batteries during times when the electricity rate is low and to discharge the batteries during times when the electricity rate is high. In this manner, a reduction in electricity cost can be expected. A concrete instance follows.

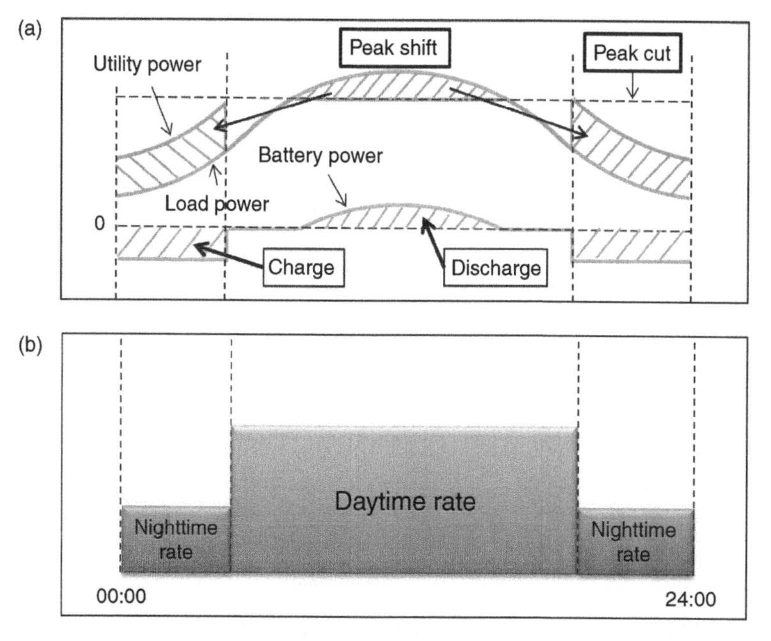

FIGURE 10.13 Daily load curve and electricity rate. (a) Day load curve of the peak cut/peak shift operation; (b) electric rate.

Fig. 10.14 shows the result of the experiment in the case of the peak cut/peak shift for the AC microgrid system. Fig. 10.14a shows the daily load curve. Fig. 10.14b shows the DC voltage curve. Between 8:30 and 15:10, the load was over 5 kW. The maximum receiving power was about 5 kW, and it was confirmed that the system was performing the peak cut operation. The batteries were charged from 23:00 to 24:00 and from 0:00 to 3:10. The batteries made up for a power shortage by discharging when the utility power exceeded 5 kW. It was confirmed that the system performed the peak shift operation. We

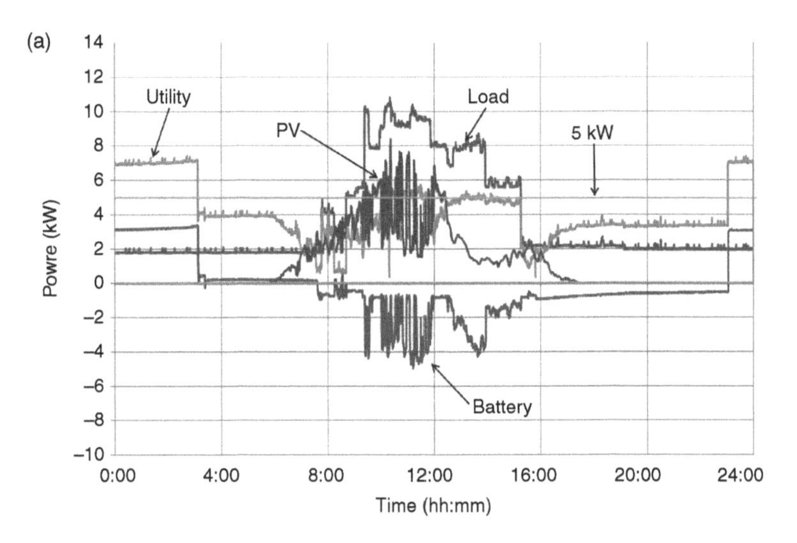

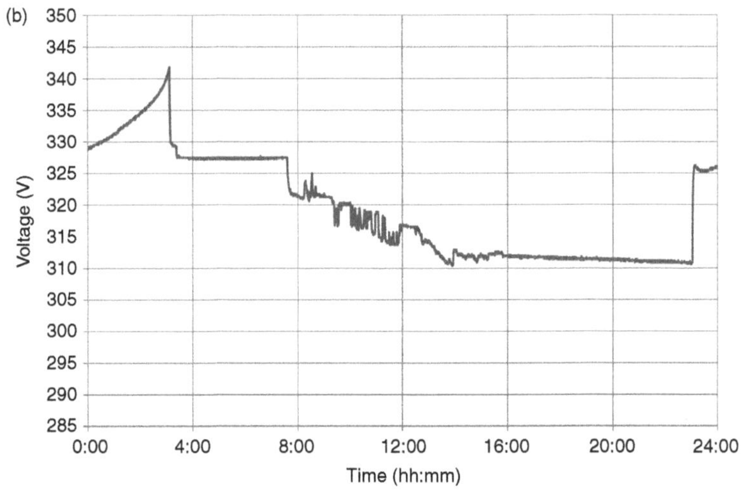

FIGURE 10.14 Daily power load curve and DC voltage curve. (a) Daily load curve; (b) DC voltage curve.

investigated whether a reduction of electric power cost and a smoothing of the received power occurred by the peak cut/peak shift operation.

10.3.2 Receiving constant power mode operation

When the receiving constant power mode operation is put into effect, an adjustment of supply and demand is put into effect by the energy storage system in the microgrid. As a result, influence on the utility power system of the microgrid can be reduced. Predictions of the amount of distributed generation and the electric load power quantity are needed for the implementation of this operation mode. The quantity of constant power received is calculated based on the prediction results.

Fig. 10.15 shows the output of solar power generation and daily load power curve when receiving a constant amount of power from the utility power system. On this day, the predicted amount of PV power was forecast to be insufficient to meet load demand, even though we charged the batteries from 7:00 to 23:00 on the previous day.

Therefore, it was possible to maintain battery capacity while reducing the amount of power received during the day. On this day, the amount of charge and discharge of the battery is a discharge of 8 kWh, the prediction errors are much better than the aforementioned results. Battery voltage on this day is shown in Fig. 10.16. It can be seen that it was possible to maintain the battery voltage between 315 and 335 V.

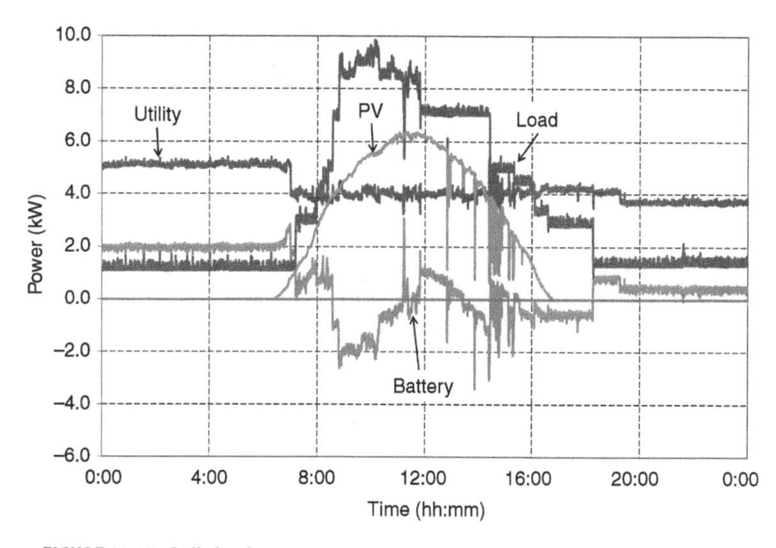

■ **FIGURE 10.15 Daily load curve.**

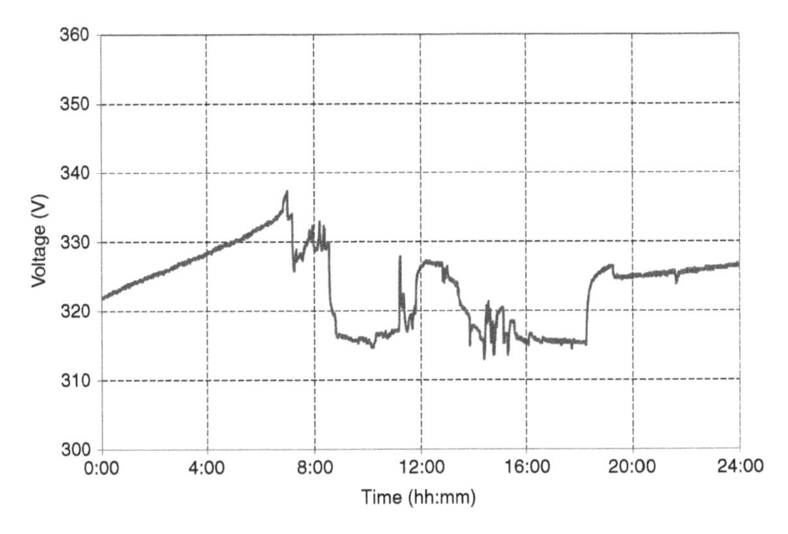

■ **FIGURE 10.16 Battery voltage.**

10.4 BASIC CONCEPT OF DC MICROGRIDS

Recently, distributed generation and energy storage technology have become a point of focus for various reasons based on consideration forwards the environment and energy. These technologies can improve energy efficiency of DC loads and electric loads that contain a DC link, such as light-emitting diode (LED) light, LED television, and air conditioning. The use of a DC power supply can be expected to reduce the number of AC/DC converters, as shown in Fig. 10.17. Because a DC distribution grid can do without a DC–AC inverting link in distributed generation, the DC load, power electronic equipment, etc. are merged into the AC grid, with the ability to improve power distribution reliability and power quality.

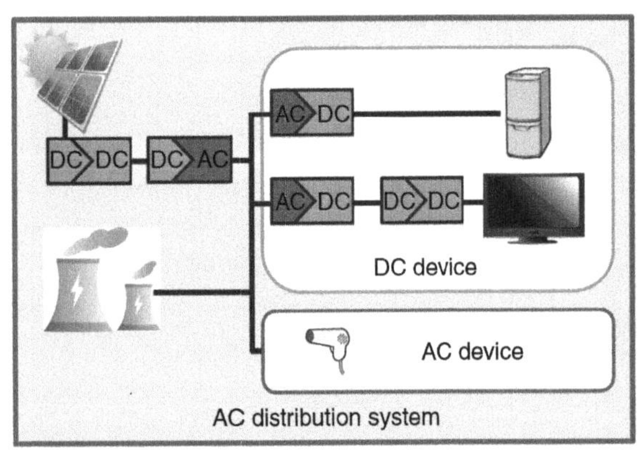

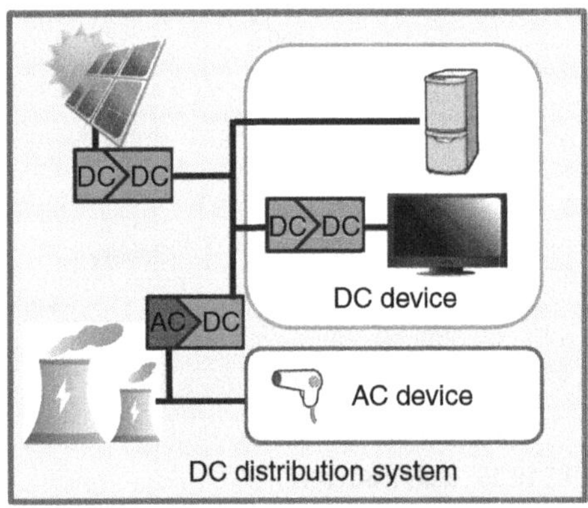

■ **FIGURE 10.17 AC and DC distributed systems.**

Table 10.5 Calculation Result of the Energy Saving Effect by the Direct Current Electricity Supply [5]

Equipment	Number of Units	Consumed Power of One (Wh/year)	Consumed Power (Wh/year)	Amount of Efficient Improvement by the Power Supply Side in DC/AC Reduction (%)	Efficiency Ratio of a Household (%)
Air conditioning (2.2 kW)	3	700	2100	3.0	2.66
Refrigerator (500 L)	1	300	300	3.0	0.24
Hot water supply	1	600	600	3.0	0.47
Induction Heating Cooking Heater	1	600	600	3.0	0.57
Microwave	1	70	70		
Rice cooker	1	100	100		
LED illumination	5	120	480	3.0	0.76
Liquid crystal television (40 in.)	2	150	300		
Recorder	1	80	80		
Other			1696		
Total			6326		4.70

Next, we consider the applicability of the DC power supply. There are two types to change electric power in home electric appliances.

Type 1: Equipment with AC motors and heaters, etc.
Type 2: Equipment with a rectifier and an inverter.

Type 1 is suitable for an AC power supply, and Type 2 is feasible by adopting an appropriate direct current voltage, showing the merit of a DC power supply. Therefore, the efficiency improvement when putting a DC power supply into use with equipment that can be put in the home was calculated as a test. Test results of electricity supply efficiency using a direct current electricity supply are shown in Table 10.5. As a result, it seems possible to reduce the power consumption by about 4.7%. The power consumption at home make up about 1/3 of the power consumption in the whole of Japan. By using DC power supply, it expect that power consumption in the whole Japan can be reduced by about 2-3%.

Figs. 10.18 and 10.19 show the main DC microgrid demonstration tests in Japan and the USA. Thus, the DC microgrid experiments aim at a more efficient use of energy.

10.4.1 **System operation of DC microgrid [7]**

Fig. 10.20 shows the configuration of a DC microgrid system. The DC power system consists of PV and WG generation systems, DC/DC converters, a bidirectional converter, storage batteries, DC load, and AC load. The DC

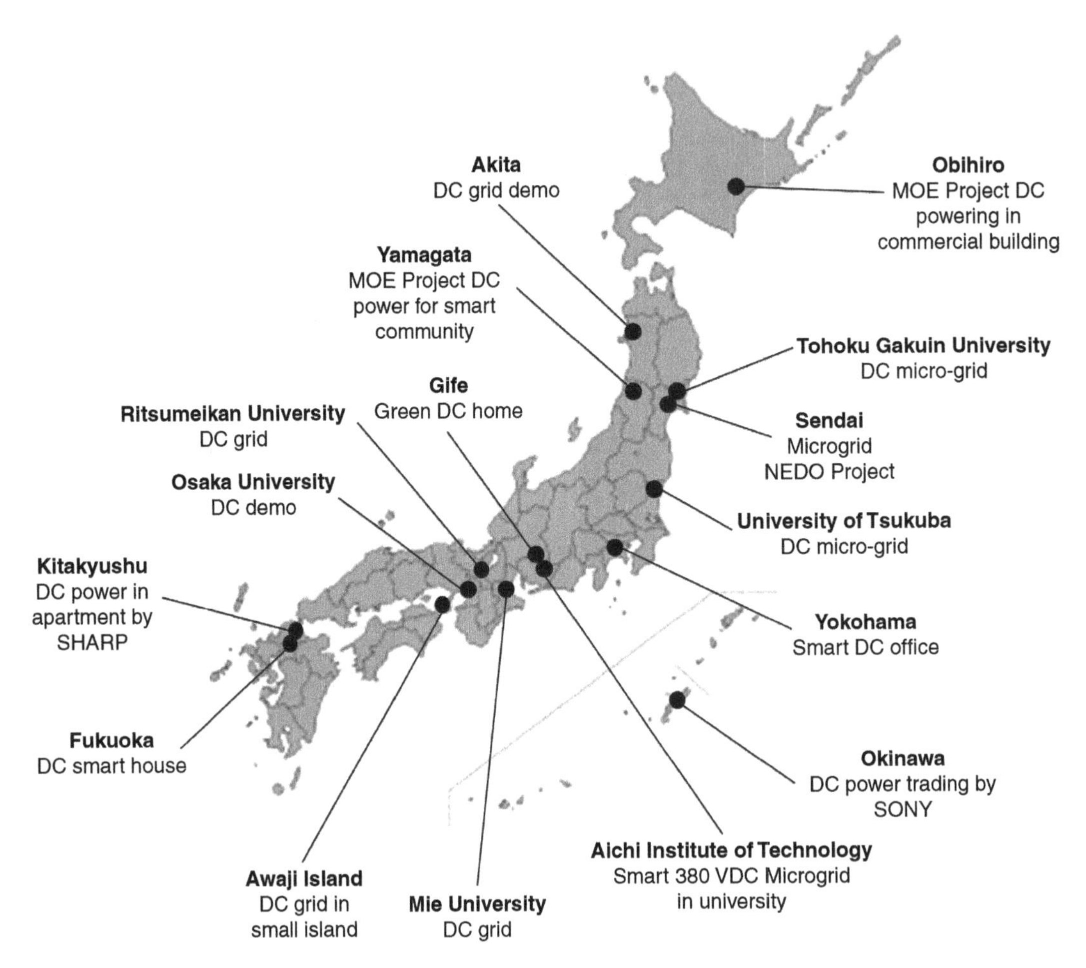

■ FIGURE 10.18 Demonstration tests in Japan [6].

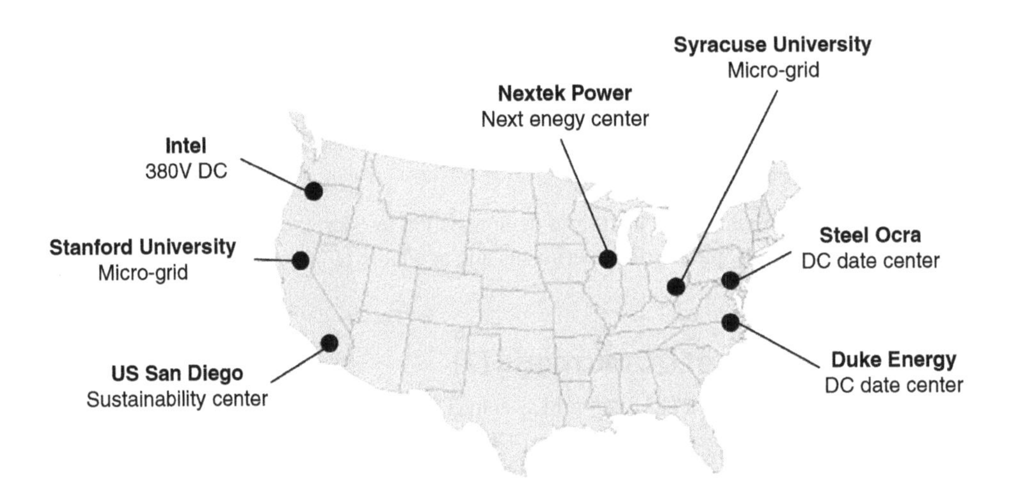

■ FIGURE 10.19 Demonstration tests in the USA [6].

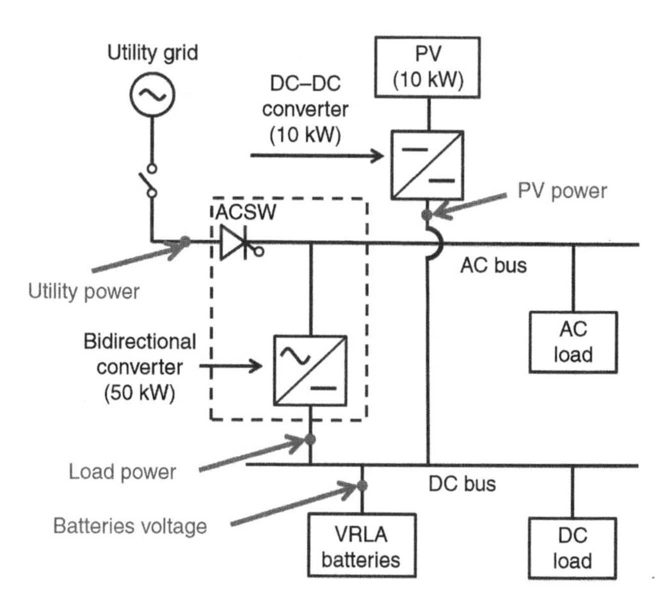

FIGURE 10.20 DC microgrid system.

power produced by PV and WG generation is connected through a DC/DC converter to the DC bus, and power is supplied through the bidirectional converter to the AC load and the DC load. At this point, if the demand for load power is larger than the amount of power supplied via PV and WG generation, an amount equal to the power shortage is supplied from the storage batteries via the bidirectional converter. If the load power demand is less than the amount of power generated by the PV and WG, the surplus power is used to charge the storage batteries. Furthermore, in situations where the discharge of power from the batteries continues until there is a shortage of stored power, the system transitions to the interconnection mode and the batteries are charged.

Fig. 10.21 shows the power flow and the load curve data for this system in a day. Fig. 10.22 shows the battery voltage. The horizontal axis is the time of day. The curves show the utility power, PV power, load power, and battery voltage. The main AC load is caused by lighting and an air conditioner. The main DC load is used for the computer servers. In Figs. 10.21 and 10.22, this means that the system is connected to the utility grid. During the hours from 0:00 to 18:30, it is not connected to the utility grid, and the power supply to the load is covered by PV and battery discharge or battery discharge alone. Then, it is proven that there is a correlation between the fluctuation of power variation of the solar PV

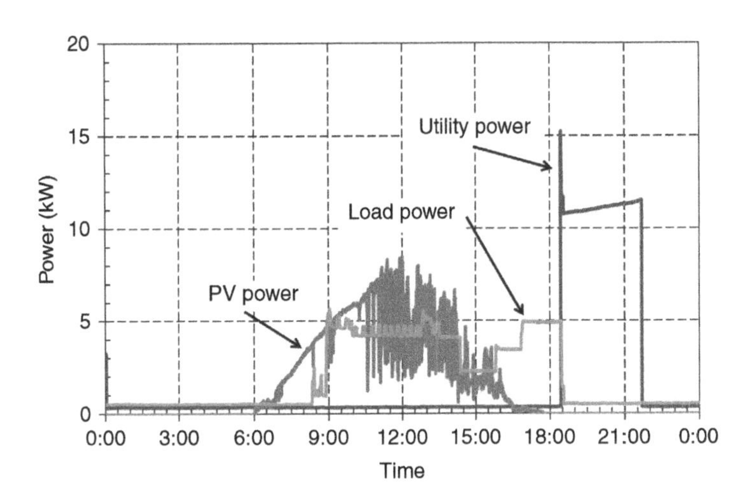

■ **FIGURE 10.21 Power flow and load curve in a day.**

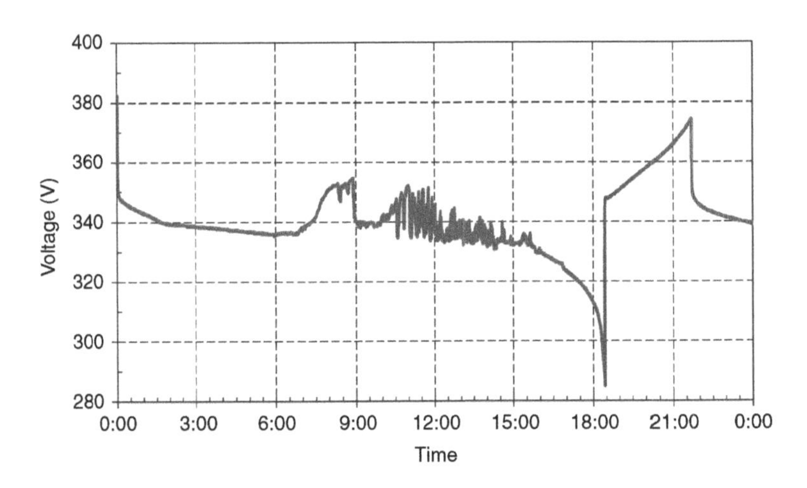

■ **FIGURE 10.22 Battery voltage in a day.**

generation and the output of the converter. In the fluctuation of an output of the converter, this adjusts the balance of power supply and power demand.

10.5 EXAMPLES OF MICROGRIDS IN THE WORLD

A microgrid may also be defined as a self-contained subset of an area electric power system with access to indigenous distributed generation sources, energy storage devices, power loads, and control system. Microgrids

Table 10.6 Some Examples of Microgrid Deployments in Different Parts of the World [8]

Location	Microgrid
North America	Fort Zed, Fort Collins. Colorado; University of San Diego. California; Santa Rita jail, Santa Rita. California; Perfect Power. Chicago, Illinois; BCiT microgrid, Vancouver, BC, Canada; Balls Gap Station, Milton, West Virginia
South America	Robinson Crusoe Island, Chile; OHagUe's microgrid, Chile; Huatacondo's microgrid. Chile
Europe	Model City of Manheim, Germany; Cell Controller Project, Denmark; CRES-Gaidouromanlra, Kythnos, Greece; Liandcr's Holiday Park at Bronsbcrgen, Zuiphcn, The Netherlands; RSE-DER test facility. Italy; TECNALIA-DRR test facility. Bilbao, Spain; PIME'S project. Dale. Norway; Szentendre, Hungary; Salburua. Spain; La Graciosa Island microgrid, Spain; Optimagrid, Spain; iSare project, Guipiizcoa, Spain
Asia	Rural PV hybrid microgrid. West Bank; Hangzhou Dianzi University, China; NbDO microgrid. Aichi, Kyotang, Elaciiinohc. Japan; NEDO Tohoku Fukushi University, Sendai. Japan; Shimi/u Corp. microgrid. Tokyo Gas microgrid, Aiclii Institute of Technology microgrid, Japan; INER microgrid, Taiwan
Africa	Diakha Madina, Senegal
Australia	CSIRO. Kings Canyon, Coral Bay, Brcmer Bay, Denhem, Esperence, Hopctoun, King Island, Roltnest Island

Note: *Information from the US Department of Energy Renewable and Distributed Systems Integration projects and C1GRE WGC6.II.*

possess several advantages, such as the potential for higher power supply availability and security for critical loads, investment deferrals in transmission and centralized generation plants, the provision of ancillary services to a business continuity plan, and opportunities for economic incentives for customers. Therefore, a multilateral match is accomplished throughout the world for microgrids. A typical match is shown in Table 10.6.

10.6 CONCLUSIONS

In this chapter, exchange electricity supply and DC electricity supply were explained along with the microgrid, which is a small-scale utility grid into which renewable energy has been introduced. I hope that the use of AC/DC microgrid systems will become increase in future.

REFERENCES

[1] Takashi T, Keiichi H, Yoshiaki O, Kazuto Y, Ichiyanagi K. Development of uninterruptible power supply system with distributed generators (DGs). INTELEC 2008;30:574–578.

[2] Murai H, Takeda T, Hirose K, Okui Y, Iwase Y, Yukita K, Ichiyanagi K. A study on charge patterns for uninterruptible power supply system with distributed generators. INTELEC 2009;31:534–538.

[3] Hiroaki M, Takashi T, Kazuto Y, Hiroshi M. Comparative study of peak cut/peak shift operation using microgrid. ICEE 2014;SPGP-2357.

[4] Tomoki K, Kazuto Y, Yasuyuki G, Katsuhiro I, Tomohito U, Keiichi H, Yoshiaki O. Operation method of micro grid using the forecasting method by neural network. ISAP 2013;929.

[5] Hiroshi K. Technology and problem of direct current power supply in the home using the 300 voltage. EMC 2013; 2.5(298): 25–38 (in Japanese).

[6] Kazuto Y, Takashi T. Technical trends on the development of DC power system. IEEJ 2015;135(6):366–369.

[7] Yukita K, Shimizu Y, Goto Y, Yoda M, Ueda A, Ichiyanagi K, Hirose K, Takeda T, Ota T, Okui Y, Takabayashi H. Study of AC/DC power supply system with DGs using parallel processing method. IPEC 2010;22A2-3:722–725.

[8] Abbey C, Cornforth D, Hatziargyriou N, Hirose K, Kwasinski A, Kyriakides E, Platt G, Reyes L, Suryanarayanan S. Powering through the storm: microgrids operation for more efficient disaster recovery. IEEE Power Energy Mag 2014;12(3):67–76.

Chapter 11

Stability problems of distributed generators

Toshihisa Funabashi*, Jinjun Liu**, Tomonobu Senjyu†
*Institute of Materials and Systems for Sustainability (IMaSS), Nagoya University, Nagoya, Japan;
**Xi'an Jiatong University, Xi'an, China; †Department of Electrical and Electronics Engineering,
University of the Ryukyus, Okinawa, Japan

CHAPTER OUTLINE

11.1 Voltage stability in distribution systems 262
11.1.1 Definitions 262
11.2 Stability problem with DGs connected to a weak power system 262
11.2.1 Voltage stability analysis 263
11.2.2 Voltage stability index 264
11.2.3 Battery control method 264
11.2.4 Simulation results 267
11.3 Stability problem with power electronics in DGs 267
11.3.1 Terminal characteristics of submodule 268
11.3.2 Stability criteria 269
11.3.3 Comparison between stability criteria 271
11.3.4 Conclusions 272
11.4 Stability problems in microgrids 273
11.4.1 Microgrid model 274
11.4.2 Inverter control method 276
11.4.3 FRT requirements 277
11.4.4 Criteria of power quality 277
11.4.5 Simulation and results 277
11.4.6 Conclusions 279
References 279

Integration of Distributed Energy Resources in Power Systems

11.1 VOLTAGE STABILITY IN DISTRIBUTION SYSTEMS

Voltage stability covers a wide range of phenomena and means different things to different engineers [1]. It is a fast phenomenon for engineers involved with induction motors and it is a slow phenomenon for other engineers. Some engineers and researchers have discussed whether voltage stability is a static or dynamic phenomenon. Voltage instability and voltage collapse are used somewhat interchangeably. Voltage stability or voltage collapse has often been viewed as a steady state problem. It was viewed suitable with static power flow type analysis. A CIGRE report [2] recommends analysis methods and power system planning approaches based on static models for voltage collapse prevention. However, the network maximum power transfer limit is not necessarily the voltage stability limit. Voltage instability or voltage collapse is a dynamic problem. Of course, the power system is a dynamic system. In contrast with rotor angle stability (synchronous stability), the dynamics mainly involves the loads and the voltage controllers. The definition of voltage stability is given in CIGRE report [3]. Voltage stability is a subset of power system stability.

11.1.1 Definitions [1]

A power system at a given operating states is small disturbance voltage stable if, following any small disturbance, voltages near loads are identical or close to the predisturbance values. (Small disturbance voltage stability corresponds to a related linearized dynamic model with eigenvalues having negative real parts. For analysis, discontinuous models for tap changers may have to be replaced with equivalent continuous models.)

A power system at a given operating state and subject to a given disturbance is voltage stable if voltages near loads approach postdisturbance equilibrium values. The disturbed state is within the region of attraction of the state post-disturbance equilibrium.

A power system at a given operating state and subject to a given disturbance undergoes voltage collapse if postdisturbance equilibrium voltages are below acceptable limits. Voltage collapse may be total (blackout) or partial.

11.2 STABILITY PROBLEM WITH DGs CONNECTED TO A WEAK POWER SYSTEM

Recent liberalization of the power market combined with growing concern about the depletion of energy resources has led to an increase in the introduction of solar power generation in the electric power grid. Moreover, the move to all-electric home systems in the interest of economic benefits results in an increase of demand, causing the power system to operate near power transmission capacity in severe situations [4]. With more efficient use of transmission lines, it is possible that the power systems can be operated near voltage stability limits. As a result the possibility of voltage collapse will increase [1,5]. Therefore, voltage stability analysis is a major consideration in the stable operation of the power system.

Voltage stability has been analyzed in a variety of ways. Some of the analysis techniques include P (Active power)–V (Voltage) analysis, which concerns the relationship of the voltage and active power in the transmission system, and Q (Reactive power)–V analysis, which concerns the relationship of the voltage and reactive power [6–11]. Indicators of voltage stability proposed in these papers include finding the change in active and reactive power with respect to the change in voltage from the P–V and Q–V characteristics, the proximity of the high and low voltage vectors from the PV characteristics, and the voltage stability margin of the active power that can be consumed by the load. There has been little work done with regards to voltage stability considering both the active and reactive power at the same time because analysis becomes difficult. However, it is expected that accurate voltage stability analysis can be performed considering both the active and reactive powers simultaneously. Because active and reactive power are both main values regarding transmission characteristics, the P–Q characteristics, should also be considered.

We have previously proposed a voltage stability limit index that takes into account the active and reactive power in the transmission [12].

In this section, a method to improve the voltage stability of the power system is presented by using active and reactive power information of power transmission lines in accordance with the voltage stability [13]. Simulation results show that from the actions of charging and discharging with active and reactive power control, the voltage stability of the system can be improved with an appropriately large capacity storage battery installed at substation.

11.2.1 Voltage stability analysis

The model assumed in this section is shown in Fig. 11.1. The Institute of Electrical and Electronics Engineers 5-bus system is used [14]. In a transmission line model, V_i is the sending end voltage and V_j is the receiving end voltage, P_k is the active power being sent to the receiving side, and Q_k is the reactive power being sent. The power flow equation of the 2-bus system is expressed by the simple equation. There is a limit to possible transmission power for any given time; that power is called the power stability limit and the voltage is called the voltage stability limit [15]. The P–Q characteristics of the power stability limit are shown in Fig. 11.2. From the P–Q characteristics the relationship between

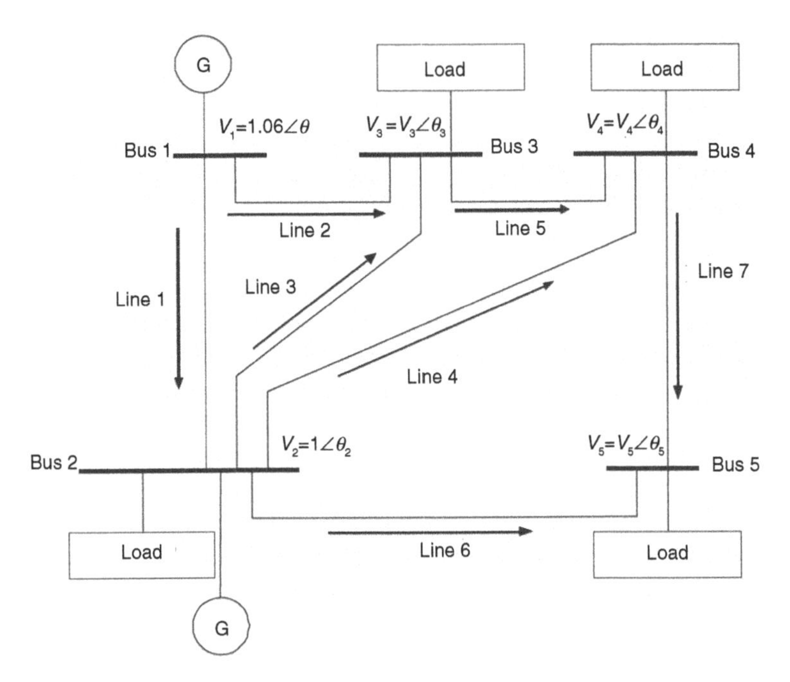

■ **FIGURE 11.1 IEEE 5-bus system model [13].**

V_i and P_j at the voltage stability limit can be shown. Then, the (P,Q)−V characteristics of a stable power limit is shown in Figs. 11.3 and 11.4.

11.2.2 **Voltage stability index**

From the current point K of P, Q an unstable point on the curve is obtained using Lagrange multiplier. The distance between the current point K and the nearest point of voltage instability D, can be obtained. Using these points, the nearest operating point is determined. Distances ΔP and ΔQ in Fig. 11.5 denote the distance to the nearest point from the operating point as the voltage stability index.

Therefore, the voltage stability index in each bus is evaluated as the largest of the voltage stability indices of the transmission lines connected to the corresponding bus. It should be noted that the voltage stability deteriorates as the index approaches 1.0 [12].

11.2.3 **Battery control method**

A method was proposed to improve voltage stability by using a large capacity battery, which is introduced to each load substation [13]. The slope of the tangent $\partial Q/\partial P$ at the closest point $D(Po,Qo)$ on the voltage stability

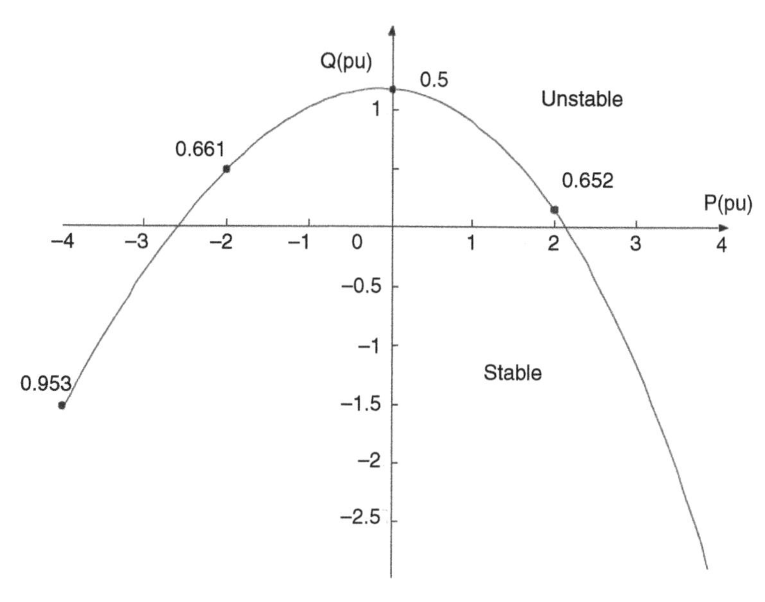

■ **FIGURE 11.2 P—Q characteristics [13].**

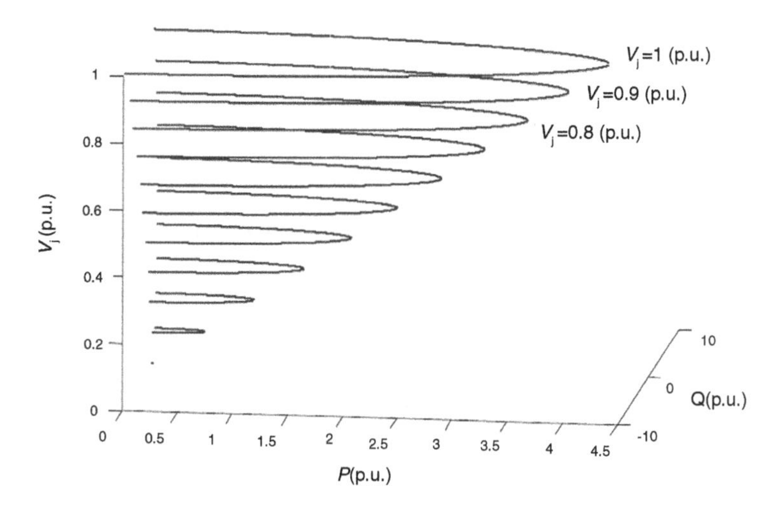

■ **FIGURE 11.3 (P, Q)—V characteristics [13].**

limit surface to the operating point $K(P_{Lk},Q_{Lk})$ is determined as shown in Fig. 11.5, where P_{Lk},Q_{Lk} are the active and reactive power flowing through the transmission line k.

It is possible to determine the slope at the nearest point $D(Po,Qo)$. Using the aforementioned idea, it is then possible to obtain an appropriate value

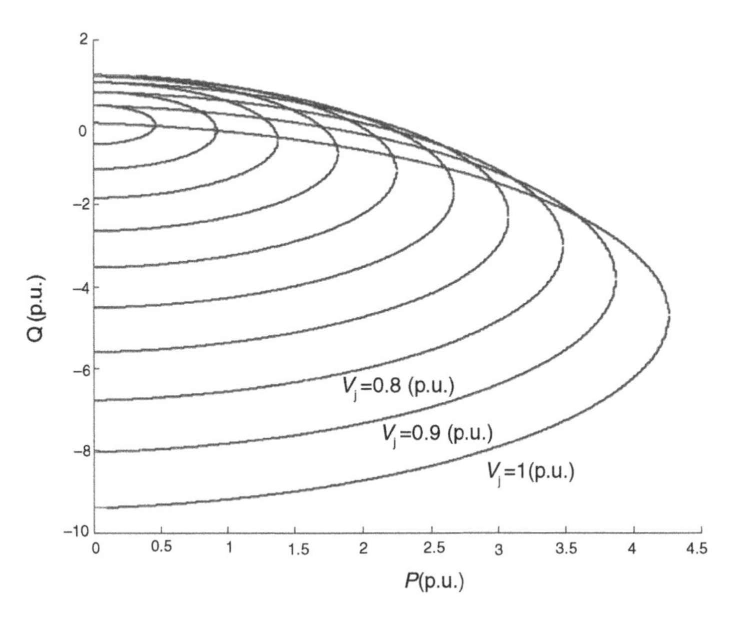

■ FIGURE 11.4 (P, Q)—V characteristics [13].

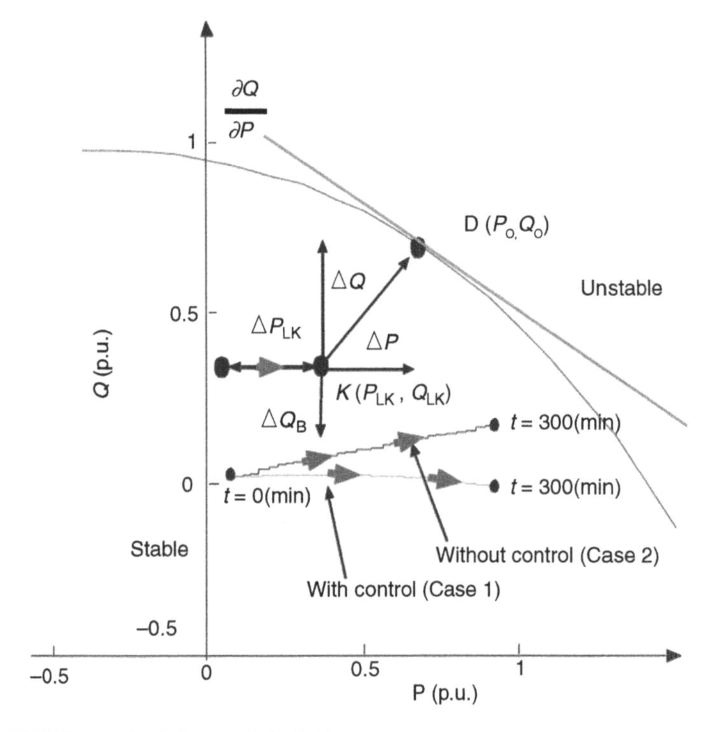

■ FIGURE 11.5 P—Q characteristics [13].

for reactive power compensation ΔQ_B for the battery with respect to the increase in active power ΔP_{Lk} flowing through the transmission line [13].

11.2.4 Simulation results

From the start of the simulation for a time interval of $t = 300$ min the load power connected to bus 5 was increased at a rate of $1 \times t$ (MW). The storage battery is connected to bus 5, the compensated reactive power were calculated for cases with/without compensating battery. The results show that the injection volume of the reactive power increases correspondingly with the increase in the load active power. It is also seen that the voltage stability index correspondingly increasing with the increase in load active power. It is possible to improve the system voltage stability by introducing a large battery in the time range from $t = 0 \sim 300$ min.

It is shown by simulation that it is possible to improve the voltage stability by controlling the reactive power of the storage battery based on the voltage stability limit indicators with respect to an increased load. Considering the control of active power of the battery, further improvement of voltage stability is possible for more severe cases, but for the sake of simplicity the conditions here only include reactive power control of the battery.

11.3 STABILITY PROBLEM WITH POWER ELECTRONICS IN DGs [32]

Compared with traditional power conversion system, the multimodule distributed system has the following benefits: (1) it can provide the possibility to coordinate several sources and loads; (2) the power and voltage rating can be easily increased by connecting different module in cascade or parallel; (3) the production and maintenance are more easily due to the standardized design. Based on these benefits, the multimodule distributed system has been widely adopted in practical applications. However, the dynamic characteristics of each submodule in the system are quite different, thus the interaction among different submodules will deteriorate the performance of the total system. Although each submodule is stable in standalone mode, the total system may be unstable due to the interaction.

The stability issues of the multimodule system have been discussed by some researchers [16–21] and also for AC systems [22–24]. Following the impedance-based stability criterion in cascade DC system [25], the multimodule system is divided into two groups: source group and load group. The interaction between source group and load group is considered. The equivalent output impedance of source group and the equivalent input impedance

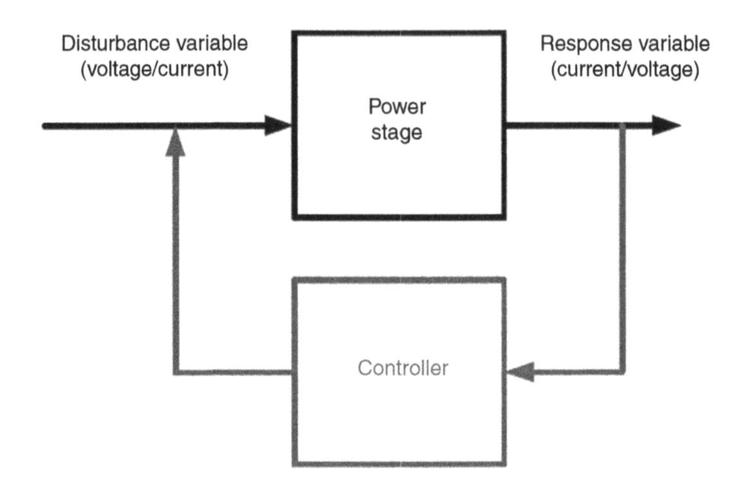

■ FIGURE 11.6 Structure of active module [32].

of load group are utilized to judge the system stability. In recent years, it has been discovered that the stability issues of cascade system are not relying on the distinction of source and load any more [26–28]. Instead, the terminal property of each submodule turns out to be the key point in the stability analysis. According to this recognition, the study about the terminal characteristics of submodule is carried out [29]. Based on the study about the behavior of submodule, two generalized stability criteria for multimodule distributed system have been proposed [30–33].

In this section, [32] two proposed stability criteria are explained, based on a paper. Based on the comparison between two proposed stability criteria, a comprehensive understanding about the stability issues of multimodule system is illustrated.

11.3.1 Terminal characteristics of submodule

When the electrical interaction is analyzed, external behavior of each submodule is of more interest than its internal loop stability. In a DC system there are only two variables at the common terminal: current and voltage. Hence, the relationship between terminal current and voltage describes the terminal characteristic of each module. The terminal impedance or admittance of the converter is the linearized model of terminal behavior around a certain operating point. When the terminal property of active module is considered, it is clear that the terminal characteristics do not only depend on power stage but also on the controller as shown in Fig. 11.6.

It is very important to distinguish disturbance variables from response variables in the controller design. In normal conditions, only one kind of ideal source is appropriate at the input or output terminal of the active converter due to the limitation in topology design. For example, if a capacitor is connected in parallel with an ideal voltage source, the effect of capacitor is attenuated by the voltage source and the capacitor may be damaged due to the overcurrent when there is a steep change in the voltage source. Therefore, the capacitor should avoid being connected with ideal voltage source in parallel directly. For a similar reason, the inductor should avoid being connected with ideal current source in series directly.

According to the type of available source at the common terminal, the active converter can be classified into two groups: current-fed (CF) converter and voltage-fed (VF). In VF converters, the terminal voltage is treated as the disturbance variable, while the terminal current is treated as the response variable. Thus, the terminal characteristic is equivalent to admittance (Y). Since the VF converter should be stable in standalone mode, no right hand pole (RHP) is allowed in terminal admittance. There are still CF converters in practical applications. The terminal current should be treated as the disturbance variable and the terminal voltage should be treated as the response variable in CF converter. The terminal characteristic of CF converter is equivalent to impedance (Z).

It is easy to draw the equivalent circuit of each module according to the basic circuit theory when the type of terminal property is fixed. The terminal property of common bus terminal is more attractive in the stability analysis, so only the characteristics of common bus terminal of each module are concerned. The Thevenin equivalent circuit is used to model the terminal characteristics of Z-type module while Y-type module is represented by the Norton equivalent circuit in general cases. This set makes the following analysis clear. Otherwise, if the Norton equivalent circuit is used for Z module where the terminal voltage is regulated tightly, an infinite current source is needed because the terminal impedance is almost 0 Hz. However, the infinite source is not acceptable in both practical application and theoretical analysis.

11.3.2 Stability criteria

In a multimodule system all the Z-type modules are represented by the Thevenin equivalent circuit and all the Y-type modules are represented by the Norton equivalent circuit. In the simplified equivalent circuit of the total system is shown in Fig. 11.7. All the Z-type modules are replaced by an equivalent Z-type module and all the Y-type modules are replaced by an equivalent Y-type module.

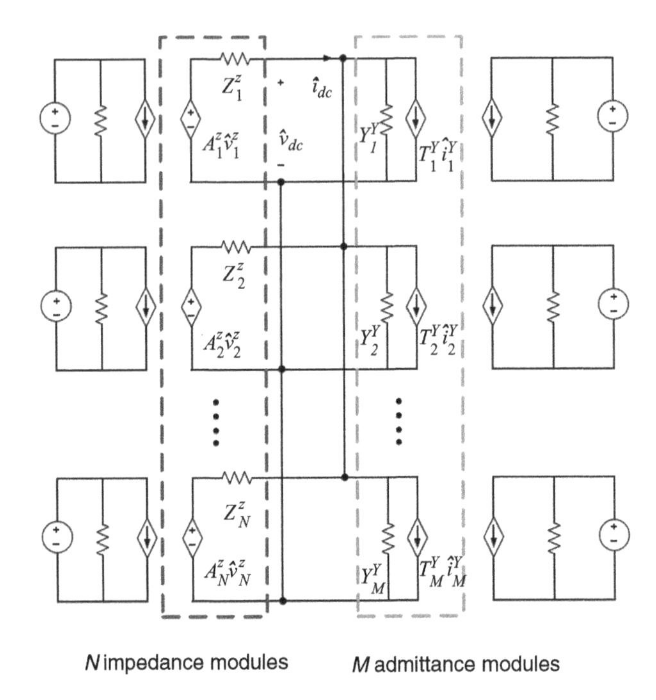

■ **FIGURE 11.7 Equivalent circuit of multimodule system [32].**

Based on the equivalent circuit, two stability criteria can be derived.

11.3.2.1 ***Stability criterion 1 [30]***

The submodules are divided into two groups and the mathematical expression of DC bus voltage can be expressed. Since each submodule is stable in standalone mode, it is clear that there should be no RHP in the numerator of DC bus voltage equation. Then, the stability of the total system depends on the numbers of right hand zero (RHZ) in the denominator of the equation.

In practical application, detailed models of some modules are unknown sometimes. In that case, it is impossible to derive the mathematical transfer function. However, the terminal characteristics of these modules can be estimated according to the measured data by network analyzer. When the measured data are analyzed, graphic stability analysis methods such as Nyquist stability criterion are more suitable.

Since all the modules are required to be stable in standalone mode, there is no RHP in the equivalent admittance of all Y-type modules. Hence, only the numbers of RHP in the equivalent impedance of all Z-type modules should be estimated. Due to the stability requirement for submodules in standalone

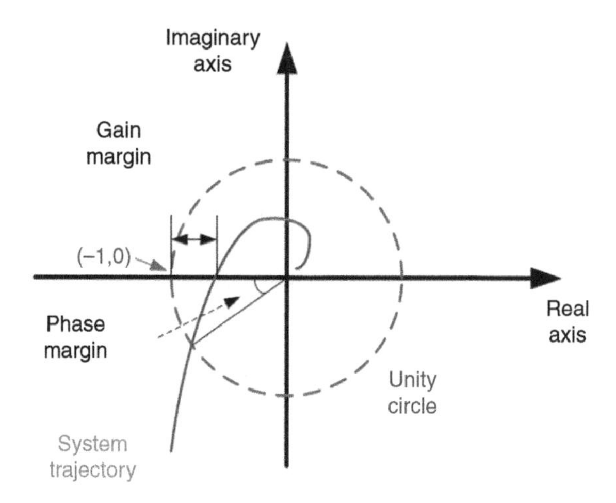

FIGURE 11.8 Definitions of gain margin and phase margin [32].

mode, there is no RHP in the numerator. Therefore, only the numbers of RHZ in the denominator need to be estimated. Similarly, the Nyquist criterion is applied to the denominator of Zeq to assess the numbers of RHZ

Based on the analysis, a two-step stability criterion is proposed. The first step can be explained as the assessment of the interaction among all the Z-type modules. If there is only one Z-type module in the system, there is no interaction. Then, the first step can be neglected and only the second step is needed. The second step is used to assess the interaction between the equivalent Z module and equivalent Y module.

11.3.2.2 *Stability criterion 2 [31]*

All the submodules can be treated as connected in parallel. Then, the small signal expression of DC bus voltage can also be expressed. Similar to the analysis procedure in criterion 1, the Cauchy theorem is employed to develop the stability criterion. Based on the aforementioned analysis, a two-step stability criterion is proposed.

11.3.3 **Comparison between stability criteria**

The validities of criterion 1 and 2 in stability assessment have been proved in [30] and [31]. In some cases, the stability criterion is used to design the parameters to realize specific gain margin and phase margin in addition to stability assessment. In classical control theory, the definitions of gain margin (GM) and phase margin (PM) relate to the distance between system trajectory and critical point $(-1, 0)$, as shown in Fig. 11.8.

It is clear that the second step of criterion 1 is based on the Nyquist criterion and the critical point is $(-1, 0)$. However, the second step of criterion 2 is based on Cauchy theorem and the critical point is $(0, 0)$. Therefore, the criterion 2 is not suitable for parameter design unless the definitions of GM and PM are modified with the point $(0, 0)$. There are also big differences between these two criteria in some applications. For example, an individual submodule A will be added to an existing system B. Both the individual submodule A and the system B are stable before integration. The stability status of integrated system needs to be assessed before the integration.

If the detailed structure of the system B is known, then both criterion 1 and 2 can be used to assess the stability. However, the effect of criterion 1 and 2 are different if the detailed structure of the system B is unknown. It should be noted that the criterion 1 is based on the concept that the total system can be divided into equivalent Z-type group and equivalent Y type group. If the detailed system is unknown, then it is impossible to divide the total system into two equivalent groups. Therefore, criterion 1 will fail in this case.

However, criterion 2 is based on another concept that all the submodules are connected in parallel. Therefore, if the system B is stable, then it can be derived that there is no RHZ in the sum of equivalent admittance of each submodule. Since the admittance is the reciprocal of impedance, it can be explained as that there is no RHP in the equivalent impedance of system B. Based on the terminal characteristics defining method in Section 2, the system B can be regarded as a Z-type module. Then, the system integration problem turns into the stability issue of a two-module system. If module A is a Y-type module, then the system can be regarded as a Z + Y-type system. If module A is a Z-type module, then the system can be regarded as a Z + Z-type system. Therefore, the stability criterion of two module system in [34] can be used to solve stability problem caused by system integration. Based on the aforementioned discussion, it is clear that criterion 1 is more advantageous in system parameter design and criterion 2 is more useful in system integration application.

11.3.4 Conclusions

Comparisons between two existing generalized stability criteria of multi-module distributed system have been made. It is clear that criterion 2 is more useful in system integration application. This is because that criterion 1 is developed based on a partial recognition about the system behavior. Meanwhile, the understanding about the system behavior in criterion 2 is clearer. In distributed multimodule system, the bus voltage for each submodule is the same, thus all the submodules should be regarded as in parallel connection.

This is the key point of system behavior study and all the other existing stability criteria are just equivalent mathematical tools. It should be noted that there are some potential problems due to the nonideality in practical application. The measuring range and resolution of the network analyzer will influence the accuracy in measurement. The noise in measurement will also affect the accuracy. These practical problems may lead to wrong stability assessment when the trajectory is very close to the critical point so that it is very hard to make sure whether the trajectory encircle the critical point or not. In the study of stability of cascade system, this problem is solved by setting different forbidden regions to keep the trajectory off the critical point.

11.4 STABILITY PROBLEMS IN MICROGRIDS

In recent years, the microgrid which is a small-scale power grid with renewable and clean DGs has attracted a lot of attention and is studied actively all over the world for exhaustion of fossil fuels and global warming. The primary advantage of microgrid is reduction of transmission loss and cost of power lines construction, improvement of the stability of energy supply for diversification of energy sources and suppression of power variation due to renewable energy. Furthermore, microgrid can be switched from grid-connected mode to isolated mode at fault events in utility grids [35,36]. Therefore, since the Great East Japan earthquake occurred, penetration of microgrid is important in Japan. However, it is difficult to maintain voltage and frequency and suppress harmonic because of small system at isolated mode [12,32]. In the microgrid composed of only inverter-based DGs, these problems are especially serious because there are no rotating machines. Also switching from the grid connected mode to the isolated mode at fault events in utility grid may have significant influence on voltage and frequency fluctuation beyond the standard permissible levels depending on the condition of supply and power [37,38]. In addition, the fluctuation ranges change by types of microgrid load. When a microgrid switches from grid connected mode to isolated mode uninterrupted, investigation of the transient characteristics is also required. Microgrid has to be constructed in consideration of these well.

In previous research, it was confirmed that microgrid composed of only inverter-based DGs can be a self-sustained operation [39,40]. In addition, transient characteristics of switching from grid-connected mode to isolated mode have been investigated in microgrid including rotating machines [41]. However, in microgrid composed of only inverter-based DGs, the transient characteristics have not been investigated. Furthermore, inverter-based DGs of the aforementioned references do not take the fault ride through (FRT) requirements into consideration.

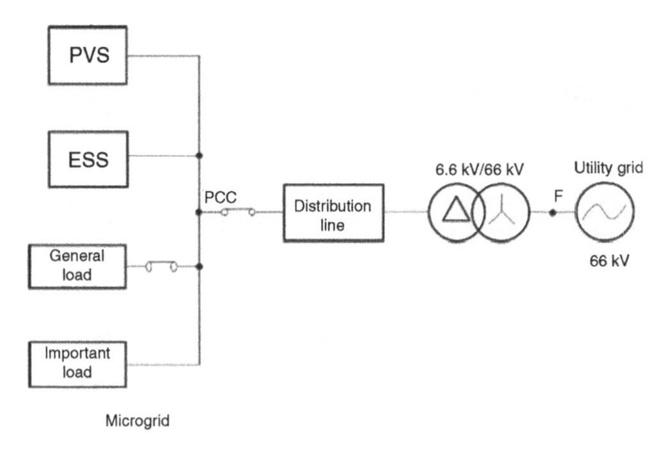

■ **FIGURE 11.9 Microgrid system.**

In this study, assuming a microgrid composed of inverter-based DGs, these transient characteristics were investigated to consider whether microgrid can be self-sustained operation without interruption at fault event.

11.4.1 Microgrid model

Fig. 11.9 shows a single-line layout of the utility grid and the microgrid system. A microgrid model, simulated on eXpandable Transient Analysis Program [42] software, is analyzed including a mix of photovoltaic system (PVS) rated 60 kW, energy storage system (ESS) rated 150 kW, and a combination of passive RL. Voltage and frequency are respectively 6.6 kV and 60 Hz in microgrid.

In grid connected mode, based on the DQ-current control strategy, power conditioning subsystems (PCSs) control real power and reactive power. Subsequent to islanding condition detection and confirmation, the DQ-current controller of the PCS of storage battery is disabled and voltage controller is made active. Thus, voltage and frequency in microgrid is decided by the PCS of storage battery.

11.4.1.1 ***PVS***

PV is represented using the equivalent circuit. This PV is interconnected to the utility grid through PCS, transformer, and filter. Fig. 11.10 shows the PVS model. The DC–DC converter in PCS performs maximum power point tracking control. Therefore PVS always outputs maximum power and power output control such as the output suppression and reactive power output are not carried out.

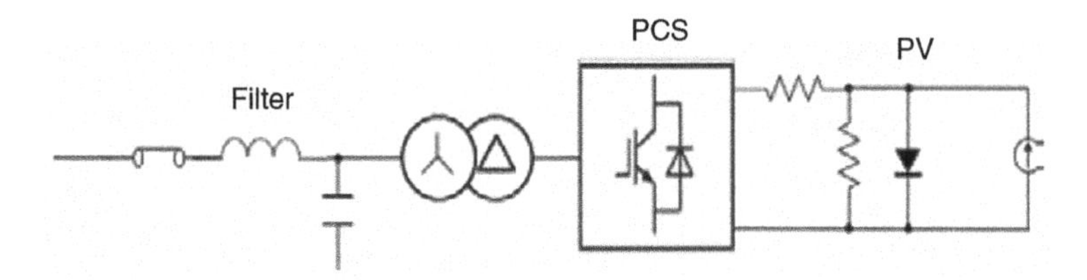

■ **FIGURE 11.10 PVS model.**

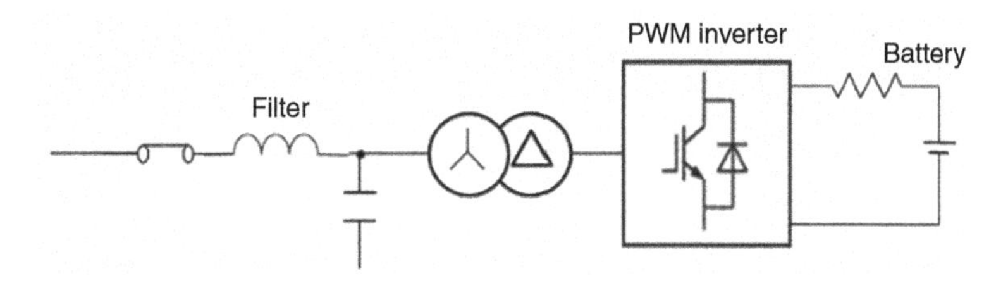

■ **FIGURE 11.11 ESS model.**

11.4.1.2 ESS

Power storage unit is presented by the DC voltage source. It is interconnected to the utility grid through PCS and transformer, filter. Fig. 11.11 shows the ESS model. This ESS model has grid connected mode and isolated mode. Usually ESS is grid connected for controlling on the basis of frequency and voltage in utility grid. When fault accident and disturbance occur, ESS switches from grid connected mode to isolated mode uninterruptedly. Then inverter is in constant voltage constant frequency control and determines voltage and frequency in the microgrid.

11.4.1.3 High-speed circuit breaker

After voltage drop detection, the utility grid and microgrid is cut off after several milliseconds. At the same time as high-speed circuit breaker operates, the information is transmitted to the ESS and the ESS switches to isolated mode automatically.

11.4.1.4 Load

There are two loads: general load (150 kVA) and important load (50 kVA). These loads are both constant impedance load (PF = 0.9). The general load is disconnected from utility grid when ESS becomes isolated mode at fault accident.

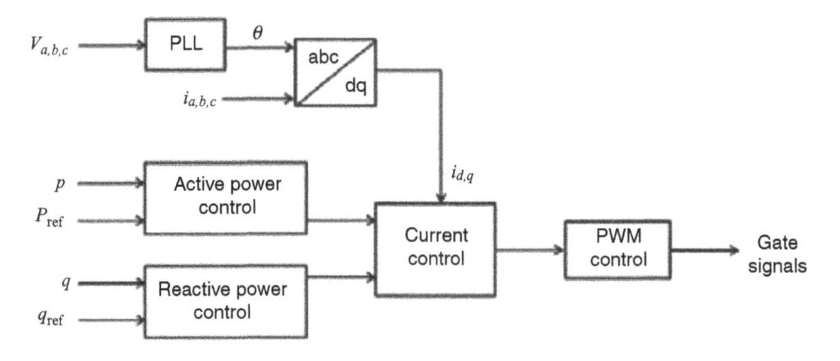

■ FIGURE 11.12 Grid connected mode.

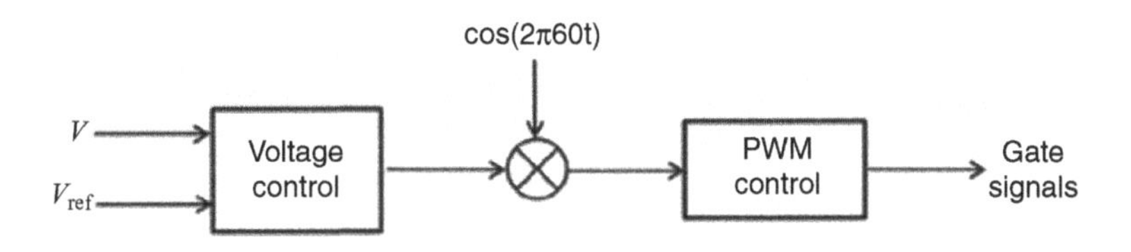

■ FIGURE 11.13 Isolated mode.

11.4.2 Inverter control method

11.4.2.1 Grid connected mode

Fig. 11.12 shows the inverter control system of grid connected mode. Here, $v_{a,b,c}$ and $i_{a,b,c}$ are the inverter output voltages and output currents. Inverter control system consists of power control using the instantaneous active power and the instantaneous reactive power and current control at grid connected mode. Instantaneous active and reactive powers p and q are determined by v and i in the d and q axes and using a dq transformation. i_{dref} and i_{qref} are determined from deviation of p and p_{ref}, q and q_{ref}, respectively, through PI control. Further PI control is executed in current control. In addition, a noninteracting control must be adopted in there.

11.4.2.2 Isolated mode

Fig. 11.13 shows the inverter control system of isolated mode. Here, V is the effective value of the inverter output line voltage. As the frequency is 60 Hz fixed, inverter control system consists of voltage control at isolated mode. Voltage amplitude signal is determined from deviation of V and V_{ref} (1 p.u.) though proportional-integral (PI) control.

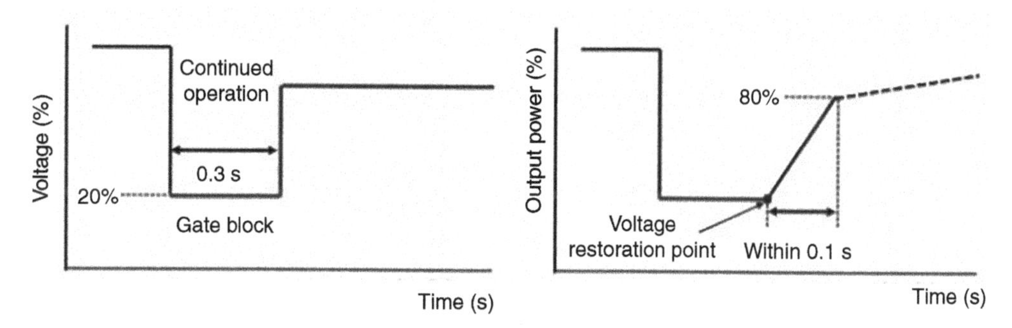

■ FIGURE 11.14 FRT requirements.

11.4.3 FRT requirements

Fig. 11.14 shows the FRT requirements which were decided recently in Japan similarly to requirement in USA and Europe. When a fault accident occurs in the utility grid, distributed generations are disconnected from there. However, voltage and frequency may be affected seriously because of imbalance between demand and supply by fault accident at mass introduction of renewable energy. In order to prevent this, FRT requirements are necessary to continue the operation.

In this study, PVS meets the FRT requirements [43]. PVS must block unwanted parallels off to continue operation of 0.3 s using FRT function when momentary terminal voltage drops. If the fault event is removed within 0.3 s and voltage in the microgrid recovers to the proper range, PVS continues to work. However, if the fault event continues more than 0.3 s, PVS is disconnected from microgrid.

11.4.4 Criteria of power quality

The targets of power quality control at grid connected mode and isolated mode are shown in Table 11.1 [44]. Voltage and frequency drop are unavoidable from fault event occurrence to the time ESS becomes isolated mode. Therefore, these control targets do not consider its reduction.

11.4.5 Simulation and results

11.4.5.1 Sensitivity analysis of cutoff delay time of circuit breaker

The simulation model is the same as that in Fig. 11.9. Islanding takes place as a result of a three-phase fault occurring at point F after 0.93 s from the start of the simulation and voltage of the utility grid is dropped to 0.3 pu. A circuit breaker is opened after several milliseconds and, at the same time, the ESS is switched to isolated mode and general load is disconnected from utility grid. By changing

Table 11.1 Criteria of Power Quality

Contents	Grid Connected Mode	Isolated Mode
Voltage	1 p.u. ± 5% (Base: 6.6 kV)	1 p.u. ± 5% (Base: 6.6 kV)
Frequency	±0.1 Hz	±0.5 Hz
Harmonic	Overall current distortion factor Less than 5%	Overall voltage distortion factor Less than 5%

the cutoff, the delay time from fault event occurs and the ESS makes isolated mode a parameter, simulation in three cases (Case 1: 3 ms, Case 2: 15 ms, Case 3: 27 ms) is carried out. In addition, the PVS output is 60 kW and ESS did not perform charge and discharge at grid connected mode for simplicity.

Various waveforms in Case 1 are calculated. In addition, the comparison result of the effective value voltage and frequency of the microgrid of each case are made and maximum values of effective value voltage and frequency are derived.

From these results, the microgrid can transfer into isolated mode without uninterruption. Furthermore, it is shown that the microgrid can be in isolated mode operation while maintaining the power quality in case1. As the cutoff delay time becomes longer, the maximum value of the effective value voltage and frequency increases, and exceeds the standard permissible levels in Cases 2 and 3. Harmonics are within it in all cases.

11.4.5.2 *Sensitivity analysis of cutoff delay time of circuit breaker with IM load*

Important load is changed to IM load (26 kW $\times$ 3) from constant impedance load and the same simulation is performed. By changing cut off delay time as a parameter, simulation in three cases (Case 1: 10 ms, Case 2: 30 ms, Case 3: 50 ms) is carried out. The comparison results of the effective value voltage and frequency of the microgrid of each case are made and maximum value of effective value voltage and frequency are derived.

From simulation results, it is shown that the microgrid can be transferred to isolated mode operation while maintaining the power quality in case1. In Case 2, although microgrid can be transferred to isolated mode operation, the frequency exceeds the criteria. In case 3, the voltage decreases and becomes 0.65 p.u. in 1.25 s and it does not recover to prefault value. This is because voltage recovery takes more time than 0.3s, and PVS is disconnected from utility grid by FRT.

11.4.6 Conclusions

In this section, it was investigated whether the microgrid can be a self-sustained operation without interruption at fault events considering FRT functions in inverter-based DGs. It can be confirmed that the microgrid can be a self-sustained operation. Furthermore, the microgrid can be a self-sustained operation while meeting criteria of power quality depending on the load. In addition, the influence that the IM load gives to the voltage is greater and the voltage cannot be restored to prefault values by disconnecting the PVS by FRT when the voltage recovery is slow. In the future, we will use fine sensitivity analysis, and make conditions that ensure the microgrid can be a self-sustained operation.

REFERENCES

[1] Taylor CW. Power system voltage stability. New York: McGraw-Hill; 1994.

[2] CIGRE Working Gropup 38.01. Planning against voltage collapse. Electra 1987;111:55–75.

[3] CIGRE Task Force 38-02-10. Modeling of voltage collapse including dynamic phenomena; 1993.

[4] Yuan-Kang W. A novel algorithm for ATC calculations and applications in deregulated electricity markets. Int J Elec Power 2007;29(10):810–21.

[5] Ajjarapu V, Lee B. Bibliography on voltage stability. IEEE Trans Power Syst 1998;13(1):115–25.

[6] Leonardi B, Ajjarapu V. An approach for real time voltage stability margin control via reactive power reserve sensitivities. IEEE Trans Power Syst 2013;28(2):615–25.

[7] Corsi S, Taranto GN. Voltage instability – the different shapes of the "nose", "bulk power system dynamics and control – VII. Revitalizing operational reliability, 2007 iREP symposium; 2007. pp. 1–16.

[8] Lin L, Wang J, Gao W. Effect of load power factor on voltage stability of distribution substation. IEEE power and energy society general meeting; 2012. pp. 1–4.

[9] Okamoto H, Tanabe R, Tada Y, Sekine Y. Method for voltage stability constrained optimal power flow (VSCOPF). IEEJ Trans Power Energ 2001;121(12):1670–80.

[10] Glavic M, Lelic M, Novosel D, Heredia E, Kosterev DN. simple computation and visualization of voltage stability power margins in real-time, transmission and distribution conference and exposition. 2012 IEEE PES 2012;1–7.

[11] Mousavi OA, Bozorg M, Ahmadi-Khatir A, Cherkaoui R. Reactive power reserve management: preventive countermeasure for improving voltage stability margin. Power and energy society general meeting 2012 IEEE; 2012. pp. 1–7.

[12] Tachibana M, Palmer MD, Senjyu T, Funabashi T. Voltage stability analysis and (P,Q)–V characteristics of multibus system. CIGRE AORC technical meeting 2014; 2014.

[13] Tachibana M, Palmer MD, Matayoshi H, Senjyu T, Funabashi T. Improvement of power system voltage stability using battery energy storage systems. 2015 International conference on industrial instrumentation and control (ICIC), College of Engineering Pune, India; May 28–30, 2015.

[14] Venkatesh P, Rajendra O, Yesuraj A, Tilak M. Load flow analysis of IEEE14 bus system using MATLAB. Int J Eng Res Technol 2013;2(5):149–55.

[15] Venkatesh B, Rost A, Chang R. Dynamic voltage collapse index – wind generator application. IEEE Trans Power Del 2007;22(1):90–4.

[16] Panov Y, Rajagopalan J, Lee FC. Analysis and design of N paralleled DC-DC converters with master-slave current-sharing control. Applied power electronics conference and exposition, 1997. APEC '97 conference proceedings 1997, Twelfth Annual, Feb 23–27, 1997, vol.1, 1997. pp. 436–442.

[17] Thottuvelil VJ, Verghese GC. Analysis and control design of paralleled DC/DC converters with current sharing. IEEE Trans Power Electron 1998;13(4):635–44.

[18] Xie X, Yuan S, Zhang J, Qian Z. Analysis and design of N paralleled DC/DC modules with current-sharing control. Power electronics specialists conference, 37th IEEE, June 18–22, 2006. PESC '06, 2006. pp. 1–4.

[19] Hou D, Liu J, Wang H, Huang W. The stability analysis and determination of multi-module distributed power electronic systems. Power Electronics for Distributed Generation Systems (PEDG), 2010 2nd IEEE international symposium, June 16–18; 2010. pp. 577–583.

[20] Liu F, Liu J, Zhang H. Comprehensive study about stability issues of multi-module distributed system. Proceedings of the 7th international power electronics conference, Hiroshima, Japan; 2014.

[21] Wang H, Liu J, Huang W. Stability prediction based on individual impedance measurement for distributed DC power systems. 2011 IEEE 8th international conference on power electronics – ECCE Asia, Jeju, Korea, May 30–June 03, 2011. pp. 2114–2120.

[22] Liu Z, Liu J, Bao W. A novel stability criterion of AC power system with constant power load. In: Proceedings of 27th annual IEEE applied power electronics conference and exposition. Orlando, FL, USA; 2012. p. 1946–50.

[23] Liu Z, Liu J, Wang H. Output impedance modeling and stability criterion for parallel inverters with average load sharing scheme in AC distributed power system. In: Proceedings of 27th annual IEEE applied power electronics conference and exposition. Orlando, FL, USA; 2012. p. 1921–6.

[24] Liu Z, Liu J, Zhao Y. Output impedance modeling and stability criterion for parallel inverters with master-slave sharing scheme in AC distributed power system. In: Proceedings of 27th annual IEEE applied power electronics conference and exposition. Orlando, FL, USA; 2012. p. 1907–13.

[25] Middlebrook RD. Design techniques for preventing input-filter oscillations in switched-mode regulators. Proceedings of Powercon 5; 1978. pp. A3-1–A3-16.

[26] Suntio T, Leppaaho J, Huusari J, Nousiainen L. Issues on solar-generator interfacing with current-fed MPP-tracking converters. IEEE Trans Power Electron 2010;25(9):2409–19.

[27] Sun J. Impedance-based stability criterion for grid-connected inverters. IEEE Trans Power Electron 2011;26(11):3075–8.

[28] Liu F, Liu J, Zhang B, Zhang H, Hasan SU, Zhou S. Unified stability criterion of bidirectional power flow cascade system. Applied power electronics conference and exposition (APEC), 2013 28th Annual IEEE, March 17–21; 2013. pp. 2618–2623.

[29] Liu F, Liu J, Zhang B, Zhang H, Hasan SU. General impedance/admittance stability criterion for cascade system. ECCE Asia Downunder (ECCE Asia), 2013 IEEE, June 3–6; 2013. pp. 422–428.

[30] Liu F, Liu J, Zhang H, Xue D. Generalized stability criterion for multi-module distributed DC system. J Power Electron 2014;14(1).

[31] Liu F, Liu J, Zhang H, Xue D. Terminal admittance based stability criterion for multi-module DC distributed system. Applied power electronics conference and exposition (APEC), 29th Annual IEEE; 2014.

[32] Liu F, Liu J, Zhang H. Comprehensive study about stability issues of multi-module distributed system. In: Proceedings of 7th international power electronics conference. Hiroshima, Japan; 2012.

[33] Liu F, Liu J, Zhang H. Stability issues of Z + Z type cascade system in hybrid energy storage system (HESS). IEEE Trans Power Electron 2014;29(11).

[34] Liu F, Liu J, Zhang B, Zhang H, Hasan SU, Zhou L. Stability issues of Z + Z or Y + Y type cascade system. Energy conversion congress and exposition (ECCE), 2013 IEEE, September 15–19; 2013. pp. 434, 441.

[35] Schwaegerl C, Tao L. Quantification of technical, economic, environmental and social benefits of microgrid operation, Chapter 7. In: Hatziargyriou N, editor. Microgrids architechtures and control. Chichester, West Sussex, UK: Wiley, IEEE Press; 2014.

[36] Gabbar H, Bower L, Pandya D, Islam F. Resilient micro energy grids with gas-power and renewable technologies, ICPERE, Bali, Indonesia; 2014.

[37] Hojo M, Ikeshita R, Yamanaka K, Ueda Y, Funabashi T. Frequency regulation in a microgrid by voltage phasor regulation of a photovoltaic generation unit, 2015 IEEJ PES – IEEE PES Thailand Joint Symposium, Thailand; 2015.

[38] Ota T, Mizuno K, Yukita K, Nakano H, Goto Y, Ichiyanagi K. Study of load frequency control for a microgrid, power engineering conference, AUPEC 2007, Australasian Universities; 2007.

[39] Sumita J, Nishioka K, Noro Y, Ito Y, Yabuki M, Kawakami N. A verification test result of isolated operation of a microgrid configured with new energy generators and a study of improvement of voltage control, IEEJ Trans, PE, vol. 129. 2009. pp. 57–65 (in Japanese).

[40] Arai J. New inverter control in an isolated microgrid composed of inverter power sources without synchronous generator, IEEE, PES, Thailand Joint Symposium; 2015.

[41] Kasem A, Zeineldin H, Kirtley J. Microgrid stability characterization subsequent to fault-triggered islanding incidents, IEEE Trans, PE, vol. 27. 2012. pp. 658–669.

[42] Noda T, Ametani A. XTAP, chapter 5. Numerical analysis of power system transients and dynamics IET; 2015.

[43] JEAC. Grid-interconnection code, JESC E0019; 2012 (in Japanese).

[44] Takano T, Kojima Y, Temma K, Simomura M. Isolated operation at Hachinohe micro-grid project. IEEJ Trans PE 2009;129:499–506 (in Japanese).

Chapter 12

Virtual synchronous generators and their applications in microgrids

Toshifumi Ise*, Hassan Bevrani**

*Graduate School of Engineering, Osaka University, Suita, Osaka, Japan; **Department of Electrical & Computer Engineering, University of Kurdistan, Kurdistan, Sanandaj, Iran

CHAPTER OUTLINE

12.1 Basic concepts of virtual synchronous generators 282
12.2 Control schemes of virtual synchronous generators 284
12.2.1 VSYNC's VSG Design 285
12.2.2 IEPE's VSG topology 286
12.2.3 KHI's VSG 287
12.2.4 VSG system of Osaka University 287
12.3 Applications for microgrids 290
References 292

12.1 BASIC CONCEPTS OF VIRTUAL SYNCHRONOUS GENERATORS

The portion of distributed generating (DG) units and renewable energy sources (RESs) in power systems with respect to its total power capacity is increasing rapidly; and a high penetration level is expected for the next two decades. In the conventional centralized power generations, in which the synchronous machine dominates, enormous synchronous generators (SGs) comprise rotating inertia due to their rotating parts. The intrinsic kinetic energy (rotor inertia) and damping property (due to mechanical friction and electrical losses in stator, field, and damper windings) of the bulk synchronous generators play a significant role in grid stability. These generators are capable of injecting the kinetic energy preserved in their rotating parts to the power grid in the case of disturbances or sudden changes of generation and load power. Moreover, the slow dynamics of the huge generators allows

Integration of Distributed Energy Resources in Power Systems

the system to dampen the transients after a change or disturbance through oscillations and thereby remain balanced between the power generation and demand within the required timescale.

The growing DG/RES units have either very small or no rotating mass (which is the main source of inertia) and damping property and therefore, the grid dynamic performance and stability is affected by insufficient inertia and damping of DGs/RESs. The most challenging issue with the inverter-based units is to synchronize the inverter with the grid and then to keep it in step with the grid even when disturbances or changes happen [1–3]. A power system with a big portion of inverter-based DGs is prone to instability due to the lack of adequate balancing energy injection within the proper time interval [4]. Instability may happen in the form of frequency variation with high rate, low/high frequency, and out-of-step of one or a group of generators. Voltage rise due to reverse power from photo voltaic (PV) generations, excessive supply of electricity in the grid due to full generation by the DGs/RESs, power fluctuations due to variable nature of RESs, and degradation of frequency regulation, especially in islanded microgrids, can be considered as negative results of these issues.

One solution towards stabilizing such a grid is to provide additional inertia, virtually. A virtual inertia can be established for DGs/RESs using short-term energy storage together with a power electronics inverter/converter and a proper control mechanism. By controlling the output of an inverter, it can emulate the behavior of a real synchronous machine. In this idea, the inverter-based interface of the DG unit is controlled in a way to exhibit a reaction similar to that of a synchronous machine to a change or disturbance. This concept is known as virtual synchronous generator (VSG) [5] or virtual synchronous machine (VISMA) [6]. This design is expected to operate like a synchronous generator, exhibiting the amount of inertia and damping properties, by controlling the amplitude, frequency, and the phase angle of its terminal voltage. Therefore, it can contribute to the regulation of grid voltage and frequency. In addition, synchronizing units, such as phase-locked loops, can be removed [7]. As a result, the virtual inertia concept may provide a basis for maintaining a large share of DGs/RESs in future grids without compromising system stability.

The objective of the VSG scheme is to reproduce the dynamic properties of a real SG for the power electronics-based DG/RES units, in order to inherit the advantages of a SG in stability enhancement. The principle of the VSG can be applied either to a single DG, or to a group of DGs. The first application may be more appropriate to individual owners of DGs, whereas the second application is more economical and easier to control from the network operator point of view [8]. The dynamic properties of a SG provides the possibility of adjusting active and reactive power, dependency of the grid frequency on

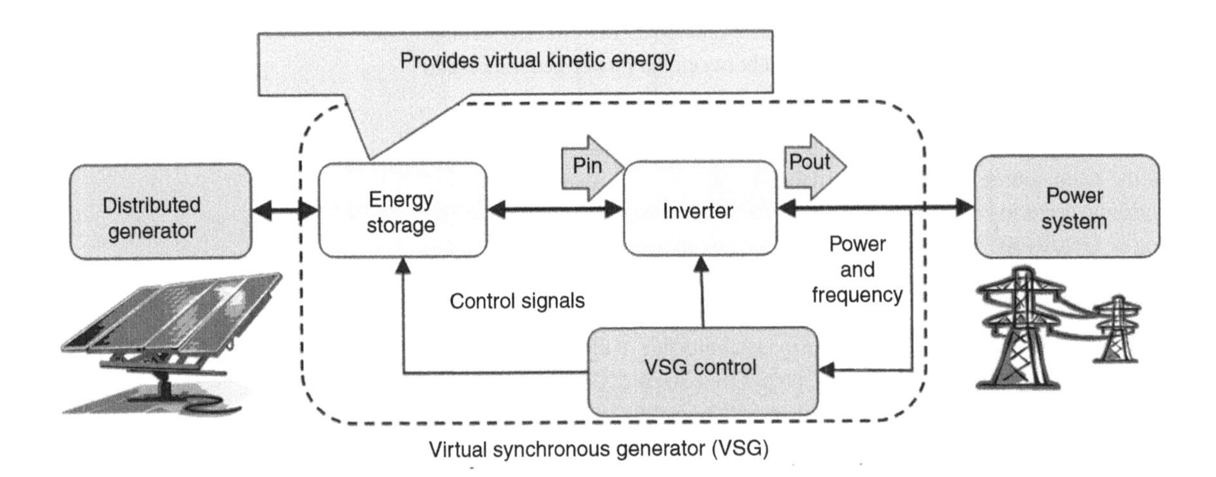

FIGURE 12.1 General structure of VSG.

the rotor speed, and highlighting the rotating mass and damping windings effect as well as stable operation with a high parallelism level [9].

The VSG consists of energy storage, inverter, and a control mechanism as shown in Fig. 12.1. In this scheme, the VSG serves as an interface between the direct current (DC) bus and the grid. The virtual inertia is emulated in the system by controlling the active power through the inverter in inverse proportion of the rotor speed. Aside from higher frequency noise due to switching of inverter's power transistors, there is no difference between the electrical appearance of an electromechanical SG and electrical VSG from the grid's point of view [10].

In the VSG control block, generally, a dynamic equation similar to the swing equation of the SGs is embedded that determines the output power based on the rate of the change of frequency and the frequency mismatch with respect to the nominal frequency.

12.2 CONTROL SCHEMES OF VIRTUAL SYNCHRONOUS GENERATORS

The idea of VSG control emerged in October 2007 and, up to now, several groups have developed various designs with the same fundamentals introduced in Section 12.1. The VSG concept and application were introduced in Refs. [11,12]. The same concept under the title of Synchronverter is described in Ref. [13]. The VSYNC project under the sixth European Research Framework program [5,8,11,14–20], the Institute of Electrical Power Eng. (IEPE) at Clausthal University of Technology in Germany [9,10,21,22], the VSG research team at Kawasaki Heavy Industries (KHI) [23], and the Osaka University [4,24–29] in Japan are explained in detail next.

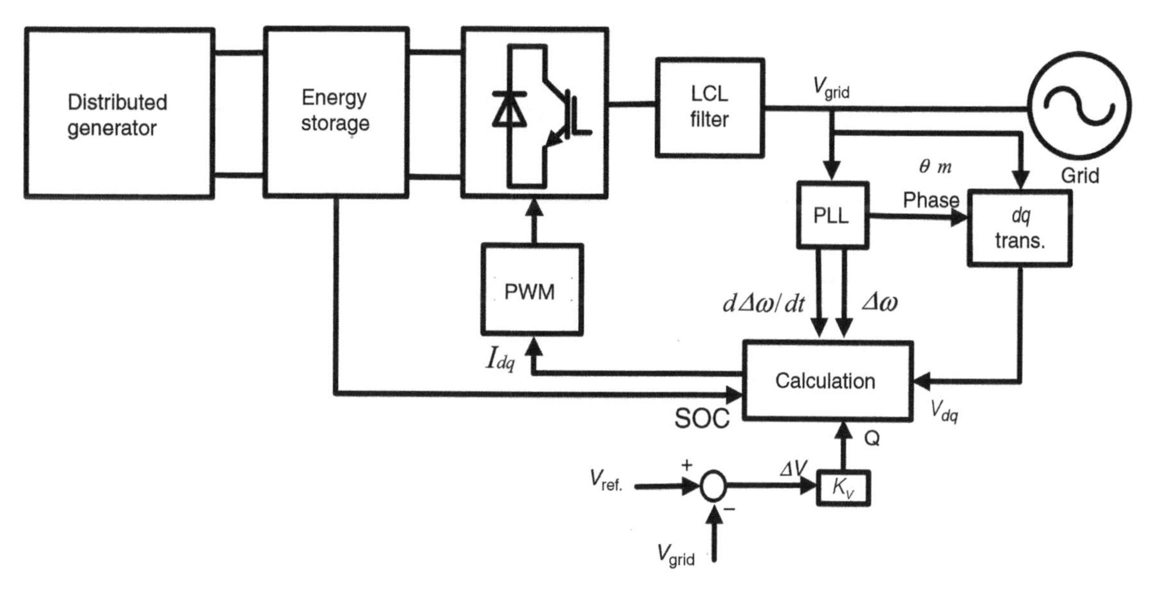

FIGURE 12.2 Structure of VSG developed by VSYNC group.

12.2.1 VSYNC's VSG Design

The initial VSG invented by this group with its full control connections is shown in Fig. 12.2. Here, energy storage unit connected to the grid through an inverter and LCL (L(Inductor), C(Capacitor) and L are combined type) filter. The VSG control produces the current reference to be used in the current controller. The phase locked loop (PLL) is used to produce the rate of frequency change (*dω/dt*) using the grid terminal voltage *Vg*. In addition, it provides the phase angle reference for rotating frame for *dq* control of inverter quantities. The "Calculation" block produces the current reference in *dq*-coordinates from the measured grid voltage, state of charge (SOC) of the storage device, $\Delta\omega$, $d\Delta\omega/dt$, and voltage reference through the following equations:

$$P = K_{SOC}\Delta SOC + K_P\Delta\omega + K_I\frac{d\Delta\omega}{dt} \qquad (12.1)$$

$$Q = K_V\Delta V \qquad (12.2)$$

$$i_d = \frac{v_d P - v_q Q}{(v_d + v_q)^2} \qquad (12.3)$$

$$i_q = \frac{v_d Q - v_q P}{(v_d + v_q)^2} \qquad (12.4)$$

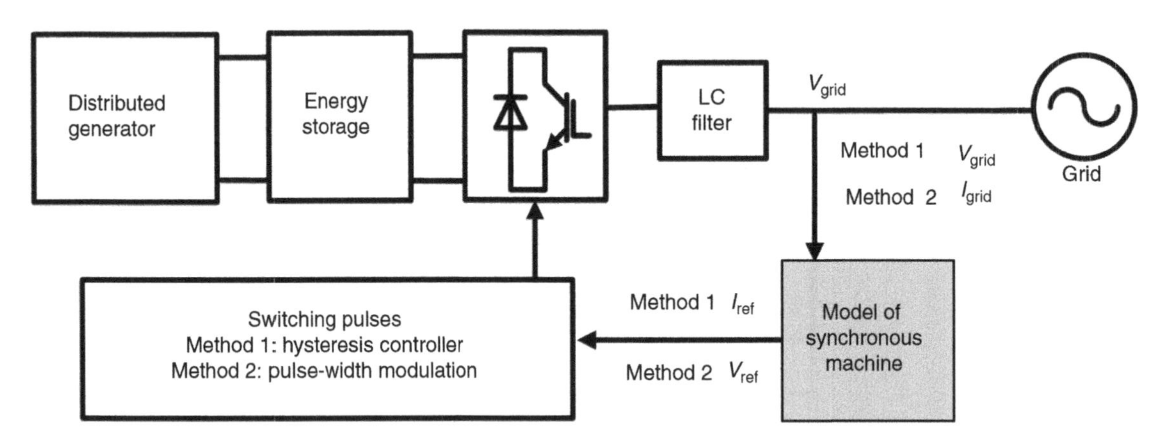

■ FIGURE 12.3 Basic structure of VISMA.

K_{SOC} must be set such that the active power P is equal to the nominal VSG output power, when the SOC deviation (ΔSOC) is at its maximum level. Similarly, the K_V must be chosen such that the VSG produce its maximum reactive power for a specified voltage deviation (eg, 10% [24]).

12.2.2 IEPE's VSG topology

This group developed a VSG design and called it VISMA [9,10,21,22]. The basic idea is shown in Fig. 12.3. The initial VISMA denoted as "Method 1" in Fig. 12.3 implements a linear and ideal model of a synchronous machine to produce current reference signals for the hysteresis controller of an inverter [9,22]. The control diagram of this scheme is shown in Fig. 12.4. The grid voltage is measured and, using a SG model, the current reference is produced to trigger the inverter switches through hysteresis controller. E_p and T_m in Fig. 12.4 are the voltage reference resembles the electromotive force of a SG and the virtual mechanical torque resembles the prime mover torque of an SG, respectively. Then the authors added an algorithm to compensate small disturbances and improve the quality of the grid voltage.

In Fig. 12.4, J is the moment of inertia, R_S is the stator resistance, and L_s is the stator inductance, K_d is the mechanical damping factor, $\varphi(s)$ is the phase compensation term with the transfer function of $1/(0.5s + 1)$, ω is the angular velocity, θ is the angle of rotation, T_m and T_e are the mechanical input and electrical torques, respectively.

The phase compensation term ensures that the virtual damping force counteracts any oscillating movement of the rotor in opposite phase. Despite simplifying the excitation winding, the induced electromotive force is given by adjustable amplitude E_P and the rotation angle θ [10].

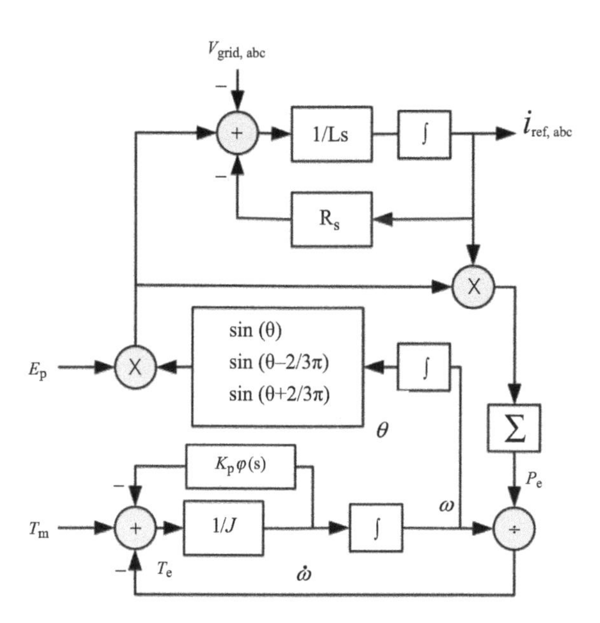

■ **FIGURE 12.4 Block diagram of the VISMA for method 1.**

12.2.3 KHI's VSG

KHI's VSG uses the phasor diagram of a SG to produce current reference [23]. The relation between voltage and current phasors of an SG is algebraic. The complete control diagram is shown in Fig. 12.5. The load angle δ is produced from the governor model that has the active power command, active power feedback signal and the reference angular velocity as inputs and uses a droop controller. The automatic voltage regulator produces the electromotive force E_f, from reactive power command, reactive power feedback signal, the voltage reference, and the voltage feedback signal through a droop controller. The δ, E_f, and grid voltage signals v_d, v_q are used to produce the current reference based on the phasor diagram with the virtual armature resistance and virtual synchronous reactance.

12.2.4 VSG system of Osaka University

The block diagram of this scheme is shown in Fig. 12.6 [24,25]. The well-known swing equation of synchronous generators is used as the heart of the VSG model:

$$P_{out} = P_{in} - J\omega_m \, {}^{d\omega_m}\!/\!_{dt} - D\Delta\omega \tag{12.5}$$

where P_{in}, P_{out}, J, ω_m, and D are the input power (as same as the prime mover power in a synchronous generator), the output power of the VSG,

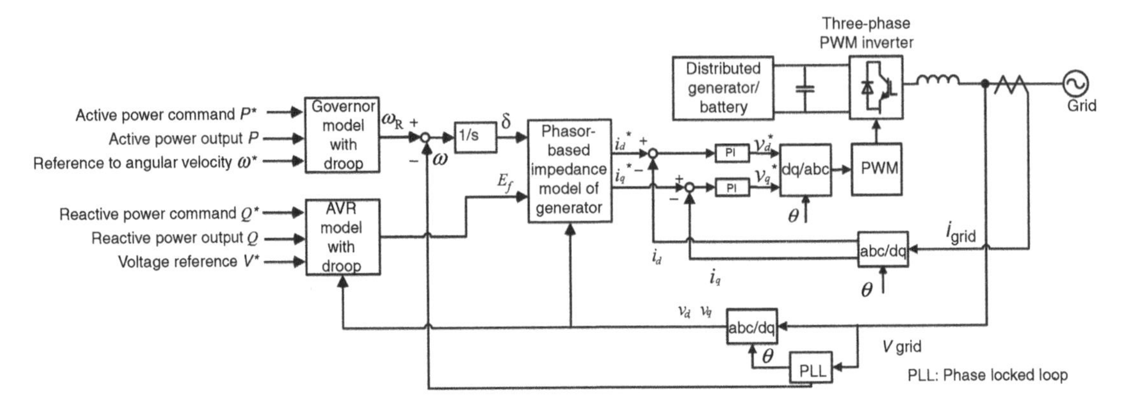

■ FIGURE 12.5 The algebraic type VSG introduced by the Kawasaki Heavy Industries.

the moment of inertia of the virtual rotor, the virtual angular velocity of the virtual rotor, and the damping factor, respectively. $\Delta\omega$ is given by $\Delta\omega = \omega_m - \omega_{grid}$, ω_{grid} being the grid frequency or the reference frequency when the grid is not available. Using voltage and current signals measured at the VSG terminals, its output power and frequency are calculated. A governor model shown in Fig. 12.7 is implemented to tune the input power command based on the frequency deviation. The grid frequency is detected by a frequency detector block that can be a PLL. Having the essential parameters, (12.5) is solved by numerical integration. By solving (12.5), the momentary ω_m is calculated and by passing through an integrator, the virtual mechanical phase angle θ_m is produced.

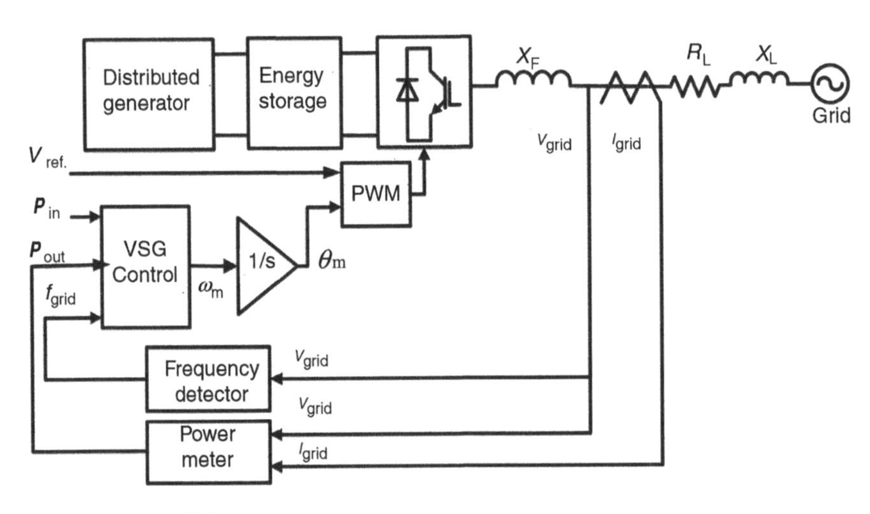

■ FIGURE 12.6 Block diagram of the VSG by the Osaka University.

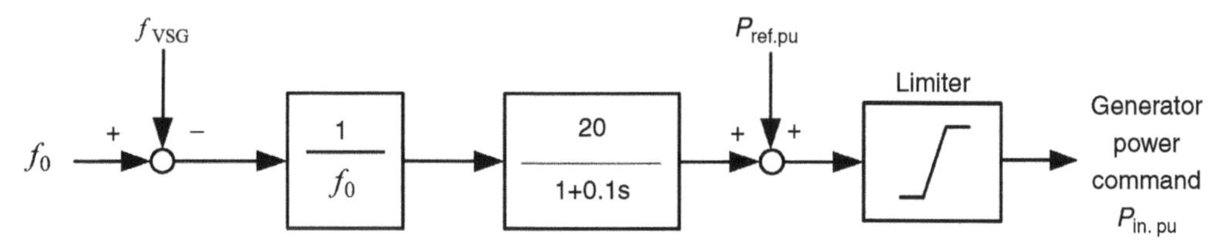

FIGURE 12.7 Governor model for the VSG of Fig. 12.6.

$V_{ref.}$ in Fig. 12.6 is the voltage reference that determines the voltage magnitude at the inverter terminal. Implementing a controller for $V_{ref.}$ results in a regulated voltage and reactive power at the VSG terminal. The phase angle and the voltage magnitude reference are used as the VSG output voltage angle and magnitude commands to generate PWM pulses for the inverter.

The value of J together with D in (12.5) determines the time constant of the VSG unit. Selecting the proper value of them is a challenging issue without a routine. Mimicking a synchronous machine, J is the inertia emulating characteristic given by $J = 2HS_0/\omega_0^2$, where H is the machine inertia constant, S_0 is the nominal apparent power of the machine, and ω_0 is the system frequency. The parameter H tells that for which period of time the machine is able to supply the nominal load based solely on the energy stored in the rotating mass. The higher H, the bigger the time constant, resulting in a slower response but smaller frequency deviation after a change or disturbance. Although it depends on the machine size and power, for typical synchronous machines H varies between 2 and 10 s.

By (1.5), the initial rate of frequency change ($d\omega/dt$) provides an error signal (with equilibrium of zero). When this signal is zero and frequency matches ω_{grid}, the output power of VSG follows the power command P_{in}. When there is a frequency variation due to a change of disturbance, the error signal will be a nonzero value and causes power oscillations. If the term $D\Delta\omega$ is neglected, power will be exchanged only during the transient state without necessarily returning back the output frequency to the nominal value. In order to cover this issue, a frequency droop part $\Delta\omega$ is added as shown in (12.5). The $D\Delta\omega$ emulates the damper windings effect in a SG, and represents the linear damping. It must be chosen so that the P_{out} to be equal with the nominal power of the VSG when the frequency deviation is at the specified maximum value [14].

Considering only the virtual inertia effect ($J\omega_m d\omega/dt$), increasing the moment of inertia, J reduces the maximum deviation of the rotor speed following a disturbance; however, the natural frequency and the damping ratio of the system may be decreased [4,30].

In summary, the virtual mass counteracts the frequency drops by injecting/extracting active power and the virtual damper suppresses the oscillation so these features are equally effective to electromechanical synchronous machines. J and D should be fixed so that the VSG exchanges its maximum active power when the maximum specified frequency variation and rate of frequency change occur. The larger J and D means that more power will be either injected or absorbed for the same amount of frequency deviation and rate of frequency change, respectively. However, oppositely, large values of J and D with specific power rating results in a small frequency excursion.

As mentioned, increasing of J provides a higher amount of equivalent inertia for the VSG, however there is a limit. This limit is mainly imposed by the inverter capacity and PLL accuracy. The inverter capacity does not have the overload capacity of a synchronous machine. Thus, a high derivative term leads to bigger power overshoots during transients (frequency deviations), and the inverter must sustain an important overload. The accuracy in frequency tracking depends on the performance of the implemented PLL. Therefore, the optimal value of derivative term in (12.5) can be obtained by a tradeoff between the virtual inertia, the inverter overload capacity, and the PLL characteristics.

This group has added reactive power control to have a constant voltage at VSG terminals [26], and evaluated the performance in various voltage sag conditions, and enhanced the voltage sag ride-through capability of the VSG [27]. Oscillation damping approaches have been developed for a DG using the VSG [28,29].

12.3 APPLICATIONS FOR MICROGRIDS

A microgrid is an interconnection of domestic distributed loads and low voltage distributed energy sources, such as microturbines, wind turbines, PVs, and storage devices. The microgrids are placed in the low voltage (LV) and medium voltage (MV) distribution networks. This has important consequences. With numerous DGs connected at the distribution level, there are new challenges, such as system stability, power quality and network operation that must be resolved applying the advanced control techniques at LV/MV levels rather than high voltage levels which is common in conventional power system control [31,32].

The VSG systems can be used as effective control units to compensate for the lack of inertia and result in the control of active and reactive power as well as microgrid voltage and frequency. A microgrid with VSG units is shown in Fig. 12.8. The VSGs can be connected between a DC bus/source

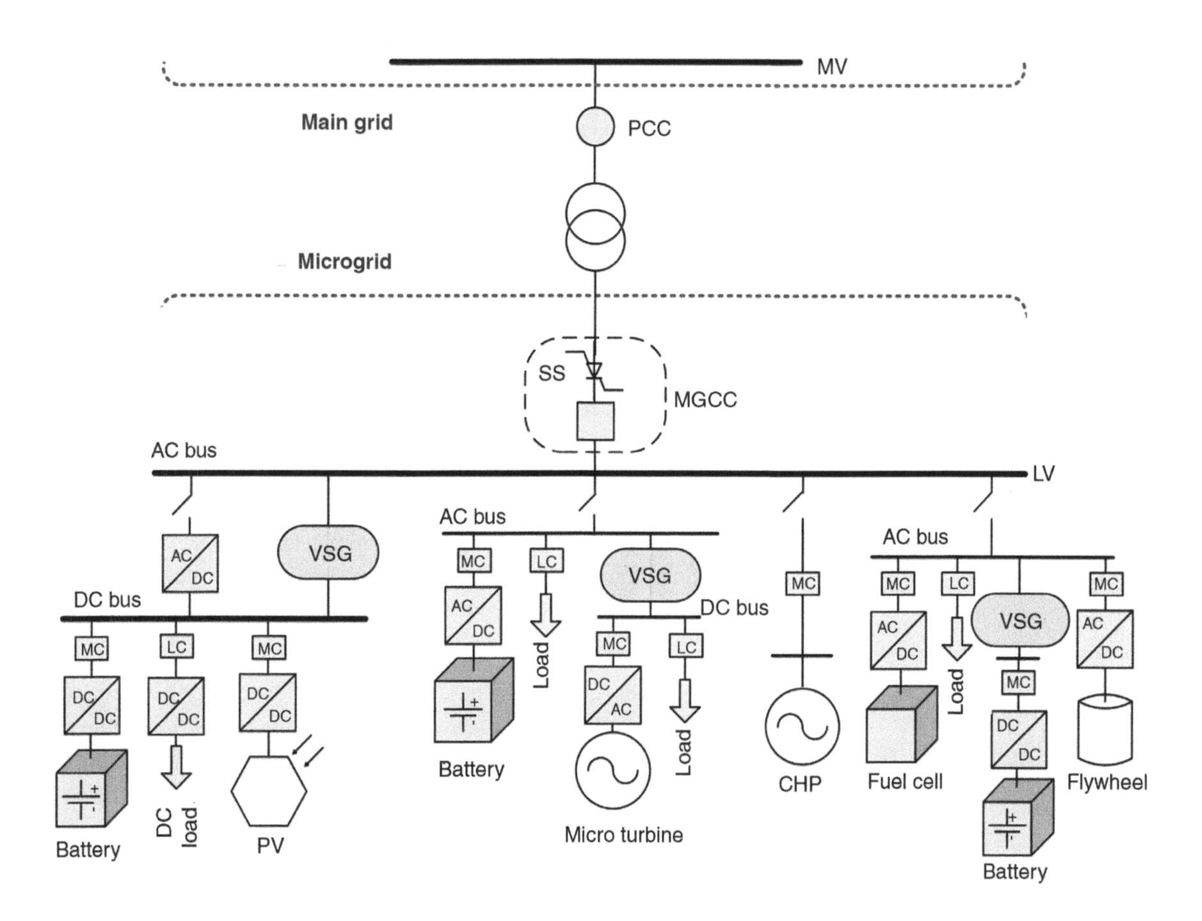

■ **FIGURE 12.8 A structure of a microgrid with multi VSG units.**

and an AC bus, anywhere in the microgrid. These systems are going to be more vital to overcome fluctuations caused in the microgrid due to integration of large number of DGs with low or no inertia [31]. Some loads can be also locally controllable using the load controllers. The load controllers are usually used for demand side management.

In microgrids with small power capacity, a change or disturbance in the system (such as a temporary imbalance between production and demand after loss of a large generating unit) results in a high rate variation in the rotating speed of generators. Conventional technologies used for power generation are not always capable of responding quickly enough to prevent unacceptably low frequency in such cases, even when the available amount of frequency control reserve exceeds the power deviation [31]. It results in relatively frequent use of load-shedding, with subsequent consequences on the economic activity, to restore the power equilibrium and prevent frequency collapse [33]. With an appropriate control strategy, the VSGs equipped with fast-acting storage devices can help microgrids to mitigate the frequency excursions caused by generation outages, thus reducing the need for load shedding [31].

During the grid connected operation, all the DGs and inverters in the microgrid use the signals of grid voltage and frequency as reference for voltage and frequency. In this mode, it is not possible to highlight the VSG contribution to the grid inertia, due to system size differences. However, in islanding, the DGs lose that reference. In this case the DGs may use the VSG units, and may coordinate to manage the simultaneous operation using one effective control techniques such as master/slave control, current/power sharing control, and generalized frequency and voltage droop control techniques [32]. The balance between generation and demand of power is an important requirement of the islanded operation modes. In the grid-connected mode, the microgrid exchanges power to an interconnected grid to meet the balance, while, in the islanded mode, the microgrid should meet the balance for the local supply and demand using the decrease in generation or load shedding [31].

During the islanded mode, if there are local load changes, local DGs will either increase or reduce their production to keep the energy balance constant as far as possible. In an islanded operation, a microgrid works autonomously, therefore must have enough local generation to supply demands, at least to meet the sensitive loads. In this mode, the VSG systems may present a significant role to maintain the active and reactive power [31].

Immediately after islanding, the voltage, phase angle, and frequency at each DG in the microgrid change. For example, the local frequency will decrease if the microgrid imports power from the main grid in a grid-connected operation, but will increase if the microgrid exports power to the main grid in the grid-connected operation [31]. The duration of islanded operation will depend on the size of storage systems. In this case they are sized to maintain the energy balance of the network for few minutes. The VSG control algorithms for islanding and grid-connected modes are different, as the islanded microgrid has to define its own frequency and voltage to maintain operation [16]. When the main grid has returned to normal operation, the frequency and voltage of the microgrid must be synchronized with and then reconnected to the main grid.

REFERENCES

[1] Zhong Q-C, Hornik T. Control of power inverters in renewable energy and smart grid integration. New York, NY, USA: Wiley; 2013.

[2] Zhang L, Harnefors L, Nee H-P. Power-synchronization control of grid-connected voltage-source converters. IEEE Trans Power Syst 2010;25(2):809–20.

[3] Blaabjerg F, Teodorescu R, Liserre M, Timbus A. Overview of control and grid synchronization for distributed power generation systems. IEEE Trans Ind Electron 2006;53(5):1398–409.

[4] Alipoor J, Miura Y, Ise T. Power system stabilization using virtual synchronous generator with alternating moment of inertia. IEEE J Emerg Select Top Power Electron 2014;PP(99):1–8.

[5] van Wesenbeeck MPN, de Haan SWH, Varela P, Visscher K. Grid tied converter with virtual kinetic storage. PowerTech, 2009 IEEE Bucharest; 2009, pp. 1–7.

[6] Beck H-P, Hesse R. Virtual synchronous machine. International conference on electrical power quality and utilisation, EPQU, Barcelona, Spain; October 2007, pp.1–6.

[7] Zhong Q-C, Nguyen P-L, Ma Z, Sheng W. Self-synchronized synchronverters: inverters without a dedicated synchronization unit. IEEE Trans Power Electron 2014;29(2):617–30.

[8] Visscher K. de Haan SWH. Virtual synchronous machines (VSGs) for frequency stabilization in future grids with a significant share of decentralized generation. Smart grids for distribution, 2008. IET-CIRED. CIRED Seminar; 2008, pp. 1–4.

[9] Chen Y, Hesse R, Turschner D, Beck HP. Comparison of methods for implementing virtual synchronous machine on inverters. International conference on renewable energies and power quality-ICREPQ'12, Spain; March 2012.

[10] Yong C, Hesse R, Turschner D, Beck HP. Improving the grid power quality using virtual synchronous machines. Power engineering, energy and electrical drives (POWERENG), 2011 international conference; 2011, pp. 1–6.

[11] Driesen J, Visscher K. Virtual synchronous generators. Proceedings IEEE power and energy society general meeting – conversion and delivery of electrical energy in the 21st Century; 2008, pp. 1–3.

[12] Loix T, De Breucker S, Vanassche P, Van den Keybus J, Driesen J, Visscher K. Layout and performance of the power electronic converter platform for the VSYNC project. Proceedings of IEEE powertech conference; 2009, pp. 1–8.

[13] Zhong Q-C, Weiss G. Synchronverters: inverters that mimic synchronous generators. IEEE Trans Ind Electron 2011;58(4):1259–67.

[14] Karapanos V, de Haan SWH, Zwetsloot KH. Testing a virtual synchronous generator in a real time simulated power system. International conference on power systems transeints (IPST2011), Delft, Netherland; June 2011.

[15] Albu M, Nechifor A, Creanga D. Smart storage for active distribution networks estimation and measurement solutions. Instrumentation and measurement technology conference (I2MTC), 2010 IEEE; 2010, pp. 1486–1491.

[16] Van TV, Visscher K, Diaz J, Karapanos V, Woyte A, Albu M, Bozelie J, Loix T, Federenciuc D. Virtual synchronous generator: an element of future grids. Innovative smart grid technologies conference Europe (ISGT Europe), 2010 IEEE PES; 2010, pp. 1–7.

[17] Albu M, Diaz J, Thong V, Neurohr R, Federenciuc D, Popa M, Calin M. Measurement and remote monitoring for virtual synchronous generator design. Applied measurements for power systems (AMPS), 2010 IEEE international workshop; 2010, pp. 7–11.

[18] Karapanos V, Yuan Z, de Haan S. SOC maintenance and coordination of multiple VSG units for grid support and transient stability. 3rd VSYNC workshop, Cheia, Romania; June 2010.

[19] Albu M, Calin M, Federenciuc D, Diaz J. The measurement layer of the Virtual Synchronous Generator operation in the field test. Applied measurements for power systems (AMPS), 2011 IEEE international workshop; 2011, pp. 85–89.
[20] Karapanos V, de Haan S, Zwetsloot K. Real time simulation of a power system with VSG hardware in the loop. IECON 2011-37th annual conference on IEEE industrial electronics society; 2011, pp. 3748–3754.
[21] Chen Y, Hesse R, Turschner D, Beck HP. Dynamic properties of the virtual synchronous machine (VISMA). International conference on renewable energies and power quality (ICREPQ́11), Las Palmas de Gran Canaria, Spanien; April 2011.
[22] Hesse R, Turschner D, Beck H-P. Micro grid stabilization using the virtual synchronous machine. Proceedings of international conference on renewable energies and power quality (ICREPQ'09), Spain; 2009.
[23] Hirase Y, Abe K, Sugimoto K, Shindo Y. A grid connected inverter with virtual synchronous generator model of algebraic type. Electr Eng Jpn 2013;184(4):10–21. Wiley.
[24] Sakimoto K, Miura Y, Ise T. Stabilization of a power system with a distributed generator by a Virtual Synchronous Generator function. IEEE 8th international conference on power electronics and ECCE Asia (ICPE & ECCE); 2011, pp. 1498–1505.
[25] Sakimoto K, Miura Y, Ise T. Stabilization of a power system including inverter type distributed generators by the virtual synchronous generator. Electr Eng Jpn 2014;187(3):7–17. Wiley.
[26] Shintai T, Miura Y, Ise T. Reactive power control for load sharing with virtual synchronous generator control. 7th IEEE international power electronics and motion control conference ECCE Asia, Harbin, China; June 2012, pp. 846–853.
[27] Alipoor J, Miura Y, Ise T. Voltage sag ride-through performance of Virtual Synchronous Generator. Proceedings of IEEE internatonal power electronics conference (IPEC-Hiroshima – ECCE-ASIA); 2014, pp. 3298–3305.
[28] Alipoor J, Miura Y, Ise T. Distributed generation grid integration using virtual synchronous generator with adoptive virtual inertia. Proceedings of IEEE energy conversion congress and exposition (ECCE); 2013, pp. 4546–4552.
[29] Shintai T, Miura Y, ise T. Oscillation damping of a distributed generator using a virtual synchronous generator. IEEE Trans Power Del 2014;29(2):668–76.
[30] Torres M, Lopes LAC. Virtual synchronous generator control in autonomous wind-diesel power systems. IEEE electrical power and energy conference; 2009, pp. 1–6.
[31] Bevrani H, Ise T, Miura Y. Virtual synchronous generators: a survey and new perspectives. Int J Elec Power 2014;54:244–54.
[32] Bevrani H, Watanabe M, Mitani Y. Microgrid controls. In: Beaty HW, Fink DG, editors. Standard handbook for electrical engineers, Section 16. 16th ed. New York: McGraw-Hill; 2012.
[33] Serban I, Marinescu C. Frequency control issues in microgrids with renewable energy sources. Advanced topics in electrical engineering (ATEE), 2011 7th International Symposium on; 2011, pp. 1–6.

Chapter 13

Application of DERs in electricity market

Yusuke Manabe
Funded Research Division Energy Systems (Chubu Electric Power), Institute of Materials and Systems for Sustainability (IMaSS), Nagoya University, Nagoya, Japan

CHAPTER OUTLINE

13.1 Basic concept of electricity market and DERs 295
13.1.1 RPS 296
13.1.2 FIT 297
13.1.3 Effect of DER's mass penetration for electricity market 297
13.2 Electricity market reform and virtual power plan 299
References 302

13.1 BASIC CONCEPT OF ELECTRICITY MARKET AND DERs

In conventional power systems, vertical integrate management is carried out by a regional monopoly company. However, since the 1990s, deregulation and electric power market opening has been done in several regions all over the world. The objects of this are economic development by open competition, and construction of resilient energy system, and so on. A fundamental component of the electric power market is spot market (day-ahead market). This market decides the market price (eg, $/MWh) and electric energy quantity (MWh) of each time period [2,8]. The time period interval is 1 h or 30 min in many markets. Supply and demand biddings are closed the previous day, and all biddings are sorted to make supply and demand curves. An equilibrium point at which the two curves meet decides the market clearing price and energy quantity. Fig. 13.1 shows an image of a spot market.

This market structure is an unfit design for distributed energy resources (DERs), which are photovoltaic generation, wind turbine generation, and so on, and forecasting output is difficult. DERs should decide each bidding price and quantity of all time periods based on the day-ahead weather

Integration of Distributed Energy Resources in Power Systems

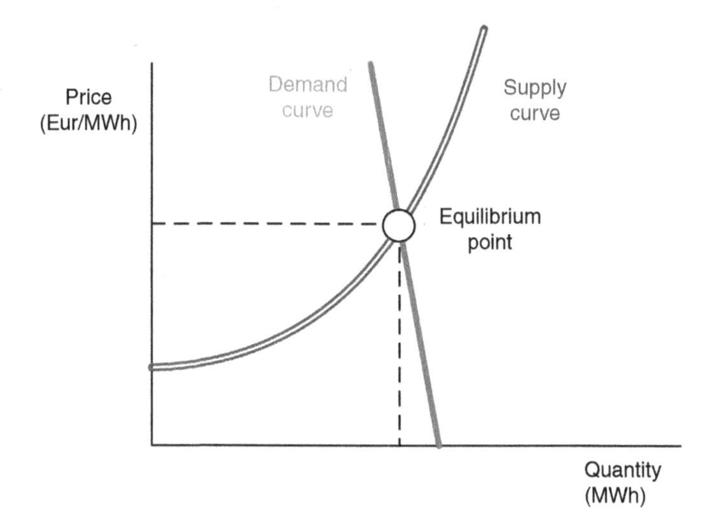

■ **FIGURE 13.1 Spot market.**

forecast. However, output forecast cannot be done perfectly, forecast error will be occurred [3]. Difference between actual output and bidding quantity give penalty to DER. This penalty reduces the economic value of DER. Additionally, the construction cost of DERs in the 1990s was higher than the present cost. The average market price which was decided by mainly thermal, large scale hydro, and nuclear generation was too low to pay for the massive initial cost of DER. However, DERs are one kind of the low carbon and renewable energy plant, it is good for security of energy source supply, and environmental protection. Therefore, the European Union, the United States, and other countries have promoted several DER policies, mainly renewable portfolio standard (RPS) and feed in tariff (FIT).

13.1.1 RPS

RPS is the regulation policy which decides the target percentage of DER output energy in the total electric energy supply, and electric power suppliers have an obligation to fulfill this target. Penalties are also imposed for nonfulfillment suppliers. RPS has been operated in majority of states in the United States (29 states), in the United Kingdom, Italy, Belgium, and Sweden, and so on [9]. In most cases, governments create renewable energy certificates (RECs) to track the fulfillment of the target percentage. These certificates is securitized the values of DER for environment protection and energy security, and allow for trading in the market. The combination of RPSs and RECs makes suppliers' businesses form more flexibly and increases the values of DER. However, RPS trading makes the RPS system more complex. Additionally, it is difficult as small-scale DERs securitize RECs.

13.1.2 FIT

FIT is the policy that decides the energy price (tariff) and purchase period of each kind of DER. All output energy is purchased without bidding to the spot market. Therefore, FIT can remove economic anxiety and increase the incentive of DER introduction. Additionally, this system is simpler than RPS. In previous research comparing FIT and RPS, FIT has been found to better promote DER development [4,7]. FIT has been operated in Germany, Spain, Italy, and some states in the United States, and so on [9]. Japan also has used FIT since 2011, and promotes the mass penetration of photovoltaic generation. However, FIT has one defect, difficult tariff setting. In general, tariffs should be set making the internal rate of return (IRR) of DERs their proper value. IRR is the discount rate value at which the net present value of future cash flow becomes same as initial investment cost. Although the tariff should be updated while considering the DER's cost due to mass production and experience accumulation in order to keep the IRR. If this is not updated, there will be excess cash flow to the DER owner. Additionally, if the penalty caused by the spot market does not occur, the DER owner will not smooth out the output fluctuation, and uncertain and intermitted output is supplied. This makes the electric power system worse and the market price becomes more unstable.

13.1.3 Effect of DER's mass penetration for electricity market

Currently, DERs mass penetrated by these promoting policies start affecting the electricity market. Specifically, the merit order of spot markets is changing. The merit order is the ranking of a power supply plant base on short-run marginal cost. In a perfect competitive situation, the best supplier's bid strategy becomes the marginal cost bidding. Therefore, the supply curve becomes the merit order naturally. If DER based on renewable energy source is bid to the spot market, zero price bidding becomes best strategy because renewable energy does not need fuel for the generation and the short-run marginal cost is almost free. Mass penetration of DER by RPS increases the bids of DER and these bids are sorted first. This moves the supply curve and equivalent points to the right and reduces the market price. Fig. 13.2 shows this price down mechanism.

DERs introduced by FIT do not bid to the spot market and DER energy is purchased by and integrated manager of the power system. Purchased DER energy is supplied on demand. As a result, the demand biding quantity to

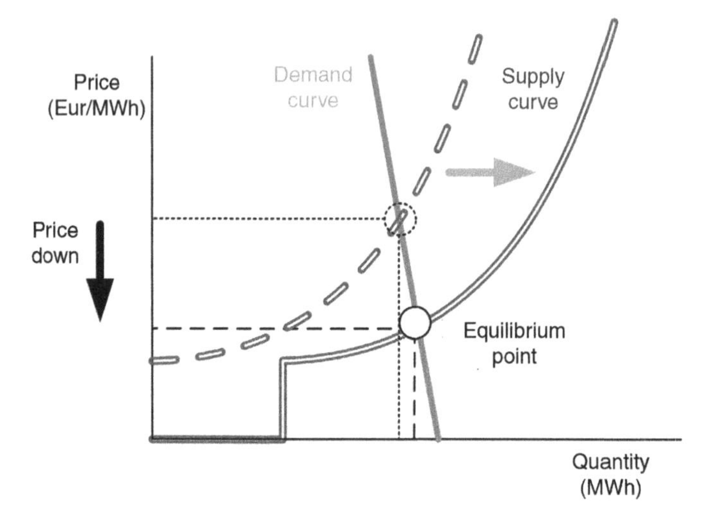

■ **FIGURE 13.2 Spot market with PRS.**

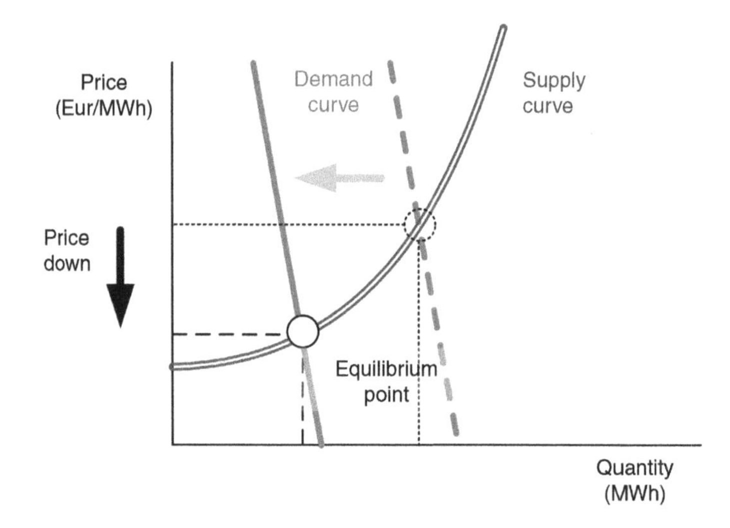

■ **FIGURE 13.3 Spot market with FIT.**

the spot market is reduced. This moves the demand curve to the left and the equivalent point to the right, and reduces the market price. Fig. 13.3 shows this price down mechanism. This is essentially the same as mass penetration of DER by RPS. Merit order change by the mass penetration of DER is merely shown by from different viewpoints.

Morthost et al. [11] researched some literature about the merit order change, and showed the effect of the price decrease in 2009 that appeared from

3–23 Eur/MWh in several European Union regions. Depreciation of the market price will bring the expected profit reduction of thermal plants with high marginal cost and low efficiency. This may promote the replacement of thermal plants and DERs. However, this effect has the following demerit: DER is forcibly penetrated by RPS or FIT. In other words, this penetration has no relation to the market principle, and it will be not happen that value reduction of DER by excess introduction. Specially, FIT will promote DER penetration until the electric power system problems happen (frequency control, distribution voltage, transmission capability, and so on). As a result, it will increase the possibility that the whole electric power system is lost and the electricity quality degrades. Additionally, it becomes a big problem as to who pays the massive construction cost of DERs. The spot market price is reduced by the RPS or FIT, but this price is wholesale. As for what will happen to retail price? The necessary cost of the achievement of the RPS target and the tariff payment of FIT are eventually charged to the customers. Therefore, the retail price will be increased. If tariff setting of FIT is mistaken, this problem will appear as an outflow of wealth from demand customers to DER owners.

Currently, the installed DER capacity is increased and the construction cost is decreased compared with the 1990s. As a result, the promoting policy's demerit which hid behind the merit when DER promoting was started is now standing out.

13.2 ELECTRICITY MARKET REFORM AND VIRTUAL POWER PLAN

To overcome the current problems of the electricity market and DERs, effective measures which do not give exceed favorable treatment to DERs and realizes a fair market competitive environment are required, and these will be introduced by renewed policies and business models of DER.

In some countries, modification of the electricity power market design has started. For example, the United Kingdom is starting the Electricity Market Reform (EMR) [6,12]. The objective of EMR is the securement of the investment needed to deliver a reliable diverse low carbon technology mix. This reform is mainly constructed following four policies [5]:

- *Feed-in tariffs with contracts for difference (FIT–CfD)*: Long-term contracts which provide revenue certainty to investors in low-carbon generation such as renewables (DERs), nuclear and carbon capture, and storage-equipped plants.
- *Capacity agreements (within a capacity market)*: Payments for reliable capacity to be available when needed, helping to ensure security of supply.

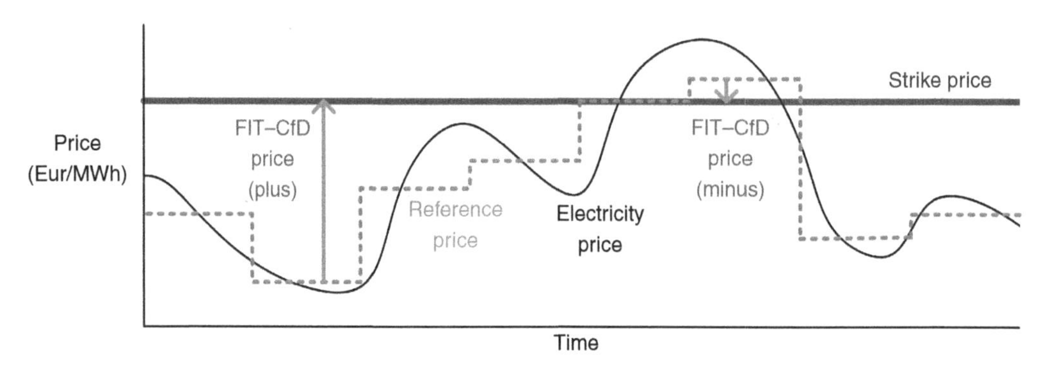

■ FIGURE 13.4 FIT–CfD.

- *Carbon price floor*: A tax to underpin the carbon price in the emissions trading scheme.
- *Emissions performance standard*: A regulatory measure which provides a backstop to limit emissions from unabated power stations.

The policy which has an especially high relationship with DERs is FIT–CfD. FIT–CfD stabilizes returns for DERs at a fixed level known as the strike price. DERs receive revenue from selling their electricity into the spot market as usual. In addition, when the market price is below the strike price they also receive a top-up payment for the additional amount. To compare with strike price, average market price in a certain period (eg, 1 month) use for reference price. Conversely if the reference price is above the strike price, the generator must pay back the difference. Fig. 13.4 shows the image of FIT–CfD.

DERs should bid to the spot market differently from normal fixed FIT. Accordingly, DERs output should be controlled in order to avoid the penalty caused by the difference of bidding quantity and actual output. FIT–CfD retains short-term market signals for efficient operation to DERs. Additionally, the strike price will be decided by auctions or use some other competitive process. Fixed FIT only improves long-term revenue certainty, but FIT–CfD makes the construction cost lower by fair competition. As a result, FIT–CfDs have more cost-effective points than other options for support, reducing the cost to consumers. It will be interesting to see the results of the FIT–CfD starting from 2017 in the United Kingdom.

Future policies for DERs such as FIT–CfD will be more competitive and DERs will be required to have high controllability. The only one kind and one

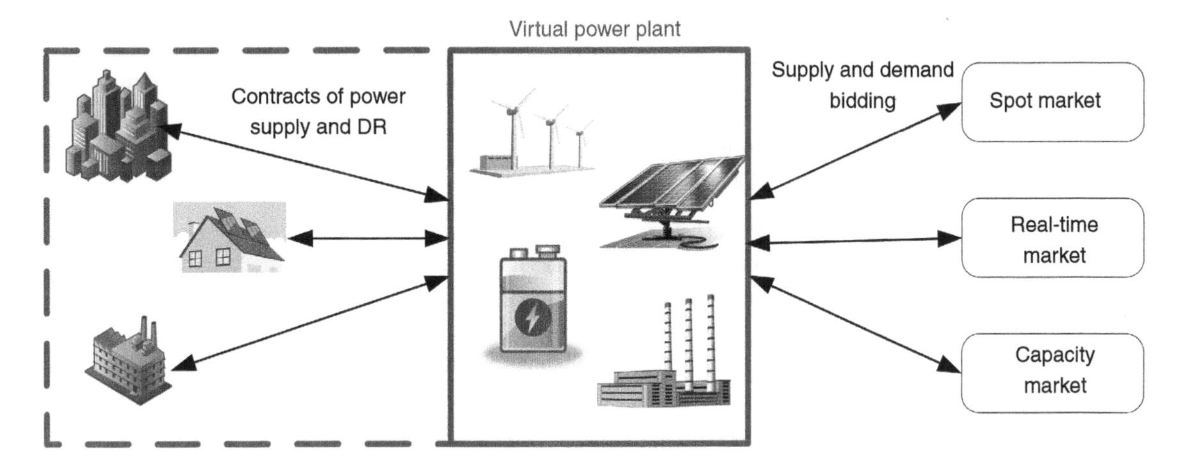

■ **FIGURE 13.5 Virtual power plant.**

plant operation will become more difficult to get high profit. Virtual power plants (VPPs) are one of the effective business models applied to this trend and maintain the economic value without FIT. VPPs combine not only several kinds of DERs but also contracts of demanded response (DR) and increases the VPP's diversity and robustness. Fig. 13.5 shows an image of VPP.

Navigant Research, a market research and consulting company specializing in green energy market, defined VPP as,

> *"A system that relies upon software and a smart grid to remotely and automatically dispatch and optimize DER via an aggregation and optimization platform linking retail to wholesale markets" [1].*

From this definition, we can take two important points. The first point is "remotely and automatically dispatch and optimize DER." The output characteristic of DER is highly dependent on where this located. In order to combine several characteristics and reduce the uncertainty effect, DERs consisting of VPPs should be widely placed and have various kinds where possible. Therefore, the remote communication environment needs to be able to operate the wide spread of DERs in cooperation. Additionally, in order to enhance the VPP's economic efficiency, entering into a real-time market or capacity market, not only the spot market, is necessary. VPPs are required for real-time operation and frequency response to enter these markets. These operations cannot be realized manually, and automatic control applied to whole VPP systems is needed. The second point is "linking retail to wholesale markets." The market price will become even more variable and unforecastable because of the mass penetration of DER with high uncertainty. Therefore, market bidding and DER operation should be optimized to maximize the VPP profit in consideration of the DER's characteristics and price variability [10]. Besides, advanced retail contracts, including DR, are useful. There is a probability that the combination of a bidding strategy and the condition of contracts which give high expected profit to VPP is proposed by taking into account the customer's and DER's characteristics.

VPP systems have already started. Next Kraftwrerk is targeting the real-time market mainly in Germany, Energy Pool is making DR an important project in France, and so on. In the future, several VPP systems will be proposed and implemented. These will have a larger effect on the electricity power system and market than simple DERs which have been introduced.

REFERENCES

[1] Asmus P, Lawrence M. Executive summary: virtual power plants – demand response, supply-side, and mixed asset VPPs: global market analysis and forecasts. Denver: Navigant Research; 2014.

[2] Belyaev LS. Electricity market reforms: economics and policy challenges. Berlin: Springer Science & Business Media; 2010.

[3] Bostan I, Bostan V, Dulgheru V, Gheorghe AV, Sobor I, Sochirean A. Resilient energy systems: renewables: wind, solar hydro. Berlin: Springer Science & Business Media; 2012.

[4] Butler L, Neuhoff K. Comparison of feed-in tariff, quota and auction mechanisms to support wind power development. Renew Energ 2008;33(8):1854–67.

[5] Department of Energy and Climate Change. Electricity market reform: policy overview. London: Department of Energy and Climate Change; 2012.

[6] Department of Energy and Climate change. Energy bill: supplementary memorandum (electricity market form). London: Department of Energy and Climate change; 2013.

[7] Dong CG. Feed-in tariff vs. renewable portfolio standard: an empirical test of their relative effectiveness in promoting wind capacity development. Energ Policy 2012;42:476–85.

[8] Gan D, Feng D, Xie J. Electricity markets and power system economics. Boca Raton: CRC Press; 2013.

[9] Japan Electric Power Information Center. Electricity industry of foreign centuries, part I. Tokyo: Japan Electric Power Information Center; 2014.

[10] Mashhour E, Moghaddas-Tafreshi SM. Bidding strategy of virtual power plant for participating in energy and spinning reserve markets – Part I: Problem formulation. IEEE Trans Power Syst 2011;26(2):949–56.

[11] Morthost PE, Munksgaard J, Ray S, Sinner AF. Wind energy and electricity prices exploring the "merit order effect". European Wind Energy Association: London; 2010.

[12] Toke D. UK Electricity Market Reform – revolution or much ado about nothing? Energ Policy 2011;39(12):7609–11.

Subject Index

A

AC. *See* Alternating current (AC)
Acknowledged mode (AM), 30
Air Force Weather Agency, 90
Air termination systems, 196, 200, 215, 217
Alternating current (AC), 2
components, 234
DC converters, 250
DC microgrids, 236, 253
average charge times, 242
basic concept of, 250
battery charge method, 242
battery voltage, 249, 254
business models, 236–242
charge pattern, 245
charge power, comparison of, 244
current/voltage during battery charge, 243
daily load curve/electricity rate, 247, 249
daily power load curve, 241, 248
distributed systems, 250
electricity rate, effect of charge times, 245, 246
examples in world, 254
Japan, demonstration tests, 251, 252
location, 254
photovoltaic (PV) and wind power generation, 235
power flow and load curve, 253
power supply, 251
supply and demand control, 246
peak cut/peak shift mode operation, 246
receiving constant power mode operation, 247
system operation of, 251
test results, 242
charge pattern, 242
test using DC power supply, 251
USA, demonstration tests, 252
electricity, 79
source, 161
AM. *See* Acknowledged mode (AM)
Ar capability, 217
Artificial neural networks (ANNs), 123
Automatic voltage regulator produces, 282
Autoregressive integrated moving average (ARIMA) model, 97
Autoregressive moving average (ARMA) model, 97
Auxiliary grounding electrodes, 211

B

Background traffic (BT), 29
Battery charge, current/voltage, 243
Battery current, 242
Battery energy storage system (BESS) technologies, 52
active and reactive powers, 67
based energy acquisition model, 137
power electronics, 232
basic configuration of, 233
Battery voltage, 249
BESS technologies. *See* Battery energy storage system (BESS) technologies
Biomass fuels, 1, 8
Bus voltages, 46

C

Capacity agreements, 294
Carbon price floor, 294
Cauchy theorem, 266, 267
Central energy management system, 21
Centralized control methods, 52
Chubu region, 85
Circuit breakers, 161
Client–server communication model, 27
Close-to-Earth-surface temperatures, 9
Cloudy weather condition, simulation results, 137
CO_2 emissions, 146
Combined heat and power (CHP) systems, 4
Constant voltage (CV) control, 242
Constructing air-termination system, 208
Conventional distributed system, 158
Conventional energy resources, 1
Conventional power grid, 52
Conventional power systems, 290
Conventional technologies, 286
Cosimulation environments, 25
Current-fed (CF) converter, 264
Current transformer (CT), 183

D

Day-ahead forecasting, 122
DC. *See* Direct current (DC)
DC–DC converter, 230
Demanded response (DR), 295
DER technology. *See* Distributed energy resource (DER) technology
DGs. *See* Distributed generations (DGs)
Direct current (DC), 2
bus voltage, 139, 242, 279
distribution system, 139
power supply strategies, 139
smart house model, 134
voltage source, 215
Distributed energy protection system, 161
Distributed energy resource (DER) technology, 15, 158, 181, 228
in distribution power systems, 15
electrical power grid, 176
electric grid, 166
in electricity market, basic concept, 290
FIT, 292
RPS, 291
electricity market reform, 294
energy management systems, 132
cloudy weather condition, simulation, 137
DC smart house model, 134
electric price, 135
home energy management systems, basic concepts of, 132–136
SC system, model of, 135
smart grid model, 134
sunny weather condition, simulation, 136
fail-to-trip, 175
feed in tariff (FIT), 290
CfD stabilizes, 295
fuse operating scheme, 190
generate DC/AC power, 228
phasor diagram, 229
power quality problems, 233
PV power generation systems, power electronics, 230
basic configuration of, 230
output current and voltages, 231
output power, characteristics of, 231
simple model of, 230

Distributed energy resource (DER) technology (*cont.*)
simplest current source model, 233
simplest model of, 229
ICT, role of, 25–27
impact on protection system, 174–175, 179
coordination, loss of, 177–179
generalized method, loss of coordination, 178
hosting capacity, 176–177
mal-trip and fail-to-trip, 176
monitoring and control, 181
MV and LV circuit breakers, 180
protection failure, 175
installation changes, 174
interconnection system, 180
internal rate of return (IRR), 292
inverter mode, 183
islanded operation, 182
market structure, 290
massive construction cost, 293
massive penetrations. *See* Massive DER penetrations
mass penetration, 292
overcurrent protection scheme, 189, 190
PAC, for distribution protection, 184
penetration limit, 177
preliminary conditions, 162
protection characteristics, 189
protection equipment for, 183
fuses, 188
overcurrent relays, 183–185
reclosers, 185–186, 187
sectionalizes, 187
recent technological trends, 188
renewable portfolio standard (RPS), 290
safety requirements, 164
spot market, 291
VC–DER unit, 191
virtual power plan, 294, 296, 297
voltage control scheme, 190
voltage-sourced converter (VSC), 190
Distributed energy sources, fluctuating power, 132
Distributed generations (DGs), 16, 52, 57, 137, 277
application of, 2
inverter, 9
maximum rating, 2
power electronics, 278
power fluctuations, 278
PV generators, 61
rating of, 2
reactive power control, 65
resources, 3
fuel cell, 5–6
microturbine generator (MTG) system, 4
reciprocating engine, 3–4
sources, 4
Distribution energy resource layout, 159
Distribution network, protection system of, 161
Distribution power systems
cellular communication network, 44
communication networks, models of, 27
cellular communication network model, 29
wired communication network model, 28
compensation cycles, number of, 44
countermeasures, 17
distributed energy resources (DERs), 15
ICT, role of, 25–27
energy resources, integration of, 15
gossip-like VVC MAS procedure, 21–25
interconnection issues, 17
mean and standard deviation values, 37, 38
mixed integer linear programming (MILP) model, 39
OLTC, power loss reductions, 41
packets, number of, 45
power distribution feeder, model, 32–33
power feeder, in black/communication network, 35, 36
TCP/UDP node models, 29
test results, 33–34
TF2 and scenario BT0-PDR0, 40
TF1, bus voltage variations, 42
TF2, bus voltage variations, 43
TF1, power loss variation, 42
TF2, power loss variation, 43
volt–VAR control, 19–21
wired communication network, 34
Distribution system operators (DSOs), 17
Distribution system, protection, 160
energization area, 165
fault conditions, 165–166
fault currents change, 166
fault detection, 168
fault isolation, 168
fault location, 168
fault location isolation and service restoration (FLISR) application
fault investigation, time line, 167
general protection, 160–164
grounding scheme, 164
IEEE standards, for protection, 164
power system stability, 168
service restoration, 168
smart protection, 166
synchronization, 164
voltage requirements, 164
Doubly fed induction generator, 6
DQ-current control strategy, 9

E

Electrical energy system, 161
Electrical equipment
lightning protection of, 197
power system, 170
Electrical power systems
complexity of, 158
network, 159
Electric double layer capacitor (EDLC), 9
Electric grid enforces, challenges, 160
Electricity demand, 78
Electricity Law in Japan, 57
Electricity market reform (EMR), 294
Electricity power company, 133
Electric power grids, 53
Electric power systems, 110, 157
end-user's level voltage, 157
Electric vehicles (EVs), 132
as storage, control strategies for, 145
Electrogeometric model (EGM), 196
Emission-free SG system, 12
Emissions performance standard, 294
EMTP-rv converts, 33
EMTP-rv model, 32
DG model, 32
Energy generating source, 163
Energy storage devices, 1
with state-of-the-art technologies, 159
Energy storage system (ESS), 9, 120, 121, 128
battery energy storage system, 10
control strategies, 137
AC grid-side inverter control system, 141
battery control system, 142
DC power system, 140
fault, DC bus constant voltage control system, 142

ground-fault, simulation, 143
rainy weather condition, 138
state of charge, 144
wind generator-side converter control system, 141
electric double layer capacitor, 10
flywheel, 11
model, 10
plug, in electric vehicle, 11
superconducting magnetic energy storage, 10
switches, 10
EPOCHS framework, 26
ESS. *See* Energy storage system (ESS)
European Center for Medium range Weather Forecasting (ECMWF) forecast model, 90, 99, 101
eXpandable Transient Analysis Program, 9
Expansion planning period, 149
External system definition/domain (ESD) model, 28
External system (Esys) module, 28

F

Faraday cage, 210
Fault clearance, 183
Faulted circuit indicator (FCI), 168
Fault location isolation and service restoration (FLISR) application, 167
intelligent electronic device (IED), 168
Fault ride through (FRT), 9
Faults, on overhead electrical power transmission, 161
Feed in tariff (FIT), 290
tariff setting, 293
Feed-in tariffs with contracts for difference (FIT–CfD), 294
Finite-difference time-domain (FDTD) method, 211
Fluctuation smoothing approaches, 120
Forecasting applications, 128
energy-storage systems
wind farms/solar power plants, scheduled generation of, 128, 129
wind output
ramp variation, suppression of, 129
Forecasting methods, 122
difficulties, 122–123
forecast/actual total output, 128
forecasted results, examples of, 125
normalized forecast errors, normalized frequency, 127
physical approach, 123
structure of, 124
probabilistic forecast, 125
regional forecasting, 124
statistic approach, 123
wind farm outputs, 126
Forecasting photovoltaic (PV) power output
annual RMSEs, 87
cloud types, classification of, 81
different forecast models, combination of, 93
Duck curve, residual electricity load, 102, 105, 106
in electric power systems, 78–79
ensemble average irradiance, fluctuation profiles of, 83
forecasted results, examples of, 98–100
forecasting methods
accuracy measures, 88–89
overview, 86
global horizontal irradiance, at single point, 81
ground-based all-sky images, 96
irradiance forecasts by different models, 94
JMA, 89
NWP models, 91
Kalman filter, time series of coefficients, 93
low-path filter applied spatial average GHI, 85
NWP models, 89–91
ensemble forecast of, 92
latitude–longitude grid, 90
physical approach, basic steps of, 88
power output fluctuation characteristics, 79
irradiance, fluctuation characteristics of, 81
smoothing effect, 82
power system operations, 80
fluctuations, 101–103
ramp events, duration time and fluctuation width, 86
RMSE
of different forecast models, 99, 101
indication of, 89
satellite cloud motion vector approach, 94
satellite image, irradiance estimation based, 95
smart house, energy management examples of, 104
United States/Japan demonstration smart grid project, 104–105
smoothing effect on accuracy, 100–101
spatial average irradiance, fluctuation characteristics of
in utility service area, 84
spatiotemporal interpolation/ smoothing, 92
statistical models, 96
postprocessing, 92
USI forecast output using all-sky image, 97
visible satellite image, 82
world cumulative installed capacity, 78
Forecasting wind power output, in electric power systems, 110
large-scale regional wind power as installed capacity percent, 116
maximum variation, 115
power output fluctuation characteristics, 112
fundamentals, 112
power spectral density (PSD), 118
relative power fluctuation frequencies for time period, 117
single wind power generator, 113
standard deviation (SD), 115
umbrella curve, 115
wind farms
in Hokkaido, Japan, 119
power curve of, 114
wind power generation, global growth of, 111
wind power generator
with energy-storage system, circuit topology of, 120
real power output, 113, 114
Fuel cell, 1, 5
Fukushima I Nuclear Power Plant, 55
Fuse-blowing schemes, 190
FW energy storage system (FESS), 11

G

Gain margin (GM), 266
Geographic information system (GIS), 15
medium voltage distribution network, 16
Geostationary Operational Environmental Satellites (GOES), 94

Geothermal systems, 9
Global Forecast System (GFS), 90
Global horizontal irradiance (GHI), 79
  multiday forecast models, 98
Global spectral model (GSM), 93
Global Wind Energy Council, 110
3GPP technical report TR 36.822, 30
Great East Japan earthquake, 268
Green environment, 158
Grid-connected mode, 287
Ground-based all-sky images, 96

H

Heat pump (HP), 132
High renewable energy penetration, 56
High voltage (HV)
  MV distribution transformers, 17
  transmission network, 16
Home energy management system (HEMS), 104, 132
  controls, storage battery, 105
HV. *See* High voltage (HV)
Hydro energy, 9

I

IEC 62305, for lightning protection design, 207
IEC international standards 62305 series, 196
IEC standard, 203, 225
IEEE 5-bus system model, 279
Independent System Operator (ISO), 101
Information and communication technologies (ICT), 17
Institute of Electrical and Electronics Engineers (IEEE) P1547-2003, 159
Institute of Electrical Power Eng. (IEPE), 279
Intelligent power grid, 1
Internet protocol (IP), 20
Inverter control method, 11, 271
  cutoff delay time, sensitivity analysis
    circuit breaker, 272
  cutoff delay time, sensitivity analysis of
    circuit breaker with IM load, 273
  FRT requirements, 272
  grid connected mode, 271
  IEEE 5-bus system model, 259
  isolated mode, 271
  power quality, criteria of, 272, 273
Irradiance forecasting, 94
  errors, smoothing effect of, 100
Island operation (IO), 182

J

Japan Meteorological Agency (JMA), 89

K

Kalman filter, 92
Kawasaki Heavy Industries (KHI), 279
Kolmogorov spectrum, 118

L

Lagrange multiplier, 285
LED. *See* Light-emitting diode (LED)
Lessening peak load demand, 55
Light-emitting diode (LED), 250
  illumination, 250
  television, 250
Lightning channel impedance, 206
Lightning current, 197
Lightning damage, mechanisms, 195
Lightning discharges, 217
Lightning electromagnetic impulse, 195
Lightning flashes, 200, 225
Lightning impulse voltages, 223
Lightning overvoltage, 195
Lightning performance, of grounding system, 210
Lightning protection levels (LPL), 207
Lightning protection system (LPS), 207
  design, 214
  for photovoltaic system, 221
Lightning protection zone (LPZ), 207
Lightning striking, 223
  distance, 200, 207
  position, 218
Lightning surge, on power cable, 216
Load ratio control transformer (LRT), 52
Low voltage (LV)
  bus, 15
  distribution systems, 3
Low-voltage ride through (LVRT) capability, 18
LPMS division, of electrical system, 210
LV. *See* Low voltage (LV)

M

Massive DER penetrations, 51
  BESS control system, 62
  control method
    constraints, 58
    objective function, 58
    objective function and constraints, 58
  control objectives, 53
    distributed generations, challenges of, 56
      high renewable energy penetration, 56
      power balancing, 56
      power losses, 57
      real-time power market, impacts of, 57
    distributed generations, importance of, 54
      energy security, improvement of, 55
      lessening peak load demand, 55
      local economics and communities, 55
      power demands, faster response, 55
      power system, enhancement of, 56
      supply reliability/power management, 55
      transmission loss, reduction, 54
    overview of, 57
  distribution loss, 65
  distribution system
    model of, 64
    parameters of, 63
  particle swarm optimization (PSO), 60
    BESS, interconnection point, 62
    plug-in electric vehicle, 63
    PV generator system, 61
  PEVs, commercialization of, 53
  proposed method, 66
  PV output power, 65
  PV, reactive power control system, 62
  renewable energies, 52
  simulation results, 63
    of comparison method, 68
      dynamic responses, 66
    of proposed method, 71
      dynamic responses, 70
    without optimization approach
      control, 67
      dynamic responses for, 66
  zero emission-based smart distributed system, 54
MATLAB®, 146
Maximum power point tracking (MPPT) control system, 61, 137

Mean bias error (MBE), 88
Medium voltage (MV)
distribution networks, 17
GIS view of, 16
distribution systems, 3
level, 15
Megasolar system, 222
Memory buffer stores, 25
Mesh type grounding electrode, 211
Metaheuristic global optimization method, 149
Microturbine generator (MTG) system
distributed generation resources, 4
system configuration of, 5
Mixed integer linear programming (MILP) model, 39
Model output statistics (MOS), 87, 123
Molten carbonate fuel cell (MCFC), 5
Multiagent system (MAS) approach, 17
Multifunctional Transport Satellite (MTSAT), 94
Multilayer perceptron network, 162
MV. *See* Medium voltage (MV)

N

Nacelle, 208
National Digital Forecast Database (NDFD), 93
Nikaho wind park, 217
Norton equivalent circuit, 264
Numerical weather prediction (NWP) models, 79, 92, 123
Nyquist criterion, 265

O

OCRs. *See* Over-current relays (OCRs)
OLTC transformer, 41
On-load tap changers (OLTCs), 18
OpenDss/ns-2 integrated tool, 26
Over-current relays (OCRs), 160
voltage control protection, 165

P

Packet discard ratio (PDR), 29
Parseval's theorem, 118
Particle swarm optimization (PSO), 60
Peak cut/peak shift mode operation, 246
Penalties, 291
Performance ratio (PR), 79
Permanent magnet synchronous generator (PMSG), 6, 137
DC distribution system, 143
generator-side converter achieves, 140
Phase locked–loop controller, 229
Phase margin (PM), 266
Phasor measurement unit (PMU) applications, 19
Photovoltaic system (PVS), 1, 9, 196, 235
battery system, demand response (DR) for, 145
energy, 3, 8
generation system, 7, 230
generator, 132
modules, insulation level of, 222
panels, 196
power generation systems, 196, 228
direct lightning flash, 223–225
lightning damage, 222
lightning protection
against lightning overvoltages, 222
principle, 221
power electronics, 230
system power output, 98, 103
Photovoltaic (PV) units, 15
Pitch angle control, 121
PLL accuracy, 285
Plug in electric vehicle (PEV), 9
Power balancing, 56
Power conditioning subsystems (PCSs) control, 9
Power generation systems, 235
Power load curve, 241
Power output smoothing control
energy-storage system
application of, 120
fluctuation smoothing, control of, 121
pitch angle control, 121
power curve with pitch angle control, 122
wind power generator, circuit topology of, 120
wind turbines, kinetic energy of, 121
Power spectrum, 118
Power supply–demand balance, 139
Power system disturbances, 169
application based three-phase fault analysis, 173, 174
electric faults, consequences of, 173
fire risk, 173
healthy systems, effect, 173
loss of apparatus, 173
threat for operator, 173
faults, 170
electrical equipment, 172
flashover, 172
mishandling, 172
surrounding weather, 172
symmetrical faults, 171
unsymmetrical faults, 171
quality issues, 169
frequency variations, 170
harmonics/transients, 170
voltage behavioral changes, 169–170
Power system operators, 78
Power system oscillations, 158
Power system protection, 160
Protection, automation, and control (PAC) system, 184
PVS. *See* Photovoltaic system (PVS)

Q

Quality of service (QoS), 19

R

Radio link control (RLC)
media access control (MAC) module, 30
Real-time power market, 57
Reciprocating engine, 3
Recloser Control M-7679, 186
Reclosers, 185
Renewable energies, 51
Renewable energy certificates (RECs), 291
Renewable energy equipment, 159
Renewable energy generation systems, 110
lightning protection principle, 197
grounding system on overvoltage, 200
IEC international standard, 207–208
lightning characteristics, 202–206
reduction, 197
shielding, 200
suppression, 198
lightning protections of, 195
mechanisms that cause, 195
wind turbine generation system, 194
Renewable energy resources, types, 1
Renewable energy sources (RESs), 1, 6, 16, 277
biomass fuels, 8
geothermal energy, 9
hydro energy, 9

Renewable energy sources (RESs) (*cont.*)
PV power systems, 7
wind energy conversion system (WECS), 6–7
RESs. *See* Renewable energy sources (RESs)
Right hand pole (RHP), 264
Right hand zero (RHZ), 265
Riverbed Modeler controller, 27
Riverbed models, 28
Root mean square errors (RMSEs), 86

S

Service data unit (SDU), 30
Small-scale power grid, 268
Smart grid (SG), 1, 11–12, 52
green system, 1
Smart house model, 1, 104, 147
Smart meter data, use of, 146–154
annual variation of total cost, 154
hot water supply, 147
investment cost, 153
operational cost for, 152
annual variation of, 153
optimal capacity/installation year, 152
optimal scheduling/expansion planning, flow chart for, 150
smart house model, 147
Solar collector (SC), 132
Solar energies, 1
Solar PV generation, 251
Solar resource knowledge management, 98
Standard deviation (SD), defined, 115
State of charge (SOC), 59, 280
deviation, 281
Static VAR compensators (SVC), 19
Sunny weather condition, simulation results, 136
Superconducting magnetic energy storage (SMES), 9
Supervisory control and data acquisition (SCADA), 20
Supervisory control systems, 188
Synchronous generators (SGs), 277
System circuit breaker (CB2), 180

T

Tabu search (TS), 135
algorithm, 146
optimization problem, 154
Terminal impedance, 263
Terminal voltage, 198
Test circuit, 242
Test feeders (TFs), 33
Time of use (ToU) electricity price, 104
Traditional power conversion system, 262
Train models, 87
Transfer hypothesis, 83
Transmission control protocol (TCP), 28
TS. *See* Tabu search (TS)

U

Universal Mobile Telecommunications System (UMTS), 29
University of California San Diego San Diego (UCSD), 96
Upward leader progression (ULP) model, 220
User datagram protocol (UDP), 28

V

VDE-AR-N 4105, 18
Vehicle-to-grid (V2G) facilities, 145
Virtual power plants (VPPs), 295, 296
economic efficiency, 296
Virtual synchronous generators, basic concepts of, 277, 278
algebraic type VSG introduced by Kawasaki heavy industries, 283
control block, 279
control schemes of, 279
general structure of, 279
IEPE's VSG topology, 281
KHI's VSG, 282
microgrid, structure of with multi VSG units, 286
VISMA, basic structure of, 281
VISMA for method, block diagram of, 282
VSG, block diagram of by Kawasaki heavy industries, 283
VSG, governor model for, 284
VSG system, of Osaka University, 282
by VSYNC group, 280
VSYNC's VSG design, 280
Virtual synchronous machine (VISMA), 278
Voltage amplitude signal, 11
Voltage collapse, 282, 284
Voltage-controlled–distributed energy resource (VC–DER), 190
Voltage-fed (VF), 264
voltage instability, 282
Voltage source inverter (VSI), 182
inverter controls, 182
Voltage stability, 166, 284
in distribution systems, 257, 258
active module, structure of, 263
battery control method, 259
definitions, 257
DGs, stability problem with power electronics, 262
IEEE 5-bus system model, 259
index, 259
P-Q characteristics, 260, 261
simulation results, 262
stability criteria, 264–266
comparison between, 266–267
terminal characteristics, 263–264
weak power system, DGs connected, 257–258
stability problems, in microgrids, 268
ESS model, 270
grid connected mode, 271
high-speed circuit breaker, 270
isolated mode, 271
load, 270
microgrid model, 269
PVS, 269
Voltage surges, 170
Voltage waveform, for lightning impulse withstand voltage test, 202
Volt-VAR control (VVC), 17
VRLA batteries, 246
VSG systems, 285
control algorithms, 287
model, 282

W

Weather data of the Japan Meteorological Agency, 146
Weather Research and Forecasting (WRF) Model, 90
WG. *See* Wind power generation (WG)
Wind energies, 1
Wind energy conversion system (WECS), 6, 7
Wind farms, 113
lightning protection, 217–220
Wind power generation (WG), 235
electromechanical characteristics of, 112
lightning protection principle

blade using receptor, 214–215
grounding resistance, 210–214
lightning damage in, 208
surge arrester/surge protective device, energy coordination of, 215–217
for wind turbine, 209–210
power electronics, 231
basic configuration of, 232
real power output, 113, 114
Wind power station, in Japan, 203
Wind turbine, 1, 196
blades, 196, 214
towers, 220
Winter lightning current, 203
World energy demand, 1
World wind energy association (WWEA), 7
WTG systems, 139

Z

Zinc oxide varistor, 198

Edwards Brothers Malloy
Ann Arbor MI. USA
April 12, 2016